ADAPTATION AND LEARNING
IN AUTOMATIC SYSTEMS

This is Volume 73 in
MATHEMATICS IN SCIENCE AND ENGINEERING
A series of monographs and textbooks
Edited by RICHARD BELLMAN, *University of Southern California*

A complete list of the books in this series appears at the end of this volume.

ADAPTATION AND LEARNING

IN AUTOMATIC SYSTEMS

Ya. Z. Tsypkin
THE INSTITUTE OF AUTOMATION AND TELEMECHANICS
MOSCOW, USSR

Translated by
Z. J. Nikolic
ESSO PRODUCTION RESEARCH COMPANY
HOUSTON, TEXAS

1971

ACADEMIC PRESS   New York and London

First published in the Russian language under the title
Adaptatsia i obuchenie v avtomaticheskikh sistemakh
Nauka, Moscow, 1968.

COPYRIGHT © 1971, BY ACADEMIC PRESS, INC.
ALL RIGHTS RESERVED
NO PART OF THIS BOOK MAY BE REPRODUCED IN ANY FORM, BY PHOTOSTAT, MICROFILM, RETRIEVAL SYSTEM, OR ANY OTHER MEANS, WITHOUT WRITTEN PERMISSION FROM THE PUBLISHERS.

ACADEMIC PRESS, INC.
111 Fifth Avenue, New York, New York 10003

United Kingdom Edition published by
ACADEMIC PRESS, INC. (LONDON) LTD.
Berkeley Square House, London W1X 6BA

LIBRARY OF CONGRESS CATALOG CARD NUMBER: 78-129017
AMS (MOS) 1970 SUBJECT CLASSIFICATION: 93-01,93C40

PRINTED IN THE UNITED STATES OF AMERICA

CONTENTS

Foreword   xiii
Preface to the English Edition   xv
Preface to the Russian Edition   xvii

Introduction   1

Chapter 1. Problem of Optimality

1.1	Introduction	5
1.2	Criteria of Optimality	6
1.3	More about the Criteria of Optimality	7
1.4	Constraints	8
1.5	A Priori and Current Information	9
1.6	Deterministic and Stochastic Processes	10
1.7	The Ordinary and Adaptive Approaches	11
1.8	On Methods for Solving Optimization Problems	12
1.9	Conclusion	14
	Comments	14
	Bibliography	15

Chapter 2. Algorithmic Methods of Optimization

2.1	Introduction	17
2.2	Conditions of Optimality	17
2.3	Regular Iterative Method	18

2.4	Algorithms of Optimization	19
2.5	A Possible Generalization	21
2.6	Various Algorithms of Optimization	22
2.7	Search Algorithms of Optimization	23
2.8	Constraints I	25
2.9	Constraints II	26
2.10	The Method of Feasible Directions	28
2.11	Discussion	28
2.12	Multistage Algorithms of Optimization	29
2.13	Continuous Algorithms of Optimization	31
2.14	Methods of Random Search	32
2.15	Convergence and Stability	33
2.16	The Conditions of Convergence	34
2.17	On Acceleration of Convergence	35
2.18	On Best Algorithms	36
2.19	Examples	37
2.20	Certain Problems	38
2.21	Conclusion	38
	Comments	39
	Bibliography	41

Chapter 3. Adaptation and Learning

3.1	Introduction	44
3.2	Concepts of Learning, Self-Learning and Adaptation	45
3.3	Formulation of the Problem	46
3.4	Probabilistic Iterative Methods	46
3.5	Algorithms of Adaptation	48
3.6	Search Algorithms of Adaptation	48
3.7	Constraints I	50
3.8	Constraints II	51
3.9	A Generalization	52
3.10	Multistage Algorithms of Adaptation	52
3.11	Continuous Algorithms	54
3.12	Probabilistic Convergence and Stability	54
3.13	Conditions of Convergence	56
3.14	Stopping Rules	57
3.15	Acceleration of Convergence	59
3.16	Measure of Quality for the Algorithms	60
3.17	The Best Algorithms	61
3.18	Simplified Best Algorithms	63

3.19	A Special Case	64
3.20	Relationship to the Least-Square Method	65
3.21	Relationship to the Bayesian Approach	65
3.22	Relationship to the Maximum Likelihood Method	66
3.23	Discussion	68
3.24	Certain Problems	68
3.25	Conclusion	69
	Comments	69
	Bibliography	72

Chapter 4. Pattern Recognition

4.1	Introduction	76
4.2	Discussion of the Pattern Recognition Problem	77
4.3	Formulation of the Problem	77
4.4	General Algorithms of Training	79
4.5	Convergence of the Algorithms	80
4.6	Perceptrons	81
4.7	Discrete Algorithms of Training	83
4.8	Search Algorithms of Training	85
4.9	Continuous Algorithms of Training	87
4.10	Comments	88
4.11	More about Another General Algorithm of Training	89
4.12	Special Cases	90
4.13	Discussion	94
4.14	Self-Learning	95
4.15	The Restoration of Probability Density Functions and Moments	96
4.16	Algorithms of Restoration	97
4.17	Principle of Self-Learning	99
4.18	Average Risk	100
4.19	Variation of the Average Risk	101
4.20	The Conditions for the Minimum of the Average Risk	101
4.21	Algorithms of Self-Learning	102
4.22	A Generalization	103
4.23	Specific Algorithms	105
4.24	Search Algorithms of Self-Learning	108
4.25	Discussion	108
4.26	Certain Problems	109
4.27	Conclusion	109
	Comments	110
	Bibliography	114

Chapter 5. Identification

5.1	Introduction	119
5.2	Estimation of the Mean Value	120
5.3	Another Approach	121
5.4	Estimation of the Variance	122
5.5	Discussion	124
5.6	Estimation of Correlation Functions	125
5.7	Estimation of the Characteristics of Nonlinear Elements	126
5.8	Estimation of the Coefficient of Statistical Linearization	128
5.9	Special Cases	128
5.10	Description of Dynamic Plants	130
5.11	Identification of Nonlinear Plants I	130
5.12	Identification of Nonlinear Plants II	131
5.13	Identification of Nonlinear Plants III	133
5.14	A Special Case	134
5.15	A Remark	135
5.16	Identification of Linear Plants I	136
5.17	Identification of Linear Plants II	137
5.18	Estimation of Parameters in Systems with Distributed Parameters	138
5.19	Noise	139
5.20	Elimination of Noise	140
5.21	Certain Problems	142
5.22	Conclusion	142
	Comments	143
	Bibliography	144

Chapter 6. Filtering

6.1	Introduction	147
6.2	Criterion of Optimality	147
6.3	Adaptive Filter	149
6.4	Special Cases	150
6.5	Adaptive Equalizer	152
6.6	Adaptive Filters Based upon Search	154
6.7	Adaptive Filter Predictor	156
6.8	Kolmogorov-Wiener Filters	157
6.9	Statistical Theory of Signal Reception	158
6.10	Criterion of Optimality in Signal Reception	159
6.11	A Decision Rule	160

6.12	Signal Detection in Background Noise I	161
6.13	Signal Detection in Background Noise II	162
6.14	Signal Extraction from Background Noise	163
6.15	Criterion of Optimal Signal Extraction	164
6.16	Algorithm of Signal Extraction	165
6.17	More about Signal Extraction from Background Noise	166
6.18	Another Criterion of Optimality	166
6.19	Optimal Receiver	167
6.20	Possible Simplifications	168
6.21	Recovery of Input Signals	169
6.22	Algorithm of Signal Recovery	170
6.23	The Influence of Noise	171
6.24	Certain Problems	171
6.25	Conclusion	172
	Comments	172
	Bibliography	173

Chapter 7. Control

7.1	Introduction	176
7.2	When Is Adaptation Necessary?	176
7.3	Statement of the Problem	177
7.4	Dual Control	178
7.5	Algorithms of Dual Control	179
7.6	Adaptive Control Systems I	181
7.7	Adaptive Control Systems II	181
7.8	Sensitivity Model	184
7.9	Adaptive Control Systems III	186
7.10	Simplified Adaptive Systems	187
7.11	Systems of Control by Perturbation	187
7.12	Algorithms of Optimal Control	189
7.13	Another Possibility	190
7.14	Extremal Control Systems	191
7.15	Algorithms of Extremal Control	192
7.16	Algorithms of Learning	193
7.17	Continuous Algorithms	194
7.18	Structural Scheme	194
7.19	Possible Simplifications	194
7.20	The Synthesis of Optimal Systems	197
7.21	Application of Adaptation Algorithms	198
7.22	The Synthesis of Optimal Systems in the Presence of Noise	199

7.23	Control and Pattern Recognition	200
7.24	Generalization of the Methods of Synthesis	201
7.25	Certain Problems	202
7.26	Conclusion	204
	Comments	204
	Bibliography	206

Chapter 8. Reliability

8.1	Introduction	210
8.2	Concept of Reliability	210
8.3	Reliability Indices	211
8.4	Definition of Reliability Indices	212
8.5	Minimization of Operating Expenditures	215
8.6	A Special Case	215
8.7	Minimization of Cost, Weight and Volume	216
8.8	Algorithms of Minimization	217
8.9	A Special Case	217
8.10	Algorithms	218
8.11	Increase of Reliability by Reserve Components	220
8.12	Increase of Reliability by Redundant Components	221
8.13	Design of Complex Systems	223
8.14	Algorithms of Optimal Productivity	223
8.15	The min-max Criterion of Optimization	225
8.16	More about the Design of Complex Systems	226
8.17	Comment	226
8.18	Certain Problems	227
8.19	Conclusion	227
	Comments	227
	Bibliography	228

Chapter 9. Operations Research

9.1	Introduction	229
9.2	Planning of Reserves	230
9.3	The Criterion of Optimality in Planning	231
9.4	Algorithm of Optimal Planning	233
9.5	More about the Planning of Reserves	234
9.6	Optimal Supply	234
9.7	Optimal Reserves	235

9.8	A Comment	237
9.9	Distribution of Production Resources	237
9.10	An Example	238
9.11	Distribution of Detection Facilities	239
9.12	Algorithm for Optimal Distribution	240
9.13	Distribution of Quantization Regions	241
9.14	Criterion of Optimal Distribution	242
9.15	Algorithms of Optimal Evaluation	243
9.16	Certain Problems	245
9.17	Conclusion	245
	Comments	245
	Bibliography	246

Chapter 10. Games and Automata

10.1	Introduction	248
10.2	The Concept of Games	249
10.3	The min-max Theorem	250
10.4	Equations of Optimal Strategies	251
10.5	Algorithms of Learning the Solution of Games	252
10.6	Games and Linear Programming	254
10.7	Control as a Game	256
10.8	Algorithms of Control	257
10.9	A Generalization	258
10.10	Threshold Elements	258
10.11	Logical Functions and Their Realizations on a Single Threshold Element	259
10.12	Criterion of Realizability	260
10.13	Algorithm of Realizability	261
10.14	Rosenblatt's Perceptron	262
10.15	Widrow's Adaline	263
10.16	Training of a Threshold Element	264
10.17	Automata	266
10.18	Description of Finite Automata	267
10.19	Stochastic Finite Automata	270
10.20	Automaton–Environment Interaction	271
10.21	The Expediency of Behavior	271
10.22	Training of Automata	272
10.23	About Markov Chains	275
10.24	Markov Learning	276
10.25	Games of Automata	277

10.26	Certain Problems	278
10.27	Conclusion	278
	Comments	279
	Bibliography	281

Epilogue 286

Subject Index 289

FOREWORD

During the past decade there has been a considerable increase of research interest in problems of adaptive and learning systems. Many different approaches have been proposed for the design of engineering systems which exhibit adaptation and learning capabilities. Preliminary results in both theoretical research and practical applications are quite promising. One of the very few books published in this area so far is the Russian edition of the present book, authored by the well-known specialist, Professor Ya. Z. Tsypkin, of the Institute of Automation and Telemechanics, Moscow, USSR. This book presents a unified treatment of the theory of adaptation and learning. The author is directly responsible for the development of the materials in a number of chapters. Applications cover pattern recognition, automatic control, operations research, reliability and communication systems.

This English edition of the book differs from the original in several details. The bibliographic material, which appeared as a section at the back of the Russian edition, has been separated, and the pertinent material is given at the end of each chapter. This material is modified to the extent that the references translated into Russian from English are replaced by the original source. In addition, English translations are substituted for the books and monographs originally published in Russian. Also, an index of basic symbols which appeared in the Russian edition has been omitted from this edition. Certain misprints and errors which occurred in the Russian version are also corrected according to the list of errata provided by the author.

Our intention was to preserve as much as possible the spirit and flavor of the original. For instance, in the expression "restoration (estimation) of the probability density functions," the word "estimation," which is commonly used in this context in the conventional English terminology, is given in parentheses. For the same reason, the words "controlled object (plant)," and "training (learning)" can also be found in the translation.

We are happy to express our gratitude to Professor Tsypkin for his kind help and the specific suggestions he has given concerning this translation.

K. S. Fu, Redactor
Z. J. Nikolic, Translator

PREFACE TO THE ENGLISH EDITION

The problem of adaptation and learning is one of the most fundamental of modern science and engineering. The complexity of many existing and developing systems, and the lack of a priori information about their properties do not permit application of the traditional methods in order to guarantee the optimal operating conditions of such systems. These problems can only be solved on the basis of certain efficient solutions of the problems of adaptation and learning. One such method is based on the adaptive approach presented in this book.

The purpose of this book is to use broad strokes in painting an attractive picture of the possibilities which the applications of the adaptive approach offer in the solutions of various problems in control theory, reliability theory, operations research, game theory, the theory of finite automata, etc.

The English translation of this book widens the audience of its readers, and this, naturally, pleases me. I will be completely satisfied if the readers of this book experience the same satisfaction which I experienced while writing it.

I have the pleasure, at this point, of expressing my gratitude to the editor of the translation, Professor K. S. Fu, and to the translator, Dr. Z. J. Nikolic, for the most pleasant and extremely useful collaboration which produced this translation.

PREFACE TO THE RUSSIAN EDITION

Many difficulties are encountered in writing about the problems of learning, self-learning and adaptation in automatic systems. Although this area is relatively new, the publications related to it are so numerous that even a short survey would be lengthy. But even if such difficulties could be overcome, we would still face another obstacle caused by a tradition which is particularly present in the theory of automatic control. Year after year, various methods are proposed for solving the same old problems, and after a certain period of competition, such methods are either forgotten, or, as it frequently happens, they continue to coexist peacefully.

Remembering that the number of problems is great, it is not difficult to imagine the situation of a specialist who attempts to understand the problems of adaptation. The author was placed in this situation when, in the summer of 1965, he started to prepare the paper " Adaptation, Learning and Self-Learning in Automatic Systems " for the Third All Union Conference on Automatic Control (Odessa, September 1965). There was a need to find a way out of the situation created, which could be found only by breaking away from tradition. Instead of treating the same problems by different methods, we attempted to treat different problems by a single method. In the beginning, hopes for success were based only on a belief that certain general laws governing adaptation should exist. But after success in discovering some of these laws, that hope became a conviction.

The probabilistic iterative methods are used as a foundation for the approach developed in this book. This approach can be applied to various problems of adaption. Shortly after the initiation of this approach, many propositions were developed and made more rigorous. It was also discovered that the results of many recent studies in adaptation and learning, as well as the results of the past studies which had not received sufficient attention, comply with the rules of the general approach.

The approach developed not only simplifies the solution of the well-known problems, but also permits the solution of many new problems.

Of course, we are far from the thought that the method presented in this book could cover all the studies on adaptation, learning and self-learning, and especially those studies in which these terms are only mentioned. It is possible that many important studies were not considered, but if this is the case, it is not because they are uninteresting and unimportant, but rather because there is not yet a general consensus which would allow us to include all of them.

The title of the book gives a sufficiently complete picture of its contents. However, there are several characteristics of the book which should be mentioned. We will not attempt to give the detailed proofs of many propositions presented. First, this would greatly increase the volume of the book and change the envisioned style; second, it seems to us that an inevitable overload of such details and refinements would prevent us from extracting and emphasizing the general ideas of adaptation and learning; finally, not all the proofs are both general and brief, and, as it will not be difficult to discover, we still do not have all but just the necessary proofs. The author has attempted to present all the questions "as simply as possible, but not simpler."

In this book, problems of varying importance, complexity and relevance are formulated. An exposition of general ideas and their applications in various problems of the modern theory of automatic control and related areas (theory of reliability, operations research, the game theory, the theory of finite automata, etc.) are also presented.

The basic text of this book contains almost no references to the literature which is related to the questions under consideration. This was done on purpose in order to preserve the continuity of presentation. For this reason, a detailed survey of the literature on adaptive and learning systems is given at the end of the book. In addition to the survey, various comments, explanations and comparisons of the obtained results with the earlier known ones are also given here.

The index of basic symbols should help the reader to become familiar with any particular chapter on application of the adaptive approach of his interest without reading the other chapters.

This book could not have been written without the support of the author's young co-workers, E. Avedyan, I. Devyaterikov, G. Kel'mans, P. Nadezhdin, Yu. Popkov and A. Propoi, who have not only taken an active role in the evaluation of the results, but have also influenced the development and the presentation of many questions considered in this book. The author is deeply grateful to them for their stimulating initiative and enthusiasm. The author is also thankful to Z. Kononova for the great help in the preparation of the manuscript of this book. Finally, the author is happy to acknowledge the great role of B. Novoseltsev. By editing the manuscript, he has helped to give this book its final form.

If, while reading this book, the reader begins to feel a desire to disagree with the author, or a tendency to improve and develop certain new results, or finally, an intention to use the ideas of adaptation in the solution of his problems, i.e., if the reader does not stay indifferent at the end, the author will be satisfied.

ADAPTATION AND LEARNING
IN AUTOMATIC SYSTEMS

INTRODUCTION

Three characteristic periods can be distinguished in the development of the theory of automatic control. For convenience, they are briefly called the periods of determinism, stochasticism and adaptivity.

In the happier days of determinism, the equations describing the states of the controlled plants* as well as the external actions (functional or perturbing) were assumed to be known. Such a complete description permitted a broad application of the classical analytic apparatus in the solution of various problems of control theory. This was especially true of the linear problems, where the powerful principle of superposition simplified their solution and gave an illusion that the principal difficulties did not exist at all. Of course, such difficulties were observed as soon as the nonlinear factors had to be considered. But even in the area of nonlinear problems, where the general regular methods of solution did not exist, significant results related to the analysis as well as the synthesis of automatic systems were obtained.

A less happy time came with the period of stochasticism. In considering more realistic operating conditions of the automatic systems, it was established that the external actions, and especially the external perturbations, were continuously varying in time and could not be uniquely defined in advance. This was also frequently true of the coefficients of the plant equations. These approaches were based on the knowledge of statistical characteristics of random functions (which had to be determined in advance), and the use of analytic methods of the times of determinism.

* Also commonly called "systems."

Introduction

The characteristic feature of these periods in the development of the theory of automatic control is that their methods and results were directly applicable to automatic systems with sufficient information, i.e., where the plant equations and the external actions or their statistical characteristics were known.

At the present "long-suffering" time (from the standpoint of automatic control theory), we become more convinced each day that in the modern complex automatic systems which operate in the most diverse conditions, the equations of the controlled plants and the external actions (or their statistical characteristics) are not only unknown, but that for certain reasons, we do not even have the possibility of determining them experimentally in advance. In other words, we are confronted with a greater or smaller initial uncertainty. Although all this makes the control of such plants more difficult, it still does not make this control impossible in principle. This is evidenced by the emergence of a new, third period in the theory of control—the period of adaptivity. The possibility of controlling the plants under incomplete and even very small a priori information is based on the application of adaptation and learning in automatic systems which reduces initial uncertainty by using the information obtained during the process of control.

One should not think that the periods of determinism, stochasticism and adaptivity have replaced each other. The following periods were always born within the preceding ones, and we now witness their existent problems.

In each of the listed periods, the basic concern in the first stage was the problem of analysis of automatic systems and an explanation of their properties. The problems of synthesis of automatic systems satisfying the specific requirements appeared later. Naturally, the desire (frequently a necessity) war born to synthesize a system which is optimal in a cetrain sense.

The problem of optimality became one of the most central in automatic control. Even if great success has not yet been achieved in selecting and formulating the performance indices, we can still be comforted by the brilliant results related to the problem of optimality which were concentrated in Pontryagin's maximum principle and in Bellman's dynamic programming method. Although raised in the field of deterministic problems, they are now solving stochastic and partly adaptive problems with definite success.

The great achievements in that direction during the period of stochasticism are the methods of Kolmogorov-Wiener and Kalman, who have considerably exhausted the linear problems of synthesis.

Unfortunately, the period of adaptivity cannot boast such brilliant results. This can be explained by the fact that the problems of adaptation and their related problems of learning and self-learning are still very new. Nevertheless, we are finding them more and more often in very diverse situations of modern automatization. In addition to the basic problem of control under incomplete a priori information or when such information is not present, i.e., in the conditions of initial uncertainty, those of adaptation take place in the estimation of the plant characteristics and actions, in learning to recognize patterns and situations, in achieving and improving the goals of control, etc.

The terms *adaptation, self-learning* and *learning* are the most fashionable terms in the modern theory of automatic control. Unfortunately, these terms, as a rule, do not have a unique interpretation and frequently have no interpretation at all. This creates a very fertile field for many unsupported exaggerated claims which can be found in the popular literature on cybernetics, and sometimes even in the pages of certain technical journals.

Nevertheless, if the numerous general discussions without much content are neglected, one can still point to many interesting approaches and results created in the solution of the above-mentioned problems.

However, it should be mentioned that until recently, these problems have been considered separately and independently from each other. The relationships between the problems were almost never noticed, although a broader look at the problems of adaptation, learning and self-learning indicates that they are so closely related that it is amazing that this was not recognized earlier.

Our main concern is to evaluate the problems of adaptation, learning and self-learning from a single point of view which would not only relate those problems considered earlier as unrelated, but also provide effective ways to their solution.

Of course, the accomplishment of this goal can be expected only if the following two conditions are fulfilled: the existence of definite concepts of adaptation, learning and self-learning, and the existence of a certain mathematical apparatus which is adequate for these concepts.

The first condition is in our hands, or at least in the hands of the various committees which determine terminology. From a number of established definitions (which sometimes contradict each other), we can either select those which are convenient for our book or we can add one more. The second condition is by far the most difficult to satisfy. But, as is frequently the case in the history of science, an adequate mathematical apparatus, although perhaps in a rudimentary stage, fortunately exists. On one hand, it consists of modern mathematical statistics, and on the other hand, it consists of a new, intensively developing branch of mathematics which is well-known under the name of mathematical programming.

Mathematical programming includes the theory and the methods of solving extremal problems, and it covers not only special topics (variational calculus, Pontyragin's maximum principle, Bellman's dynamic programming, and linear and nonlinear programming), but also the methods of stochastic approximation. The latter methods, which are unfortunately not used very extensively outside the field of mathematical statistics, play an important role in the area of our interest.

It is important to emphasize that mathematical programming does not require an analytic and formal description of the conditions of the problems. For this reason, it can cover a much wider circle of problems than the methods in which the solutions are sought in analytic form. The algorithmic form for solving extremal problems provides the possibility for using the devices of modern computing techniques while still not imposing the constraints of the analytic approach which would otherwise keep us far from solving those realistic problems which we would actually like to consider.

The algorithms of learning and adaptation must provide the optimum, in a certain sense, under the conditions of minimal a priori information. First of all, we have to become familiar with the problem of optimality and with the algorithmic methods of solving these problems. Then we can evaluate the concepts and the methods which are characteristic of the problems of adaptation and learning. Only after this and by using a single point of view and approach, can we be in a position to solve the variety of problems that interest us here.

1

PROBLEM OF OPTIMALITY

1.1 Introduction

One can say without exaggeration that the problem of optimality is a central problem of science, engineering and even everyday life. Everything that a man does, he tries to do in the best possible way. Any accepted conclusions, actions or produced devices can be considered as optimal from a certain viewpoint since they have been given a preference over other possible conclusions, actions or devices, i.e., they have been considered as the best.

In an attempt to reach a goal, three problems immediately arise. The first one is to select and formulate the goal. Something that is best in certain circumstances may be far from the best in other circumstances. The selection and the formulation of a goal depend upon many factors, and they are frequently accomplished with great difficulties. Very often we know what we want, but unfortunately, we cannot precisely formulate our wishes.

Once the goal is chosen, the second problem is to match it with the available resources, i.e., to take the constraints into consideration. Even a clearly formulated goal is not a guarantee of its own fulfillment. Goals, like dreams, are not always realized.

Finally, after selecting the goal and considering the constraints, we have to solve the third problem—to realize the goal under existent constraints, by which the real value of different mathematical methods of optimization and their power or weaknesses are actually clarified.

Therefore, the solution of an optimization problem is reduced to a sequential solution of the above-mentioned problems. In this chapter, we shall examine the necessary details of the first two problems. The third problem will be treated in the other chapters of this book.

1.2 Criteria of Optimality

Any optimization problem can be reduced to one of finding the best policy (in a certain sense) from a large number of policies. Each one of these policies is characterized by a set of numbers (or functions). The quality of a policy is measured by an index, quantitative characteristic, which defines the proximity to the set goal for the chosen policy.

The best policy corresponds to the extremum of a performance index, i.e., to the minimum or the maximum, depending on the particular problem. The performance indices are usually some functionals, which can be considered as the functions in which the curves or the vectors characterizing the policies play the role of independent variables.

A functional which depends upon a vector is simply a function of many variables. We shall further consider only those which depend upon the vectors. Such functionals can be obtained from those that depend upon the functions by using direct methods of variational calculus.

In general, the performance index can be given in the form of expectation

$$J(\mathbf{c}) = \int_X Q(\mathbf{x}, \mathbf{c}) p(\mathbf{x}) \, d\mathbf{x} \tag{1.1}$$

or briefly by

$$J(\mathbf{c}) = M_\mathbf{x}\{Q(\mathbf{x}, \mathbf{c})\} \tag{1.2}$$

where $Q(\mathbf{x}, \mathbf{c})$ is the functional of the vector $\mathbf{c} = (c_1, \ldots, c_N)$ which also depends on a vector of random sequences or processes $\mathbf{x} = (x_1, \ldots, x_N)$ with probability density function $p(\mathbf{x})$; X is the space of vectors $\mathbf{x}$. Here and in the other chapters, all vectors represent column matrices.

In expression (1.2), a possible dependence of the functionals on known vectors which will be found in the studies of specific problems is not emphasized. A number of performance indices which differ only in their form correspond to equation (1.2). For instance, a very popular criterion in statistical decision theory is the average risk (Bayesian criterion) which is defined as

$$R(d) = \int_\Lambda \sum_{\nu, \mu = 1}^N P_\nu w_{\nu\mu} d_\mu(\mathbf{x}, \mathbf{c}) p_\nu(\mathbf{x}) \, d\mathbf{x} \tag{1.3}$$

where P_ν is the probability that an observed element $\mathbf{x}$ belongs to the subset Λ_ν of Λ; $p_\nu(\mathbf{x})$ is the conditional probability density function defined over the subset Λ_ν; $d_\mu(\mathbf{x}, \mathbf{c})$ is the decision rule which depends upon an unknown vector of the parameters $\mathbf{c}$ such that

$$d_\mu(\mathbf{x}, \mathbf{c}) = \begin{cases} 1 & \text{if it was decided that } \mathbf{x} \in \Lambda_\mu \\ 0 & \text{if it was decided that } \mathbf{x} \notin \Lambda_\mu \end{cases} \qquad (1.4)$$

and $w_{\nu\mu}$ ($\nu, \mu = 1, \ldots, N$) is an element of the cost matrix W which defines the cost of incorrect decisions.

Let us write the formula for $R(d)$ as

$$R = \sum_{\nu, \mu = 1}^N w_{\nu\mu} \int_\Lambda P_\nu p_\nu(\mathbf{x}) \, d_\mu(\mathbf{x}, \mathbf{c}) \, d\mathbf{x} \qquad (1.5)$$

It follows that R can be considered as a conditional mathematical expectation of a random variable $w_{\nu\mu}$ with a certain probability distribution

$$\int_\Lambda P_\nu p_\nu(\mathbf{x}) \, d_\mu(\mathbf{x}, \mathbf{c}) \, d\mathbf{x}$$

It is sometimes more convenient to use the performance index which defines the probability that a certain variable lies between the limits $\varepsilon_1 \leq Q(\mathbf{x}, \mathbf{c}) \leq \varepsilon_2$, i.e.,

$$J(\mathbf{c}) = P\{\varepsilon_1 \leq Q(\mathbf{x}, \mathbf{c}) \leq \varepsilon_2\} \qquad (1.6)$$

By introducing the characteristic function

$$\theta(\mathbf{x}, \mathbf{c}) = \begin{cases} 1 & \text{if } \varepsilon_1 \leq Q(\mathbf{x}, \mathbf{c}) \leq \varepsilon_2 \\ 0 & \text{otherwise} \end{cases} \qquad (1.7)$$

we can transform (1.6) to the form

$$J(\mathbf{c}) = M_\mathbf{x}\{\theta(\mathbf{x}, \mathbf{c})\} \qquad (1.8)$$

which corresponds to (1.2).

The goal is to find either the minimum (for instance in the case of (1.3)) or the maximum (for instance in the case of (1.6)) of a functional. Therefore, these functionals are also frequently called the criteria of optimality.

1.3 More about the Criteria of Optimality

In addition to the criteria of optimality which are given in the form of conditional expectation (1.2) based on the ensemble average, the criteria of optimality based on the averaging of $Q(\mathbf{x}, \mathbf{c})$ with respect to time can also be used, depending on whether $\mathbf{x}$ is a random sequence $\{\mathbf{x}[n]; n = 0, 1, 2, \ldots\}$ or a random process $\{\mathbf{x}(t); 0 \leq t < \infty\}$. The criteria of optimality can be written, respectively, either as

$$J(\mathbf{c}) = \lim_{N \to \infty} \frac{1}{N} \sum_{n=0}^N Q(\mathbf{x}[n], \mathbf{c}) \qquad (1.9)$$

or as

$$J(\mathbf{c}) = \lim_{T \to \infty} \frac{1}{T} \int_0^T Q(\mathbf{x}(t), \mathbf{c}) \, dt \qquad (1.10)$$

For the ergodic stationary sequences and processes, the criteria of optimality (1.9) and (1.10) which differ from the criterion of optimality (1.2) in the averaging procedure (time or ensemble averaging), are equivalent. This means that the expressions of these functionals will always coincide if we can obtain them in an explicit form. In any other case, the criteria of optimality (1.9) and (1.10) differ from (1.2). However, this fact should not prevent us from using the criterion of optimality (1.9) or its generalization even in the cases when at each step (at each instant of time) the form of the function Q is varying, i.e., when

$$J(\mathbf{c}) = \lim_{N \to \infty} \frac{1}{N} \sum_{n=0}^{N} Q_n(\mathbf{x}[n], \mathbf{c}) \qquad (1.11)$$

or

$$J(\mathbf{c}) = \lim_{T \to \infty} \frac{1}{T} \int_0^T Q(\mathbf{x}(t), \mathbf{c}, t) \, dt \qquad (1.12)$$

The criteria of optimality have, or at least should have, a certain definite physical or geometrical meaning. For instance, in the automatic control systems the criterion of optimality represents a certain measure of deviation from the desired or prescribed system. In the problems of approximations, the criterion of optimality characterizes a certain measure of deviation of the approximating function from the approximated one. The selection of a specific criterion of optimality is always related to a tendency to find a compromise between a desire to describe the posed problem more accurately and the possibility of obtaining a simpler solution for the corresponding mathematical problem.

1.4 Constraints

If the vectors $\mathbf{x}$ and $\mathbf{c}$ in the performance indices were not constrained by certain conditions, the problem of optimality would not have any meaning. An optimization problem is created when mutually exclusive constraints exist, and the optimality is achieved by satisfying these constraints in the best manner, i.e., by selecting a policy in which the criterion of optimality reaches the extremum. The constraints, which are given in the form of inequalities, equalities, or logical relationships, isolate so-called permissible policies (variants) from a set of policies within which the optimal policy is sought.

The laws of nature which describe various events and which define the behavior of different systems also represent certain constraints. These laws, expressed in the form of algebraic differential and integral equations govern the behavior of the vectors **x** and **c**. The form of these equations depends on the character and the nature of the problem under consideration, and we shall thus find various types of equations describing the constraints called constraints of the first type. Other constraints may be caused by limited resources, energy or any other variables which the systems cannot violate due to their physical nature; these are called constraints of the second type. They are imposed on the components of the vector **c**, and are expressed on the form of equalities

$$g_v(\mathbf{c}) = 0 \qquad (v = 1, \ldots, M < N) \qquad (1.13)$$

or inequalities

$$g_v(\mathbf{c}) \leq 0 \qquad (v = 1, \ldots, M_1) \qquad (1.14)$$

where $g_v(\mathbf{c})$ is a certain function of the vector **c**.

Frequently, the constraints may refer not to the instantaneous, but rather to the average values. Then (1.13) and (1.14) are replaced by the equations or the inequalities of the mathematical expectations of the corresponding functions:

$$g_v(\mathbf{c}) = M_\mathbf{x}\{h_v(\mathbf{x}, \mathbf{c})\} = 0 \qquad (v = 1, \ldots, M) \qquad (1.15)$$

or

$$g_v(\mathbf{c}) = M_\mathbf{x}\{h_v(\mathbf{x}, \mathbf{c})\} \leq 0 \qquad (v = 1, \ldots, M_1) \qquad (1.16)$$

Therefore, the constraints are based upon the equations of the processes and upon the limits in the variations of certain functions characterizing these processes.

In automatic systems, such constraints are the equations of motion and the limits in the variations of controlled variables and control actions. In the problem of approximation, the constraints are determined by the character of the approximated function. Unfortunately, the number of constraints in realistic physical problems exceeds that for which the statement of an optimization problem is still possible and logical.

Since all these constraints reduce the number of possible solutions, it would seem that the problem of finding the optimal policy is simplified. However, the solution of this problem becomes more complicated, and the well-known classical methods of solutions become inapplicable.

1.5 A Priori and Current Information

A priori information, or, as it is frequently called, initial information, represents a collection of facts about the criterion of optimality and the constraints which are known in advance. The criterion of optimality is used to express conditions which have to be satisfied in a best manner, while the constraints describe resources. Therefore, a priori information includes the demands imposed upon the process, the character of the process equations, the values of the parameters in these equations, the properties of the external actions, and finally the process alone.

A priori information can be obtained by preliminary theoretical or experimental investigations. Such information is the starting point in solving all engineering problems and in particular, optimization problems.

Any description of a real system inevitably leads to the idealization of its properties. But, regardless of the tendency to take into consideration all the basic characteristic features of this system, we cannot hope to obtain complete a priori information, which implies absolute, exact knowledge. If we also take into consideration the presence of various disturbances which are the source of uncertainty, we have to conclude that all the cases encountered in practice are characterized by incomplete a priori information. Of course, the degree of this incompleteness can vary.

The information can be either sufficient or insufficient for the information and solution of an optimization problem. When sufficient information exists, all necessary facts about the criterion of optimality and the constraints are known, i.e., they can be written in explicit form. In the case of insufficient a priori information, the necessary facts about the criterion of optimality and/or about the constraints are not complete.

Since random variations always exist in realistic situations, a priori information becomes unreliable and inaccurate. The degree of completeness, i.e., the amount, of a priori information plays a substantial role in the formulation and the solution of an optimization problem.

On the other hand, current information is based on the observation of the process or on the results of the experiment. Therefore, the current information is renewed at each instant. The current information obtained during a preplanned experiment can be used for gathering corresponding a priori information. However, the most important role of the current information consists in compensating the insufficient volume of a priori information.

A priori information is the basis for the formulation of any optimization problem, but the current information provides the solution for that problem.

1.6 Deterministic and Stochastic Processes

The deterministic processes are characterized by the following property: If the process is known within a certain interval of time, one can completely determine its behavior outside of that interval. The criterion of optimality for a deterministic process is given in advance, and the constraints of the first and second kind are known. In stochastic processes, one can only predict the probabilistic characteristics of the behavior of these processes outside an interval of time by knowing the process within that interval.

If these probabilistic characteristics (for instance, probability distributions) are given in advance, we can again define the criterion of optimality and the constraints in an explicit form through certain conditional expectations.

The deterministic processes can be considered as special cases of stochastic processes in which the probability density function is a Dirac function, i.e., δ function: $p(\mathbf{x}) = \delta(\mathbf{x})$. In such a case, the conditional expectations in the criterion of optimality (1.2) and in the constraints are simply transformed into the deterministic functions which do not depend upon the random vector $\mathbf{x}$.

For instance, when $p(\mathbf{x}) = \delta(\mathbf{x})$, one can easily obtain the following relationships for a deterministic process from (1.2), (1.15) and (1.16) by using (1.1):

$$J(\mathbf{c}) = Q(0, \mathbf{c}) \tag{1.17}$$

$$g_v(\mathbf{c}) = h_v(0, \mathbf{c}) = 0 \quad (v = 1, \ldots, M < N) \tag{1.18}$$

$$g_v(\mathbf{c}) = h_v(0, \mathbf{c}) \leq 0 \quad (v = 1, \ldots, M_1) \tag{1.19}$$

which correspond to the constraints of type (1.13) and (1.14).

It should be obvious by now that the stochastic processes differ from one another, and in particular, from the deterministic processes, but only by the form of probabilistic characteristics—probability density functions. The volume of a priori information for deterministic processes is usually larger than for stochastic processes since the probability density functions for deterministic processes are known in advance, while for stochastic processes they are to be determined. If the probability distribution is determined in advance, and if we can manage to write the functional and the constraint equations in an explicit form, then regardless of the basic differences between the deterministic and stochastic processes, it is difficult to establish any prominent dissimilarities in the formulation and the solution of optimization problems for these processes.

We can often obtain the optimality for deterministic processes for every individual process; for instance, this was the case of the optimal systems with minimal response time. At the same time, we can only guarantee the optimality on the average for the statistical processes, but this is related more closely to the differences in ideas than to the differences in the formulation and the solution of an optimization problem.

1.7 The Ordinary and Adaptive Approaches

A problem of stochastic optimization does not differ from a deterministic problem if the probability functions of the stochastic processes are known and the performance indices $J(\mathbf{c})$ are explicitly written. Dynamic programming is then equally applicable to both deterministic and stochastic problems.

A careful look at the results obtained during the periods of determinism and stochasticism leads to the conclusion that the approaches to the solution of optimization problems are identical. The best examples are the following: the stochastic problem of Wiener-Kolmogorov filtering, i.e., the synthesis of a linear system which is optimal in the least square error sense; and the problem of analytical design of the regulator, i.e., the synthesis of the linear systems which are optimal in the least integral square error sense. If $J(\mathbf{c})$ and its distribution are assumed to be known, the approach is called ordinary.

A substantially different situation occurs when the probability distribution is not known a priori. Here, the ordinary approach becomes ineffective and we need another approach which can permit the solution of the optimization problem under insufficient a priori information without first having to determine the probabilistic characteristics. Such an approach will be called adaptive. In the adaptive approach, the current information is actively used instead of a priori information which does not exist. This method can also be used in the cases where an application of the ordinary approach requires a large effort in an advance estimation of the probability distribution.

If it is not clear whether the process is deterministic or stochastic, it is logical to try the adaptive approach, i.e., to solve the problem by using learning and adaptation during the process of experimentation.

1.8 On Methods for Solving Optimization Problems

After an optimization problem is formulated, i.e., after the criterion of optimality is selected and the constraints of the first and the second kind are specified, one must attempt to solve the problem. Although it is frequently said that the formulation of the problem means 50–80% success (depending on the temperament of the person who claims it), the remaining percentage is often large enough to preclude any success.

The solution for the problem of optimality is reduced to finding a vector $\mathbf{c} = \mathbf{c}^*$ (it is proper to call it an optimal vector) which satisfies the constraints and for which the functional

$$J(\mathbf{c}) = M_{\mathbf{x}}\{Q(\mathbf{x}, \mathbf{c})\} \tag{1.20}$$

reaches an extremum. Note that in the majority of cases, we have to determine the functions over which certain functionals reach their extremal values.

The procedures for defining such functions are frequently followed by great difficulties. In order to circumvent these difficulties, one can replace the sought

1.8 On Methods for Solving Optimization Problems

function over which the extrema of the functional are defined by a linear combination of certain linearly independent functions with unknown coefficients. In such a case, a functional which depends on a function is replaced by a functional which depends on a vector. Certain convenient examples will be provided later.

When sufficient a priori information exists, we know both the explicit expressions for the functional $J(\mathbf{c})$ and the constraints, not only for deterministic but also for stochastic processes. The ordinary approaches can then be applied to $J(\mathbf{c})$.

The ordinary approaches are very numerous, and they encompass analytic and algorithmic methods. At first glance, the analytic methods appear as the most attractive since they lead to an explicit formal solution of the problem, but their use severely limits the area of application. These methods are convenient for the solution of simple problems which frequently could only be obtained by an extreme idealization of the actual problem. For instance, the formulas for the solution of algebraic equations of the first- and second-order equations are simple. Such formulas can also be written for the equations of the third and fourth order, although they are more complicated to use. Finally, similar formulas are simply nonexistent for the equations of higher than the fourth order. How can we now be satisfied by solving the second-order equation instead of the equations of a higher order? Various approximations of analytical methods (for instance, asymptotic approximations) extend the limits of applicability, but not very much.

Algorithmic methods, which until recently have not received great attention, do not provide an explicit formal solution of the problem, but only specify the algorithm, i.e., the sequence of actions and operations which lead to the particular sought solution. The algorithmic methods are based on the numerical solutions of various equations, and now, in connection with a broad application of digital computers, they have become dominant.

The algorithmic methods provide not only the solution, but also a way to find the solution on the basis of recursive formulas. Even in the cases when the analytic methods are applicable, it is sometimes preferable to use the algorithmic methods since they offer a faster and more convenient method of obtaining the sought result. A system of a large number of linear algebraic equations is more efficiently solved by one of many iterative methods than by using Cramer's rule. If the functional is not explicitly known, the ordinary approaches are not applicable, and in order to remove the uncertainty caused by insufficient a priori information, one should use the adaptive approach. The adaptive approach is mainly related to the algorithmic, or more accurately, iterative methods. A general discussion of algorithmic methods of optimization will be found in the next two chapters.

1.9 Conclusion

In this chapter we have attempted to give a general description of the problem of optimality, its formulation and solution. We have also tried to point out the differences, and especially the similarities, of the optimization problems for deterministic and stochastic processes. We have indicated that the approach for solving an optimization problem depends on the degree of a priori information.

When sufficient a priori information exists, one can use the ordinary approach; the adaptive approach has to be used under insufficient a priori information. As we shall see later, it is sometimes more convenient to use the adaptive approach even in cases when the ordinary approach is also applicable. Such a situation occurs when a priori information can be obtained experimentally by preprocessing the available data. The restoration (estimation) of probability density functions and correlation functions used in the solution of optimization problems can serve as an example. Is it not better in such a case to solve the problem of optimality by an adaptive approach? This would not require a priori information and thus, would considerably reduce the amount of computations.

In order to avoid excessive idealization and simplification of realistic problems, we select the algorithmic method as a basic one. This permits us to obtain efficient algorithms for the solution of optimization problems even in complex cases by using various computing devices.

COMMENTS

1.1 Much literature devoted to the problem of optimality cannot be surveyed here in a brief and clear form. For our purpose, this is not necessary. Therefore, we shall only mention several basic books. First, we mention the book by Feldbaum (1965) which presents various formulations of the problem of optimality and possible approaches to its solution.

In the books by Bellman (1957, 1961) this problem is considered on the basis of his method of dynamic programming, and in the book by Pontryagin *et al.* (1962) it is presented on the basis of their maximum principle. The reader who wishes to expand his understanding of various aspects of the problem of optimality should become familiar with the following books: Boltyanskii (1966), which very simply and clearly presents the problem of optimality; Butkovskii (1965), which is devoted to the problems of optimality in the systems with distributed parameters; and Chang (1961), which also covers the problems of optimality in linear continuous and discrete systems. Perhaps the best way to become familiar with the problem of optimal control is to study a short book by Lee (1964). All of these have rather complete bibliographies.

One should not think that the problem of optimality is simple and clear. Frequently the question "What is optimality?" can cause despair and pessimism (see the papers by Zadeh (1958) and Kalman (1964)). The author, however, does not share such pessimism completely.

1.2 Here we use the notation adopted by Gantmacher (1959). All the vectors represent column matrices and the symbol $\mathbf{c} = (c_1, \ldots, c_N)$. The transpose of the column matrix is a row matrix $\mathbf{c}^T = [c_1, \ldots, c_N]$.

The Bayesian criterion is broadly used in communication theory and radar engineering and recently in control theory (see the books by Gutkin (1961), Helstrom (1960) and Feldbaum (1965)).

In formula (1.3), instead of $d\Omega(\mathbf{x}) = dx_1 \cdot dx_2 \cdots dx_m$, which defines an elementary volume in the space X, we simply write $d\mathbf{x}$. For further information on mathematical expectation of the characteristic function, see Gnedenko (1962) or Stratonovich (1966).

1.3 Concerning ergodicity, see a small but very interesting book by Halmos (1956).

1.4 The role of deterministic constraints was clearly emphasized many times by Feldbaum (1965) (see also Lee (1964) and Bellman (1957)).

1.5 Such a characteristic of a priori and current information was described by Krasovskii (1963).

1.6 The presented classification of the deterministic and stochastic processes is almost universally accepted (Bellman, 1961). Here, we have attempted to emphasize the generality of the formulation and the solution of the problem of optimality for these processes, and thus their differences were not extensively discussed.

1.7 A slightly different treatment of the adaptive approach can be found in Bellman (1961). The treatment found here is identical to one presented earlier in a paper by the author (Tsypkin, 1966). The relationship between the stochastic problem of synthesizing a linear system which is optimal with respect to the variance of the error, and the deterministic problem of synthesizing a linear system which is optimal according to the integral square error was shown by Kalman (1961), who formulated this relationship as the principle of duality.

1.8 In addition to the books mentioned above, the methods of solving the problems of optimality related to the ordinary approach are described in the books by Leitman (1962) and Merriam (1964).

BIBLIOGRAPHY

Bellman, R. (1957). "Dynamic Programming." Princeton University Press, Princeton, N.J.
Bellman, R. (1961). "Adaptive Control Processes." Princeton University Press, Princeton, N.J.
Boltyanskii, V.G. (1966). "Mathematical Methods of Optimal Control." Nauka, Moscow. (In Russian.)
Butkovskii, A.G. (1965). "Optimal Control of the Systems with Distributed Parameters." Nauka, Moscow. (In Russian.)
Chang, S.S.L. (1961). "Synthesis of Optimum Control Systems." McGraw-Hill, New York.
Feldbaum, A.A. (1965). "Optimal Control Systems." Academic Press, New York.
Gantmacher, F.R. (1959). "The Theory of Matrices," 2 Vols. Chelsea, New York.
Gnedenko, B.V. (1962). "Theory of Probability." Chelsea, New York.
Gutkin, L.S. (1961). "Theory of Optimal Radio Receivers." Gosenergoizdat, Moscow. (In Russian.)
Halmos, P.R. (1956). "Lectures on Ergodic Theory." Tokyo.

Helstrom, C.W. (1960). "Statistical Theory of Signal Detection." Pergamon Press, New York.

Kalman, R. (1961). On the general theory of optimal control, "Proceedings of the First IFAC, Automatic and Remote Control," Vol. I, pp. 481–492. Butterworths, London.

Kalman, R. (1964). When is a linear control system optimal? *Trans. ASME* **86D**.

Krasovskii, A.A. (1963). "Dynamics of Continuous Self-Organizing Systems." Fizmatgiz, Moscow. (In Russian.)

Lee, R.C.K. (1964). "Optimal Estimation, Identification and Control." MIT Press, Cambridge, Mass.

Leitman, G.E. (ed.) (1962). "Optimization Techniques with Applications to Aerospace Systems." Academic Press, New York.

Merriam, C.W. (1964). "Optimization Theory and the Design of Feedback Control Systems." McGraw-Hill, New York.

Middleton, D. (1960). "An Introduction to Statistical Communication Theory." McGraw-Hill, New York.

Pontryagin, L.S., Boltyanskii, V.G., Gamkrelidze, R.V., and Mishchenko, E.F. (1962). "The Mathematical Theory of Optimal Processes." Wiley (Interscience), New York.

Stratonovich, R.L. (1966). "Selected Problems of Noise Theory in Radio Engineering." Sovyetskoe Radio, Moscow. (In Russian.)

Tsypkin, Ya.Z. (1966). Adaptation, learning and self-learning in automatic systems, *Automat. Remote Contr.* **27** (1), 23–61. (English transl.)

Zadeh, L.A. (1958). What is optimal?, *IRE Trans. Inform. Theory* **IT-4** (1), 3.

2

ALGORITHMIC METHODS OF OPTIMIZATION

2.1 Introduction

In this chapter we shall consider recursive algorithmic methods of solving optimization problems. These methods encompass various iterative procedures related to the application of sequential approximations. Similar methods, which were first used in the solution of algebraic equations and then extended to differential and integral equations, were also developed with the help of the ideas found in functional analysis. Our immediate goal is not only to give a systematization and a comparison of sufficiently well-developed recursive methods, but also to explain their physical meaning, or more accurately, their meaning from the viewpoint of control specialists.

It will be assumed throughout this chapter that sufficient a priori information exists. Therefore, in the solution of optimization problems, we can employ the regular approach. The presented results will be used to develop the adaptive approach by analogy. We shall see that regardless of the variety of recursive methods, all of them can be reduced to relatively simple schemes.

2.2 Conditions of Optimality

In deterministic and stochastic processes under sufficient a priori information (such a case is considered only in this chapter), the criterion of optimality, i.e., the functional $J(\mathbf{c})$, is explicitly known. The constraints are also known. In the beginning, we shall assume that the constraints of the second kind do not exist, and that the constraints of the first kind are included in the functional. The initial dimension of the sought vector $\mathbf{c}$ is thus reduced.

If the functional $J(\mathbf{c})$ is differentiable, its extremum (maximum or minimum) is reached only for those values $\mathbf{c} = (c_1, \ldots, c_N)$ for which N partial derivatives $\partial J(\mathbf{c})/\partial c_v$ ($v = 1, \ldots, N$) are simultaneously equal to zero, or, in other words, for which the gradient of the functional

$$\nabla J(\mathbf{c}) = \left(\frac{\partial J(\mathbf{c})}{\partial c_1}, \ldots, \frac{\partial J(\mathbf{c})}{\partial c_N}\right) \tag{2.1}$$

is equal to zero.

The vectors $\mathbf{c}$ for which

$$\nabla J(\mathbf{c}) = 0 \tag{2.2}$$

are called the stationary or singular vectors, but all stationary vectors are not optimal, i.e., they do not correspond to the desired extremum of the functional. Therefore, condition (2.2) is only a necessary condition.

We could also write the sufficient conditions of the extremum in the form of an inequality based on the determinant which contains partial derivatives of the second order of the functional with respect to all components of the vector. However, this is not worth doing even in the case when the required computational effort is not very large.

Frequently, starting directly from the physical conditions of the problem for which the functional is defined, we can find whether the stationary vector corresponds to the minimum or to the maximum. This can be established easily in the frequent cases of our interest when there is only one extremum.

The conditions of optimality only define local extrema, and if the number of such extrema is large, the problem of finding the absolute or the global extremum becomes a very complex one. Certain possible solutions of this problem will be evaluated later.

We shall now be concerned only with the case when the optimal value of the vector $\mathbf{c}^*$ is unique. In order to be more specific, we shall also consider that the extremum of the functional is the minimum.

2.3 Regular Iterative Method

The condition of optimality (2.2) in the general case represents a set of nonlinear equations, and an analytic solution of such equations is almost never possible. However, in the case of quadratic performance indices and linear constraints of the first kind, the nonlinear equation (2.2) becomes linear, and we can use Cramer's rule to solve it. These beautiful mathematical ideas and not practical problems were, until recently, responsible for the development of the theory of optimality.

If we also mention that the linear analytic methods are applicable only for simple problems of small dimensionality, the need for developing and applying the algorithmic methods (or more accurately, iterative methods which do not require such stringent constraints) becomes obvious.

The basic idea of solving equation (2.2) using regular iterative methods consists of the following: Let us write equation (2.2) in the form

$$\mathbf{c} = \mathbf{c} - \gamma \nabla J(\mathbf{c}) \tag{2.3}$$

where γ is a certain scalar, and then seek the optimal vector $\mathbf{c} = \mathbf{c}^*$ using the method of sequential approximations or iterations.

$$\mathbf{c}[n] = \mathbf{c}[n-1] - \gamma[n] \nabla J(\mathbf{c}[n-1]) \tag{2.4}$$

The value $\gamma[n]$ defines the length of the step, and it depends on the vectors $\mathbf{c}[m]$ ($m = n-1, n-2, \ldots$). When the corresponding conditions of convergence are satisfied (this will be briefly discussed later),

$$\lim_{n \to \infty} \mathbf{c}[n] = \mathbf{c}^* \tag{2.5}$$

for any initial condition $\mathbf{c} = \mathbf{c}[0]$.

The methods of obtaining $\mathbf{c}^*$ are based on relationship (2.4) and are called the iterative methods. Since the choice of the initial vector $\mathbf{c}[0]$ uniquely defines the future values of the sequence $\mathbf{c}[n]$, these iterative algorithms will be called regular. The probabilistic algorithms, which differ from the regular algorithms, will be discussed in Chapter 3. Different forms of the regular iterative methods differ according to the choice of $\gamma[n]$.

There is an enormous number of publications devoted to regular algorithms. Unfortunately, many of them use different terminology. Here we have to find the optimal value of the vector, since we are interested in the problem of optimality. Therefore, we shall use the terminology most closely related to the problem under consideration and the corresponding algorithms of optimizations.

2.4 Algorithms of Optimization

Relationship (2.4) defines the sequence of actions which have to be performed in order to determine the vector $\mathbf{c}^*$. Therefore, it is appropriate to call (2.4) an algorithm of optimization. This algorithm of optimization can be observed as a recursive equation. By introducing the difference of the first order

$$\Delta \mathbf{c}[n-1] = \mathbf{c}[n] - \mathbf{c}[n-1] \tag{2.6}$$

2 Algorithmic Methods of Optimization

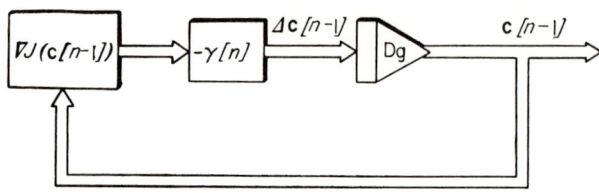

Fig. 2.1

we can easily write the algorithm of optimization in the form of a difference equation

$$\Delta \mathbf{c}[n-1] = -\gamma[n]\nabla J(\mathbf{c}[n-1]) \tag{2.7}$$

Finally, by summing both parts of this equation from 0 to n, we obtain the algorithm of optimization in the form of the sum equation

$$\mathbf{c}[n] = \mathbf{c}[0] - \sum_{m=1}^{n} \gamma[m]\nabla J(\mathbf{c}[m-1]) \tag{2.8}$$

which includes the initial value $\mathbf{c}[0]$. Therefore, the algorithms of optimization can be represented in three forms: recursive, difference and sum.

Recursive, difference and sum equations correspond to certain discrete feedback systems, which are shown in Fig. 2.1. The block diagrams include a nonlinear transformer $\nabla J(\mathbf{c})$, an amplifier with time-varying gain $\gamma[n]$ and a discrete integrator–digrator (designated by Dg in Figs. 2.1 and 2.2). The delay line is designated by TD in Fig. 2.2b. The output of the digital integrator (digrator) is always $c_v[n-1]$ (Fig. 2.2). Double lines in Fig. 2.1 indicate vector relationships. This discrete feedback system is autonomous. All necessary a priori information is already present in the nonlinear transformer.

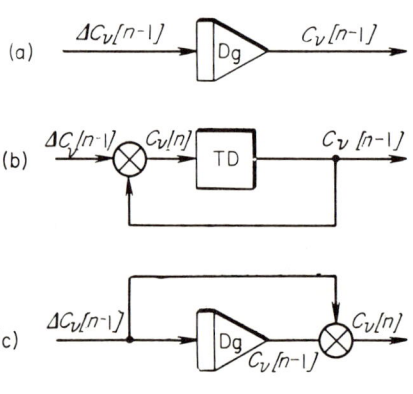

Fig. 2.2

2.5 A Possible Generalization

When $J(\mathbf{c}) = $ const has "ridges" (Fig. 2.3), the rate of convergence to the optimal point $\mathbf{c}^*$ is slow. In such cases, instead of the scalar, it is better to use the matrix

$$\Gamma[n] = \|\gamma_{\nu\mu}[n]\| \qquad (\nu, \mu = 1, \ldots, N) \quad (2.9)$$

in the algorithm of optimization.

Algorithm (2.7) is then replaced by a more general algorithm,

$$\Delta\mathbf{c}[n-1] = -\Gamma[n]\nabla J(\mathbf{c}[n-1]) \qquad (2.10)$$

In this case, the number of steps related to various components are different and mutually dependent.

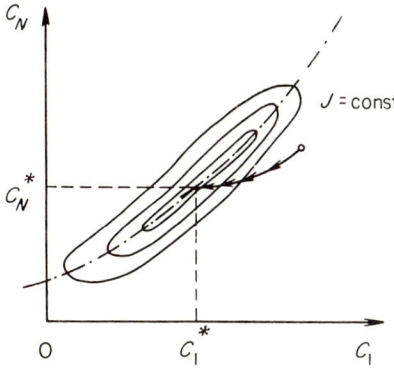

Fig. 2.3

The matrix $\Gamma[n]$ can be chosen so that in the neighborhood of the optimum point the trajectory of the parameters can be a certain straight line, i.e.,

$$\begin{aligned}\Delta J(\mathbf{c}[n]) &= -kJ(\mathbf{c}[n-1]) \qquad (k = \text{const})\\ \Delta J(\mathbf{c}[n]) &= J(\mathbf{c}[n+1]) - J(\mathbf{c}[n])\end{aligned} \quad (2.11)$$

The matrix $\Gamma[n]$ provides a linear transformation of the coordinate system under which the equidistant lines of $J(\mathbf{c})$ become concentric spheres (Fig. 2.4).

The block diagram which corresponds to the general algorithm of optimization differs from the one shown in Fig. 2.1. Instead of simple amplifiers, it has "matrix" amplifiers where all the inputs and outputs are interconnected.

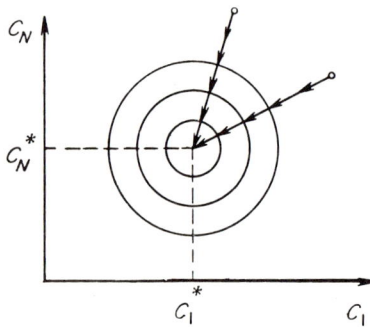

Fig. 2.4

2.6 Various Algorithms of Optimization

The selected gain coefficients of matrix or ordinary amplifiers define the type of the algorithms of optimization and their corresponding discrete systems. For instance, if $\Gamma[n] = \Gamma$, which does not depend on n, we obtain the algorithm of adaptation with constant steps and the corresponding discrete system with constant gain coefficient. If $\Gamma[n]$ depends on n, we obtain the algorithm with variable steps, and the corresponding discrete system with variable gain coefficients. In particular, the matrix $\Gamma[n]$ can be periodic, $\Gamma[n + n_0] = \Gamma[n]$. In the numerical methods of solving the system of linear equations, the algorithms discussed above are called, respectively, stationary, nonstationary and cyclic.

In general, $\Gamma[n]$ can depend on the vectors $c[m]$ ($m = n - 1, n - 2, \ldots$). In this case we have algorithms with "nonlinear" steps and the corresponding discrete systems with nonlinear gain coefficients. The relaxation algorithms, in which $\Gamma[n]$ is chosen to minimize a certain function of the error $c[n] - c^*$ at each step n, belongs to this type of algorithm.

In the coordinate relaxation algorithms, $\Gamma[n]$ can be such that only one or several components of $c[n]$ are changing at each step. In the gradient relaxation algorithms,

$$\Gamma[n] = I\gamma[n] \qquad (2.12)$$

where I is an identity matrix and $\gamma[n]$ is a scalar which also depends on the coordinates of the vector c. For instance, the well-known Newton's algorithm

$$\Delta c[n - 1] = -[\nabla^2 J(c[n - 1])]^{-1} \nabla J(c[n - 1]) \qquad (2.13)$$

belongs to the class of algorithms with nonlinear steps. Here,

$$\Gamma[n] = [\nabla^2 J(c[n - 1])]^{-1} \qquad (2.14)$$

A modification of Newton's algorithm,

$$\Delta \mathbf{c}[n-1] = -[\nabla^2 J(\mathbf{c}[0])]^{-1} \nabla J(\mathbf{c}[n-1]) \tag{2.15}$$

where $\mathbf{c}[0]$ is a certain initial value, represents an algorithm with constant steps.

The algorithm of steepest descent is

$$\Delta \mathbf{c}[n-1] = -\gamma[n] \nabla J(\mathbf{c}[n-1]) \tag{2.16}$$

which is also a relaxation algorithm. Here, $\gamma[n]$ is selected at each step by minimizing the function

$$\Psi(\gamma) = J(\mathbf{c}[n-1] - \gamma \nabla J(\mathbf{c}[n-1])) \tag{2.17}$$

Therefore, by selecting the corresponding $\Gamma[n]$ or $\gamma[n]$, we obtain various well-known algorithms.

2.7 Search Algorithms of Optimization

It is not always possible to compute the gradient of the functional (2.1) in an explicit form, and thus we cannot use the algorithms discussed above. Such a situation exists when the functional $J(\mathbf{c})$ is discontinuous or nondifferentiable, or when its dependence on $\mathbf{c}$ cannot be expressed explicitly. This last case is characteristic for the functionals of the type (1.5) and (1.8), and for the functionals formed by logical operations.

The only possible solution of the optimization problem under such conditions is probably related to the search methods of finding the extremum. If we cannot compute the gradient in advance, we must estimate it by measurements. In the search methods, the quantities which can be used for indirect estimation of the gradient are measured. There is a large variety of search methods which were mainly developed in the design of various optimum control systems. We shall not attempt to describe all the methods of search here. Only the simplest classical methods will be considered in order to clarify certain basic questions. After such explanations, the reader can easily master different methods of search which are not considered here.

Let us introduce the vectors

$$\begin{aligned} J_+(\mathbf{c}, a) &= (J(\mathbf{c} + a\mathbf{e}_1), \ldots, J(\mathbf{c} + a\mathbf{e}_N)) \\ J_-(\mathbf{c}, a) &= (J(\mathbf{c} - a\mathbf{e}_1), \ldots, J(\mathbf{c} - a\mathbf{e}_N)) \end{aligned} \tag{2.18}$$

when a is a scalar, and $\mathbf{e}_\nu$ ($\nu = 1, \ldots, N$) represents the unit vectors. In the simplest case,

$$\mathbf{e}_1 = (1, 0, \ldots, 0); \ \mathbf{e}_2 = (0, 1, \ldots, 0); \ \ldots; \ \mathbf{e}_N = (0, 0, \ldots, 1) \tag{2.19}$$

2 Algorithmic Methods of Optimization

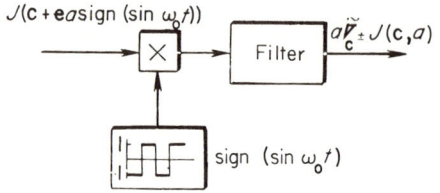

Fig. 2.5

The gradient is then approximately defined by the formula

$$\frac{J_+(\mathbf{c}, a) - J_-(\mathbf{c}, a)}{2a} = \tilde{\nabla}_{\mathbf{c}\pm} J(\mathbf{c}, a) \tag{2.20}$$

Using (2.20) instead of the gradient in (2.4), we obtain the search algorithm of optimization

$$\mathbf{c}[n] = \mathbf{c}[n-1] - \gamma[n]\tilde{\nabla}_{\mathbf{c}\pm} J(\mathbf{c}[n-1], a[n]) \tag{2.21}$$

This algorithm of optimization, which can also be written in the form of a difference or a sum, corresponds to discrete extremal systems. The estimate of the gradient (2.20) can be obtained by using a synchronous detector (Fig. 2.5) if the search oscillations are square-waves of sufficiently high frequency and an amplitude $y(t) = a \, \text{sign}(\sin \omega_0 t)$.

Actually, it is simple to verify that

$$\tilde{\nabla}_{\mathbf{c}\pm} J(\mathbf{c}, a) = \frac{\omega_0}{2\pi a} \int_0^{2\pi/\omega_0} J(\mathbf{c}, a \, \text{sign}(\sin \omega_0 t)) \, \text{sign}(\sin \omega_0 t) \, dt \tag{2.22}$$

Therefore, the block diagram of the extremal discrete system can be given in the form shown in Fig. 2.6. Usually, $a[n] = \text{const}$. This scheme differs from the ordinary (nonsearch) system (Fig. 2.1). Here we have an additional generator of search oscillations, a synchronous detector and the commutators which sequentially define the arguments $\mathbf{c} \pm a\mathbf{e}_k$ and the components $\tilde{\nabla}_{\mathbf{c}\pm} J(\mathbf{c}, a)$.

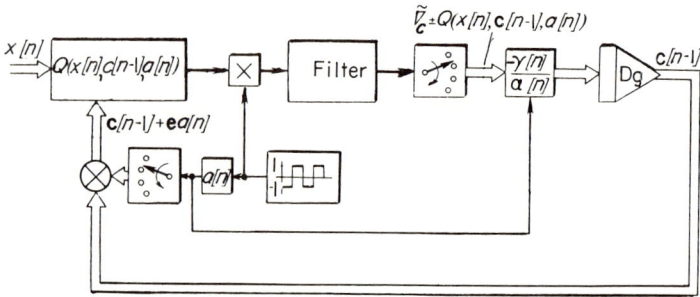

Fig. 2.6

The algorithms of optimization can also be based upon various types of searches which are extensively used in extremal systems. In particular, it may sometimes be more convenient not to vary $\mathbf{c}$ in both directions ($\mathbf{c} + a\mathbf{e}_k$ and $\mathbf{c} - a\mathbf{e}_k$), but in one direction only (either $\mathbf{c} + a\mathbf{e}_k$ or $\mathbf{c} - a\mathbf{e}_k$). In this case, instead of (2.20), we can use either

$$\tilde{\nabla}_{\mathbf{c}+} J(\mathbf{c}, a) = \frac{\mathbf{J}_+(\mathbf{c}, a) - \mathbf{J}_0(\mathbf{c})}{a}$$

or (2.23)

$$\tilde{\nabla}_{\mathbf{c}-} J(\mathbf{c}, a) = \frac{\mathbf{J}_0(\mathbf{c}) - \mathbf{J}_-(\mathbf{c}, a)}{a}$$

where $\mathbf{J}_0(\mathbf{c}) = (J(\mathbf{c}), \ldots, J(\mathbf{c}))$ is a vector with identical components. In computing $\tilde{\nabla}_{\mathbf{c}+} J(\mathbf{c}, a)$ or $\tilde{\nabla}_{\mathbf{c}-} J(\mathbf{c}, a)$, we perform one step only.

2.8 Constraints I

It is not difficult to determine the optimal vector $\mathbf{c}^*$ for the functional $J(\mathbf{c})$ when the constraints of the equality type (1.15) only are present. Using the method of Lagrange multipliers, we obtain the problem considered earlier. Let us form the functional

$$J(\mathbf{c}, \lambda) = J(\mathbf{c}) + \lambda^T \mathbf{g}(\mathbf{c}) \qquad (2.24)$$

where $\lambda = (\lambda_1, \ldots, \lambda_M)$ is still an unknown vector of Lagrange multipliers; T indicates the transport of a vector, and $\mathbf{g}(\mathbf{c}) = (g_1(\mathbf{c}), \ldots, g_M(\mathbf{c}))$ is a vector function.

The minimum of the functional $J(\mathbf{c})$ under the constraints (1.15) is found from the following system of equations:

$$\begin{aligned} \nabla_{\mathbf{c}} J(\mathbf{c}, \lambda) &= \nabla J(\mathbf{c}) + G(\mathbf{c})\lambda = 0 \\ \nabla_{\lambda} J(\mathbf{c}, \lambda) &= \mathbf{g}(\mathbf{c}) = 0 \end{aligned} \qquad (2.25)$$

where

$$G(\mathbf{c}) = \left\| \frac{\partial g_\nu(\mathbf{c})}{\delta c_\mu} \right\| \quad (\nu = 1, \ldots, M; \mu = 1, \ldots, N) \quad (2.26)$$

is an $N \times M$ matrix.

In analogy to the algorithm (2.4) for solving the system of equations (2.2), we can solve the system (2.25) by using the algorithms

$$\begin{aligned} \mathbf{c}[n] &= \mathbf{c}[n-1] - \gamma[n] \nabla_{\mathbf{c}} J(\mathbf{c}[n-1], \lambda[n-1]) \\ \lambda[n] &= \lambda[n-1] + \gamma_1[n] \nabla_{\lambda} J(\mathbf{c}[n-1], \lambda[n-1]) \end{aligned} \qquad (2.27)$$

or

$$\begin{aligned}\mathbf{c}[n] &= \mathbf{c}[n-1] - \gamma[n]\nabla J(\mathbf{c}[n-1] + G(\mathbf{c}[n-1]\lambda[n-1]) \\ \lambda[n] &= \lambda[n-1] + \gamma_1[n]\mathbf{g}(\mathbf{c}[n-1])\end{aligned} \quad (2.28)$$

The existence of equality constraints slightly complicates the structure of the system which corresponds to the algorithms of optimization. The special loops which define the Lagrange multipliers are introduced (Fig. 2.7). Many

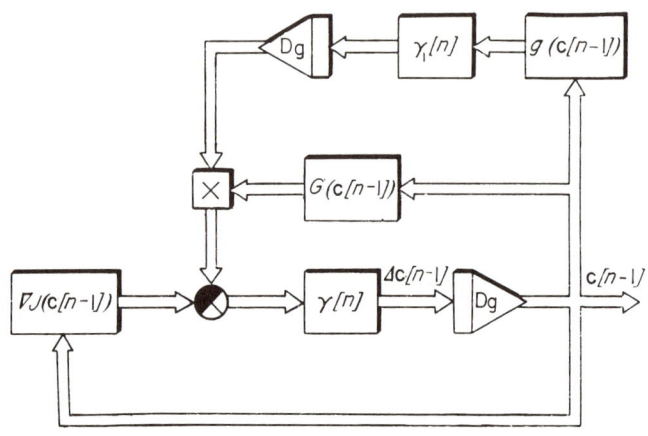

Fig. 2.7

other algorithms of optimization which differ only according to the chosen Lagrange multipliers exist, but we shall not treat them here.

2.9 Constraints II

The inequality constraints cannot be treated by the classical methods discussed thus far. In order to consider constraints of this type, we have to rely on a new mathematical apparatus—mathematical programming—which has been developed recently. The conditions of optimality are given in this case by the Kuhn-Tucker theorem, which represents a generalization of the Lagrange method for the case of inequality constraints. This theorem states that the optimal vector $\mathbf{c}^*$ satisfies the following conditions:

$$\begin{aligned}\nabla_c J(\mathbf{c}, \lambda) &= \nabla J(\mathbf{c}) + G(\mathbf{c})\lambda = 0 \\ \mathbf{g}(\mathbf{c}) + \delta &= 0 \\ \lambda^T \delta &= 0 \\ \lambda &\geq 0 \\ \delta &\geq 0\end{aligned} \quad (2.29)$$

2.9 Constraints II

The conditions are written in the vector form, $\lambda = (\lambda_1, \ldots, \lambda_{M_1})$, $\delta = (\delta_1, \ldots, \delta_{M_1})$; the inequalites $\lambda \geq 0$ and $\delta \geq 0$ indicate that all the components of these vectors are nonnegative. Moreover, it is assumed that the constraints (1.14) are such that there exists a vector $\mathbf{c}$ for which

$$g_\nu(\mathbf{c}) < 0 \qquad (\nu = 1, \ldots, M_1) \qquad (2.30)$$

This is Slater's regularity condition which is well-known in the theory of nonlinear programming. The conditions (2.29) have a simple meaning: if a certain constraint is not essential for the optimal vector $\mathbf{c}^*$, i.e., if $g_\nu(\mathbf{c}^*) < 0$ for a certain ν, the corresponding $\lambda_\nu = 0$; if $\lambda_\nu > 0$, then $\delta_\nu = g_\nu(\mathbf{c}^*) = 0$. Therefore, the Lagrange multipliers can be interpreted as certain estimates of the constraints (1.14) on the optimal value of the vector. It should be noticed that the Kuhn-Tucker theorem provides the necessary and sufficient conditions of optimality if the functionals $J(\mathbf{c})$ and $g_\nu(\mathbf{c})$ ($\nu = 1, \ldots, M_1$) are convex.

By applying the algorithms of optimization to (2.29), it is not difficult to obtain the following relationship:

$$\begin{aligned}
\mathbf{c}[n] &= \mathbf{c}[n-1] - \gamma[n][\nabla J(\mathbf{c}[n-1]) + G(\mathbf{c}[n-1])\lambda[n-1]] \\
\lambda[n] &= \max\{0, \lambda[n-1] + \gamma_1[n]\mathbf{g}(\mathbf{c}[n-1])\} \\
\lambda[0] &\geq 0
\end{aligned} \qquad (2.31)$$

The block diagram of the system which corresponds to this algorithm is shown in Fig. 2.8. The half-wave rectifier in this scheme is introduced to realize the inequality constraints.

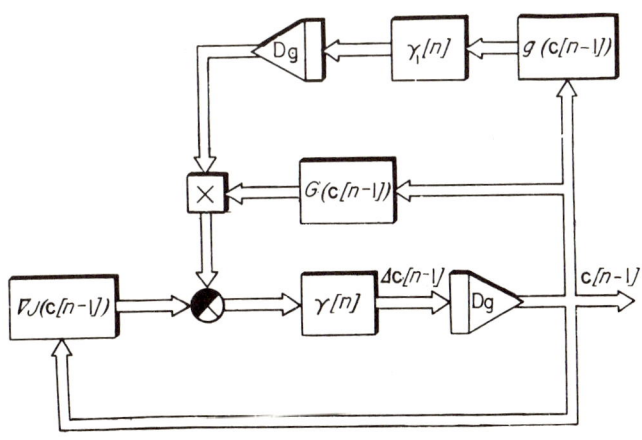

Fig. 2.8

2.10 The Method of Feasible Directions

In the preceding sections we have considered the algorithms which can be used to determine the vector **c*** under the additional inequality or equality constraints. The optimal vector **c*** and the optimal Lagrange multiplier λ* are found by these methods. We can also use the idea of motion along the gradient in the solution of these problems. These methods are called the methods of feasible directions. Their essence consists in the following.

An arbitrary point **c**$[n-1]$ is selected so that the constraint (1.14) is satisfied. For this point, we define a direction **z**$[n-1]$ and a step of length $\gamma[n]$ which makes the value of the functional smaller, but does not lead outside of the permissible region. The length of the step $\gamma[n]$ is then determined, and thus a new value of the vector **c**$[n]$ is found:

$$\mathbf{c}[n] = \mathbf{c}[n-1] - \gamma[n]\mathbf{z}[n-1] \tag{2.32}$$

The value of the functional at the new point **c**$[n]$ must be smaller than at the preceding point:

$$J(\mathbf{c}[n]) < J(\mathbf{c}[n-1]) \tag{2.33}$$

Therefore, at each step the problem of determining the new value of the vector **c**$[n]$ consists of two stages: first, the selection of the direction, and second, the length of the step in that direction.

In order to satisfy the inequality (2.33), the vector **z** has to form an oblique angle with the gradient of the functional at that point, i.e.,

$$\mathbf{z}^T[n-1]\nabla J(\mathbf{c}[n-1]) > 0 \tag{2.34}$$

A direction which satisfies (2.34) is called a feasible direction, which explains the name of the methods.

The length of the step $\gamma[n]$ can be determined as in the case of the steepest descent method, i.e.,

$$J(\mathbf{c}[n-1] - \gamma[n]\mathbf{z}[n-1]) = \min_{\gamma} J \tag{2.35}$$

Of course, we must not violate the constraints (1.14).

For some special cases, the methods of feasible directions can define the extremum in a finite number of steps. It should be noticed that many effective algorithms of mathematical programming (for instance, the simplex method in linear programming) can be treated as special cases of the methods of feasible directions.

2.11 Discussion

The algorithmic methods discussed thus far have a very simple everyday character. It is convenient to describe them by discussing the behavior of a speleologist who explores a desert and tries to find its lowest point. He can examine or follow the local character only in the vicinity of his present position. How should he behave under these conditions? Obviously, he has to determine the direction of the steepest descent (i.e., the direction of the gradient). Then, he has to move in that direction as long as it continues to descend. The points of rest are actually the local minima. If the equality constraints, i.e., narrow passages, exist, the tpeleologist has to follow them until he reaches the lowest point. If the inequality constraints exist, i.e., when the speleologist encounters a rock, he has to move around it to the place where all directions lead to the top.

Such behavior of the speleologist illustrates the basic idea of the gradient methods which is expressed in the algorithm of optimization. It must be emphasized that these methods have a local character. The speleologist who finds a certain low place in the desert cannot be certain that a lower place does not exist close by.

2.12 Multistage Algorithms of Optimization

All the algorithms of optimization considered thus far are related to single-stage algorithms. They are described by a vector difference equation of the first order, and therefore they can be called the algorithms of the first order.

If the functional $J(\mathbf{c})$ has several extrema, the single-stage algorithms can determine only local extrema. In order to determine global extrema, we can try multistage algorithms of optimization:

$$\mathbf{c}[n] = \sum_{m=1}^{s} \alpha_m \mathbf{c}[n-m] - \sum_{m=1}^{s_1} \Gamma_m[n] \nabla J(\mathbf{c}[n-m]) \tag{2.36}$$

or

$$\mathbf{c}[n] = \sum_{m=1}^{s} \alpha_m \mathbf{c}[n-m] - \Gamma[n] \nabla J\left(\sum_{m=1}^{s_1} \bar{\alpha}_m \mathbf{c}[n-m]\right) \tag{2.37}$$

Algorithms (2.36) and (2.37) are closely related to each other; one is obtained from the other by a simple substitution of variables. In these algorithms, the coefficients α_m and $\bar{\alpha}_m$ are not arbitrary but satisfy the conditions

$$\sum_{m=1}^{s} \alpha_m = 1 \qquad \sum_{m=1}^{s_1} \bar{\alpha}_m = 1 \tag{2.38}$$

which stem from the condition that $\nabla J(\mathbf{c}^*) = 0$ when $\mathbf{c}[m]$ is substituted by $\mathbf{c}^*$. The structural schemes which realize algorithms (2.36) and (2.37) are shown in Figs. 2.9 and 2.10.

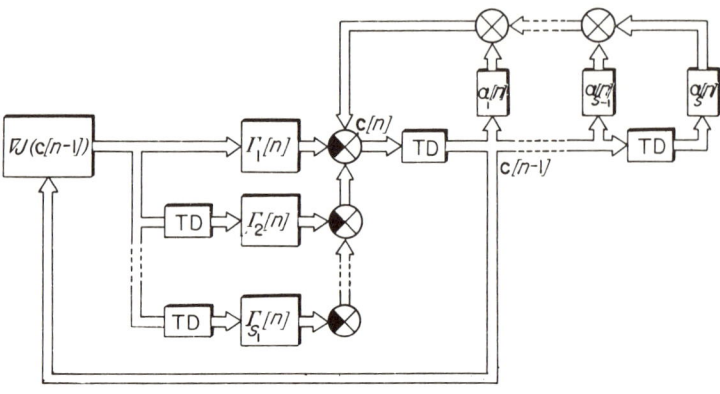

Fig. 2.9

Multistage algorithms can be given not only in recursive form but also in the difference form. In order to accomplish this, it is sufficient to use the Gregory-Newton formula which is known in the theory of finite series:

$$\mathbf{c}[n-m] = \sum_{v=0}^{m} (-1)^v \binom{m}{v} \Delta^v \mathbf{c}[n-v] \qquad (2.39)$$

where it is assumed that $\Delta^0 \mathbf{c}[n] = \mathbf{c}[n]$. After a substitution of (2.39) into (2.36) and (2.37), we obtain

$$\sum_{m=0}^{s} \beta_m \Delta^m \mathbf{c}[n-m] - \sum_{m=1}^{s_1} \Gamma_m[n] \nabla J \left(\sum_{v=1}^{m} \binom{m}{v} \Delta^v \mathbf{c}[n-v] \right) = 0 \qquad (2.40)$$

and

$$\sum_{m=1}^{s} \beta_m \Delta^m \mathbf{c}[n-m] - \Gamma[n] \nabla J \left(\sum_{m=1}^{s_1} \bar{\beta}_m \Delta^m \mathbf{c}[n-m] \right) = 0 \qquad (2.41)$$

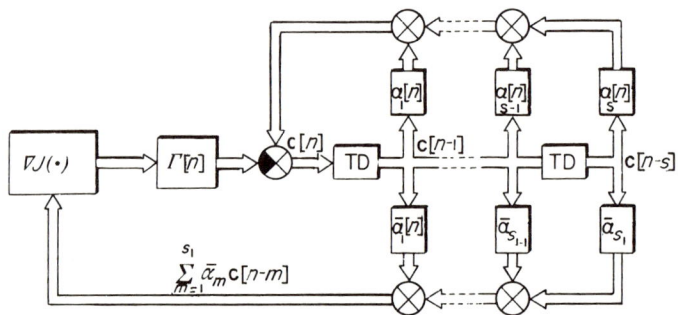

Fig. 2.10

Obviously, there is a definite relationship between the coefficients of the multistage algorithms in recursive form, i.e., α_m and $\bar{\alpha}_m$, and in difference form, i.e., β_m and $\bar{\beta}_m$. We shall not write such a relationship, since we will not be using it.

We can obtain the single-stage algorithms from the multistage algorithms for $s = s_1 = 1$. If in (2.36) and (2.37) or (2.40) and (2.41) we replace $\nabla J(\mathbf{c})$ by the estimates $\tilde{\nabla}_{\mathbf{c}\pm} J(\mathbf{c})$, defined in (2.20), or $\tilde{\nabla}_{\mathbf{c}+} J(\mathbf{c})$, we obtain the corresponding multistage search algorithms.

For a certain choice of the coefficients $\alpha_m = \alpha_m[n]$ and the matrices $\Gamma_m[n]$, the algorithms are not very sensitive on local extrema, and they also can be accelerated when there is only a single extremum $\mathbf{c}^*$ of the functional $J(\mathbf{c})$. This is due to the memory for the past values of the vectors $\mathbf{c}[n-m]$ and the gradients $\nabla J(\mathbf{c}[n-m])$, i.e., to a better extrapolation and filtering than in the case of single-stage algorithms.

A physical interpretation of such properties of the multistage algorithms of optimization will be given in the next section. Unfortunately, we still do not have a general method for selecting the coefficients α_m and $\Gamma_m[n]$.

2.13 Continuous Algorithms of Optimization

The continuous algorithms of optimization can be obtained by a limiting process from the difference equations describing the corresponding discrete algorithms of optimization discussed thus far. For instance, from (2.7) and (2.41) with $s_1 = 1$, we obtain the continuous algorithms of optimization after substituting the continuous time t for n, and the derivatives for the differences. The single-stage algorithm is then defined by the equation

$$\frac{d\mathbf{c}(t)}{dt} = -\Gamma(t)\nabla J(\mathbf{c}(t)) \tag{2.42}$$

and the multistage algorithm by

$$\sum_{m=0}^{s} \beta_m \frac{d^{s-m}\mathbf{c}(t)}{dt^{s-m}} = -\Gamma(t)\nabla J(\mathbf{c}(t)) \tag{2.43}$$

Due to their nature, the continuous algorithms cannot be written in the recursive form. Only the differential and the integral forms can exist. We shall not write the integral forms which correspond to the differential forms (2.42) and (2.43).

Since the concept of a step for the continuous algorithms is meaningless, it is better to call them the algorithms of the first and the higher orders than the single-stage and multistage algorithms. The continuous algorithms can be realized on analog computers, and can be used in the solution of finite (algebraic, transcendental) equations. If the conditions of optimality are considered as a system of finite equations, then many methods of solving finite equations can be considered as continuous algorithms of optimization.

In order to explain the properties of the algorithms of the higher order and give them a certain physical meaning, we consider the algorithm of optimization of the second order which is obtained from (2.43) for $s = 2$:

$$\beta_0 \frac{d^2\mathbf{c}(t)}{dt^2} + \beta_1 \frac{d\mathbf{c}(t)}{dt} + \Gamma(t)\nabla J(\mathbf{c}(t)) = 0 \qquad (2.44)$$

If $\beta_0 = 0$ and $\beta_1 = 1$, we have an algorithm of the first order (2.42).

Equation (2.44) describes the motion of a body ("heavy ball") with mass β_0, the coefficient of friction β_1, and the variable coefficient of elasticity $\Gamma(t)$ in the potential field. By choosing the proper parameters (β_0, β_1), the "heavy ball" can on the one hand, jump over small local minima, and on the other hand, reach the global minimum faster. Of course, this conclusion is also valid for discrete multistage algorithms of optimization, but the physical interpretation is slightly different. The reader will probably not have any difficulties in obtaining the corresponding continuous search algorithms of adaptation.

2.14 Methods of Random Search

In all regular iterative methods of search for the minima of $J(\mathbf{c})$, the current estimate $\mathbf{c}[n]$ of the parameter $\mathbf{c}$ is based on a step which is uniquely determined either by the value of the gradient $\nabla J(\mathbf{c})$ for $\mathbf{c} = \mathbf{c}[n]$, or by the value of the function $J(\mathbf{c})$ alone.

In the methods of random search, a transition from $\mathbf{c}[n-1]$ to $\mathbf{c}[n]$ is based on a random step $\gamma \xi$, where ξ is the unit random vector which is uniformly distributed in an n-dimensional unit sphere and γ is the length of the step. In this case

$$\begin{aligned} \mathbf{c}[n] &= \mathbf{c}[n-1] - \gamma \xi[n] & \text{if} & \quad J(\mathbf{c}[n-1] - \gamma \xi[n]) < J(\mathbf{c}[n-1]) \\ \mathbf{c}[n] &= \mathbf{c}[n-1] & \text{if} & \quad J(\mathbf{c}[n-1] - \gamma \xi[n]) \geq J(\mathbf{c}[n-1]) \end{aligned} \qquad (2.45)$$

There are many modifications of algorithm (2.45). In one algorithm the random step is performed only when $J(\mathbf{c}[n]) < J(\mathbf{c}[n-1])$. Otherwise, the system remains in the same state, $\mathbf{c}[n] = \mathbf{c}[n-1]$. In another algorithm, the random step may be performed only when the preceding step was not successful, etc. Finally, if in the regular gradient algorithm the quantity γ is random, we also obtain an algorithm of random search,

$$\mathbf{c}[n] = \mathbf{c}[n-1] - \gamma[n]\, \nabla J(\mathbf{c}[n-1]) \qquad (2.46)$$

where $\gamma[n]$ is a realization of a random variable γ.

2.15 Convergence and Stability

It should be mentioned that a random step which does not depend on the absolute value of the gradient, or of the function, can help in evading certain shallow local extrema of the function.

2.15 Convergence and Stability

The algorithms of optimization can be realized only if they converge, i.e., if $c[n]$ in discrete algorithms and $c(t)$ in the continuous algorithms converge to the optimal value of the vector c^*. Therefore, it is very important to establish the conditions of convergence. Only under the conditions of guaranteed convergence can we consider applying the algorithms of optimization. Since each algorithm of optimization has its corresponding autonomous feedback system, the convergence of the algorithm, and thus its realizability, is equivalent to the stability of that autonomous system. During investigations of stability, we can use the methods which are sufficiently developed in mechanics and in the theory of automatic control.

We shall now present certain possibilities for investigating the stability of closed-loop discrete systems of special structure, and thus the convergence of the algorithms of optimization. First of all, we shall use an approach which is analogous to the one used in the theory of nonlinear systems, and which can be considered as a discrete analog of Lyapunov's method.

Let us form the variational equation. We write

$$c[n] = c^* + \eta[n] \qquad (2.47)$$

where $\eta[n]$ is the deviation from the optimal vector. By substituting this value into the recurrent form of the algorithm of optimization (2.4), it is simple to obtain

$$\eta[n] = \eta[n-1] - \Gamma[n]\, \nabla J(c^* + \eta[n-1]) \qquad (2.48)$$

This difference equation has a trivial solution $\eta = 0$, since by the definition of c^*, we have $\nabla J(c^*) = 0$. The stability of the trivial solution then corresponds to the convergence of the algorithm of optimization (2.4). As it is known, two types of stability are distinguished when all the coordinates of the vector $\eta[n]$ are small. One is the local stability, and the other is global stability (for any $\eta[n]$). In order to investigate the local stability, the gradient $\nabla J(c^* + \eta)$ must first be approximated by a linear function and the obtained linear difference equation is then tested for stability. When $\Gamma[n] = \text{const}$, i.e., in the stationary case, this problem is reduced to one of finding the conditions under which the roots of the characteristic equation lie within the unit circle. Since the linear approximation is only valid for sufficiently small values of $\eta[n]$, the "stability in small" corresponds to the convergence of the algorithms of optimization when the initial vector $\eta[0]$ belongs to a certain small sphere with an unknown center. This case is neither useful nor interesting. It is much more interesting to explore the stability for any initial conditions $\eta[0]$, i.e., the "stability in large." This type of stability can be studied by a discrete analog of Lyapunov's second method.

Let us choose the norm of the vector $\eta[0]$ to be a Lyapunov function:

$$\mathscr{V}(\eta[n-1]) = \|\eta[n-1]\| \geq 0 \qquad (2.49)$$

According to (2.48), the first difference is

$$\Delta\mathscr{V}(\eta[n-1]) = \mathscr{V}(\eta[n]) - \mathscr{V}(\eta[n-1])$$
$$= \|\eta[n-1] - \Gamma[n]\,\nabla J(\mathbf{c}^* + \eta[n-1])\| - \|\eta[n-1]\| \qquad (2.50)$$

The condition of the stability in large requires the first difference to be negative. After some simple transformations, we obtain

$$\|I - \Gamma[n]\Phi(\theta\mathbf{c}^*)\| < 1 \qquad (2.51)$$

where I is a unit matrix, and

$$\Phi(\theta\mathbf{c}^*) = \left\|\frac{\partial^2 J(\theta\mathbf{c}^*)}{\partial c_\nu\,\partial c_\mu}\right\| \qquad (\nu, \mu = 1, \ldots, N),\ 0 \leq \theta \leq 1 \qquad (2.52)$$

Here we have used the principle of contraction mapping which becomes identical to the direct method of Lyapunov when a stationary point is in the origin of the coordinate system.

The principle of contraction mapping can also be applied directly to the algorithm of optimization in the recursive form (2.4) for which the stationary point corresponds to the value of the optimal vector. Of course, we obtain the same results, but perhaps in a slightly different form. Generally speaking, the conditions obtained in this fashion are only sufficient.

2.16 The Conditions of Convergence

Let us decipher condition (2.51) for the simplest case of the stationary algorithm of optimization when $\Gamma[n] = I\gamma$. We learn that it is equivalent to the following conditions:

(a) $\gamma < \gamma_0 = \text{const}$
(b) $\eta^T\,\nabla J(\mathbf{c}^* + \eta) > 0$ when $\varepsilon < \|\eta\| < 1/\varepsilon,\ \varepsilon > 0$ (2.53)
(c) $\|\nabla J(\mathbf{c})\| \leq R\|\mathbf{c}\|$ when $R = \text{const}$

Condition (a) defines the maximal gain coefficient. A particular value γ depends on the selection of a norm. Condition (b) describes the behavior of the surface $\nabla J(\mathbf{c}) = z$ in the neighborhood of $\mathbf{c}^*$, which corresponds to the fact that the system has a negative feedback loop. Finally, condition (c) defines the variations of the norm of the gradient.

Is it necessary to mention that conditions (2.53) are actually the sufficient conditions for convergence? The conditions listed above guarantee the convergence for a sufficiently broad class of **VJ(c)**, for any value **c*** and for any initial conditions **c**[0]. If we limit ourselves only to finite values of **c**[0], we can obtain various conditions of convergence for the iterative methods. Many such conditions can be found in the literature on computational mathematics.

In practical applications of the described methods we are faced with some definite difficulties; the expressions which describe the conditions of stability in small or in large always contain the unknown optimal vector **c***. In the classical theory of stability, it is always assumed that the value **c*** is a priori given, but here such an assumption is not justified, since the iterative algorithms are the tools for finding **c***. However, this difficulty can be overcome if we employ the concept of absolute stability which is broadly used by Lurie in control theory.

The conditions of absolute stability guarantee the convergence of the algorithms of optimization for any initial value **c**[0] and any a priori unknown optimal vector **c***. We would prefer to call these conditions the conditions of absolute convergence, but we fear objections, since this term is already used in the theory of convergent series.

2.17 On Acceleration of Convergence

We have already mentioned in Section 2.12 that a transition to multistage algorithms can lead to an acceleration of convergence. We will now illustrate such a possibility in a very simple example. Let us consider algorithm (2.37) for $s = 1$, $s_1 = 2$ and $N = 1$. We shall specify

$$\alpha_1 = 1 \qquad \bar{\alpha}_1 = 1 - \alpha \qquad \bar{\alpha}_2 = \alpha \qquad 0 \leq \alpha \leq 1 \qquad \Gamma[n] = I\gamma_0 \qquad (2.54)$$

Then

$$c[n] = c[n-1] - \gamma_0 \, \nabla J((1-\alpha)c[n-1] + \alpha c[n-2]) \qquad (2.55)$$

where $\nabla J(\cdot) = dJ(\cdot)/dc$. For $\alpha = 0$, this algorithm is reduced to an ordinary algorithm of the type (2.4). The difference is that in (2.55) the argument is not simply $c[n-1]$, but a weighted average of two preceding values, $c[n-1]$ and $c[n-2]$. If this weighted average is closer to the optimal value c^* than $c[n-1]$, the rate of convergence is increased. Such a situation indeed occurs when $c[n]$ converges to c^* in an oscillatory fashion. By selecting a proper $\alpha = \alpha_n$ at each step, we can considerably accelerate the convergence. It is then obvious that α_n has to depend on the difference of $c[n]$. We can set

$$\alpha_n \approx \frac{\Delta c[n-1]}{\Delta^2 c[n-2]} = \frac{c[n] - c[n-1]}{c[n] - 2c[n-1] + c[n-2]} \qquad (2.56)$$

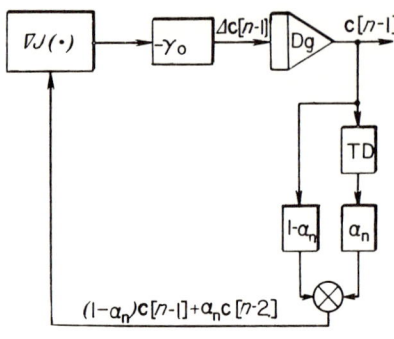

Fig. 2.11

The structural scheme described by algorithm (2.55) is shown in Fig. 2.11. Its characteristic is that it has two loops, and that the gain coefficients depend upon the current and the preceding state. We could probably accelerate the convergence by varying γ according to Δc.

2.18 On Best Algorithms

We have become familiar with the various algorithms of optimization. It is now natural to ask which algorithm we should consider to be the best for a given situation. An attempt to find the answer for a general case would be futile. However, if we specify the type of the algorithm (single-stage, discrete or continuous), the best algorithm is then found by selecting its parameters (for instance, in the case of algorithm (2.4), the matrices $\Gamma[n]$; and for algorithm (2.21), $\gamma[n]$ and $\alpha[n]$). The specific choice of these parameters depends on our definition of the best algorithms. The parameters in the algorithms of relaxation and many gradient methods are optimal with respect to the pre-selected criteria. Therefore, we obtain the algorithms which are "locally" best at each step. Of course, this does not mean that the algorithm is the best overall.

The problem of designing the best algorithms of optimization is very similar to the problems of synthesis for discrete or continuous systems which realize these algorithms. Unfortunately, we cannot yet use the powerful apparatus of the theory of optimal systems for obtaining the best algorithms. This is due to the fact that in the modern theory of optimal systems it is assumed that the initial vector **c**[0] and the final vector **c*** are known. In our case, **c*** is unknown, and moreover, our goal is to find **c***. Therefore, we are forced to consider only the "locally" best algorithms. In order to clearly explain what can be accomplished by following this route, we shall give a few examples.

2.19 Examples

If we ask that the algorithm minimizes a certain function of the error $\mathbf{\eta}[n] = \mathbf{c}[n] - \mathbf{c}^*$ at each step n, the best algorithm of optimization can be obtained by using the well-known relaxation methods or the methods of steepest descent. The coefficient $\gamma[n] = \gamma_{opt}[n]$ is then defined as the smallest positive root of the equation

$$\frac{\partial}{\partial \gamma} J(\mathbf{c}[n]) = \frac{\partial}{\partial \gamma} J(\mathbf{c}[n-1] - \gamma \nabla J(\mathbf{c}[n-1])) = 0 \quad (2.57)$$

Unfortunately, we cannot use iterative methods to determine the optimal values of γ at each stage, since the algorithms for obtaining γ would also contain an unknown parameter which has to be chosen. Sometimes these difficulties can be bypassed by a judicious choice of γ. For instance, γ can be changed until for a certain γ_1, the inequality

$$J(\mathbf{c}[n-1] - \gamma_1[n] \nabla J(\mathbf{c}[n-1])) < J(\mathbf{c}[n-1] - \gamma[n] \nabla J(\mathbf{c}[n-1])) \quad (2.58)$$

is violated. This approach is actually used in relaxation methods. Another possibility consists in defining γ for a linear approximation of (2.57). Then

$$\gamma_{opt}[n] \approx \frac{\nabla^T J(\mathbf{c}[n-1]) \nabla J(\mathbf{c}[n-1])}{\nabla^T J(\mathbf{c}[n-1]) \nabla^2 J(\mathbf{c}[n-1]) \nabla J(\mathbf{c}[n-1])} \quad (2.59)$$

or, according to algorithm (2.7),

$$\gamma_{opt}[n] \approx \frac{\Delta \mathbf{c}^T[n-1] \Delta \mathbf{c}[n-1]}{\Delta \mathbf{c}^T[n-1] \nabla^2 J(\mathbf{c}[n-1]) \Delta \mathbf{c}[n-1]} \quad (2.60)$$

where $\nabla^2 J(\mathbf{c}) = \|[\partial^2 J(\mathbf{c})]/[\partial c_\nu \, \partial c_\mu]\|$ ($\nu, \mu = 1, \ldots, N$) is a matrix of second derivatives.

It is more customary to consider minimization of a mean-square-error criterion in defining the best algorithm. Let us designate

$$V^2[m] = \sum_{n=0}^{m} \|\mathbf{\eta}[n]\|^2 \qquad V^2[m-1] = \sum_{n=1}^{m} \|\mathbf{\eta}[n-1]\|^2 \quad (2.61)$$

where $\|\mathbf{\eta}\|$ is the euclidean norm of the vector $\mathbf{\eta}$. We shall set $\Gamma[n] = I\gamma = \text{const}$ in the algorithm of optimization (2.48). Then, after taking the square of both sides in (2.48), and summing it in n from 1 to m, we obtain

$$V^2[m] = V^2[m-1] - 2\gamma_0 \sum_{n=1}^{m} \mathbf{\eta}^T[n-1] \nabla J(\mathbf{c}^* + \mathbf{\eta}[n-1])$$

$$+ \gamma_0^2 \sum_{n=1}^{m} \|\mathbf{\eta}[n-1]\|^2 \frac{\|\nabla J(\mathbf{c}^* + \mathbf{\eta}[n-1])\|^2}{\|\mathbf{\eta}[n-1]\|^2} \quad (2.62)$$

Let $\nabla J(\mathbf{c}^* + \mathbf{\eta}[n-1])$ for any $\mathbf{c}^*$ and $\mathbf{\eta}$ satisfy the conditions

$$0 < A < \frac{\|\nabla J(\mathbf{c}^* + \mathbf{\eta})\|}{\|\mathbf{\eta}\|} < B \qquad (2.63)$$

$$A\|\mathbf{\eta}\|^2 \leq (\mathbf{\eta}^T \nabla J(\mathbf{c}^* + \mathbf{\eta}))$$

Then, by using the upper and lower bounds in (2.62), we obtain the inequality

$$V^2[m] \leq V^2[m-1](1 - 2\gamma_0 A + \gamma_0^2 B^2) \qquad (2.64)$$

The right-hand side of the inequality reaches a minimum for

$$\gamma_0 = A/B^2 \qquad (2.65)$$

Therefore,

$$V^2[m] \leq V^2[m-1](1 - A^2/B^2) \qquad (2.66)$$

or

$$V^2[m] \leq V^2[0](1 - A^2/B^2)^m \qquad (2.67)$$

Thus, for the best value of γ, an upper bound on $V^2[m]$ is minimized at each step. This approach is related to the optimization of automatic systems on the basis of Lyapunov's direct method.

2.20 Certain Problems

The most important problems faced in the development and application of the algorithms of optimization are to guarantee the convergence and to improve the methods of accelerating the convergence. We could perhaps accelerate the convergence if the computed values of $\gamma[n]$ are used to direct the variations in $\mathbf{c}[n]$. It would be important to determine how $\gamma[n]$ should depend upon $\mathbf{c}[n]$. It would also be very useful to develop a certain principle for a comparison of various optimization algorithms. This problem might be solved if we could find a general expression for the functional of the vector $\mathbf{c}[n]$ such that its minimization would lead to a uniquely defined step $\gamma[n]$.

For an effective application of the multistage algorithms, it is important to find the rules for choosing their parameters. The question of optimal algorithms is still open, and we believe that it will stay open for a long time to come.

2.21 Conclusion

We have become acquainted with the various regular algorithms of optimization which eventually define the optimal vector with or without any constraints. The main efforts in the further development of these algorithms have to be directed toward an improvement of the rates of convergence. The algorithms should be capable of finding not only local but also global extrema.

Although the regular algorithms are important in themselves, we have studied them here only in order to provide certain guidelines for the development of the corresponding algorithms of adaptation. These algorithms have to replace the regular algorithms of optimization when we do not have sufficient a priori information.

COMMENTS

2.2 We use the vector description of the necessary conditions of an extremum (Lee, 1964).

2.3 The regular iterative method goes back to Cauchy. We shall resist the temptation to present the history of these methods and will only suggest the excellent article by Polyak (1963), and also the books on computational mathematics by Demidovich and Maron (1963) and Berezin and Zhidkov (1960). In these books, the traditional presentation of the iterative methods applicable to linear and nonlinear algebraic equations is presented. The book by Traub (1965) is especially devoted to the iterative methods, and the applications of these methods for the solution of boundary value problems is described in the book by Shimanskii (1966). See also the survey by Demyanov (1965).

2.4 The term digrator was proposed by I. L. Medvedev to describe a discrete integrator.

2.5 This generalization is most clearly described by Bingulac (1966).

2.6 A unique classification of the algorithms (or, as they are sometimes called, recursive schemes) does not exist. A very complete classification of the systems of linear algebraic equations is given in the book by Fadeev and Fadeeva (1963). The reader has perhaps noticed that we have not deviated much from such a classification. For the particular algorithms, see the book by Demidovich and Maron (1963). The relaxation algorithms are described in an interesting paper by Glazman (1964). (See also Glazman and Senchuk (1966).) It seems to us that the conjugate gradient methods have a great perspective (Mitter *et al.*, 1966). A review of the iterative methods of solving functional equations is given in the paper by Kantorovich (1956).

2.7 The idea of search algorithms was originated by Germansky (1934). In the mathematical literature, similar search algorithms have not been given much attention. On synchronous detection, see books by Degtyarenko (1965) and Krasovskii (1963). In the latter book, the methods of obtaining the gradient using a synchronous detector or a correlator (using noise) are broadly employed.

2.8 See the book by Lee (1964). Other types of approaches for obtaining Lagrange multipliers can be found in the paper by Polyak (1963).

2.9 The inequality constraints were treated broadly in the literature on mathematical and, particularly, nonlinear programming. We recommend books by Arrow *et al.* (1958) and Künci and Krelle (1965), and the paper by Levitin and Polyak (1966). In these works, the reader can find rigorous formulations of the Kuhn-Tucker theorem and Slater's regularity conditions. A very interesting approach to the solution of similar problems was developed by Dubovitskii and Milyutin (1965). The schemes which realize these algorithms with constraints were described by Dennis (1959).

2.10 The most complete exposition of the methods of feasible directions was given in the book by Zoutendijk (1960), who is the author of these methods.

2.11 We have paraphrased an example given by Lee (1964).

2.12 Here we have used the well-known relationship in the theory of finite differences (Tsypkin, 1963). Multistage algorithms were treated in a relatively small number of works, among which we recommend the paper by Polyak (1963).

2.13 On continuous algorithms, see the papers by Gavurin (1958, 1963), and also the paper by Alber and Alber (1966) devoted to mathematical questions. See also the papers by Ribashov (1965a, b), which describe the applications of analog computers in the solution of finite difference equations. The idea of a "heavy ball" apparently belongs to Cumada (1961); see also the paper by Polyak (1963).

2.14 The methods of random search, at one time described by Ashby (1956), have been greatly developed in the numerous papers by Rastrigin, which have been collected in his book (Rastrigin, 1965). Certain applications of the method of random search in estimating linear decision rule were described by Wolff (1966). For further study, see also the papers by Matiash (1965) and Munson and Rubin (1959).

2.15 A presentation of the direct method of Lyapunov for discrete systems can be found in the book by Bromberg (1967), and in the paper by Kalman and Bertram (1960). Different formulations of the contraction mapping principle can be found in the books by Lyusternik and Sobolev (1961) and Krasnoselskii (1956).

2.16 On absolute stability, see the already classic book by Lurie (1957). Unfortunately, the conditions of almost certain convergence have not been given much attention in the literature. The conditions of convergence for the iterative methods in general have been treated in a considerably larger number of publications. We shall mention the books by Traub (1965) and Ostrowski (1960), and also a paper by Yakovliev (1965). In these, the reader can find a comprehensive bibliography on the convergence of iterative methods. See also the book by Mikhlin (1966).

2.17 This method of accelerating the convergence belongs to Wegstein (1958). It is also described in the book by Ledley (1962).

2.18 When the author was studying this question, he was very happy to find a reference to a paper by Bondarenko (1966) with a very promising title. However, after becoming familiar with the paper, he discovered that his hope to learn how to define the concept of the "best algorithm" was not justified.

2.19 On relaxation methods, see the papers by Glazman (1964), Lyubich (1966) and Tompkins (1956).

2.20 A survey of nonlocal methods of optimization and the ways of solving multi-extremal problems can be found in the paper by Yudin (1965). An interesting method of solving multiextremal problems, the so-called method of "holes," was proposed by Gelfand and Tsetlin (1962).

BIBLIOGRAPHY

Alber, S.I., and Alber, Ya.I. (1966). An application of the methods of differential descent to the solution of nonlinear systems, Zh. Vychisl. Mat. Mat. Fiz. **7** (1). (In Russian.)
Arrow, K.J., Hurwitz, L., and Uzawa, H. (1958). "Studies in Linear and Nonlinear Programming." Stanford University Press, Stanford, Ca.
Ashby, W.R. (1956). "An Introduction to Cybernetics." Wiley, New York.
Berezin, I.S., and Zhidkov, N.P. (1960). "Computational Methods," Vol. II. Fizmatgiz, Moscow. (In Russian.)
Bingulac, S.P. (1966). On the improvement of gradient methods, IBK-450, *Electronics and Automation*. Boris Kidrič Institute of Nuclear Sciences, Beograd-Vinča.
Bondarenko, P.S. (1966). The principles of optimal computational algorithms, *Vychisl. Mat.* **1966** (3). (In Russian.)
Bromberg, P.V. (1967). "Matrix Methods in the Theory of Relay and Sampled-Data Control." Nauka, Moscow. (In Russian.)
Cumada, I. (1961). On the golf method, *Bull. Electrotech. Lab. Tokyo* **25** (7).
Degtyarenko, P.I. (1965). "Synchronous Detection in Instrumentation and Automation." Naukova Dumka, Kiev. (In Russian.)
Demidovich, B.P., and Maron, I.A. (1963). "Fundamentals of Computational Mathematics," 2nd ed. Fizmatgiz, Moscow. (In Russian.)
Demyanov, V.F. (1965). On minimization of functions over convex bounded domains, *Kibern. Dok.* **1965** (6). (In Russian.)
Dennis, J.R. (1959). "Mathematical Programming and Electrical Networks." Wiley, New York.
Dubovitskii, A.Ya., and Milyutin, A.A. (1965). The problems of optimization under constraints, *Zh. Vychisl. Mat. Mat. Fiz.* **5** (3). (In Russian.)
Fadeev, D.K., and Fadeeva, V.N. (1963). "Computational Methods in Linear Algebra." Freeman, London.
Fletcher, R., and Powell, M.J.D. (1963). A rapidly convergent descent method for minimization, *Computer J.* **6** (2).
Gavurin, M.K. (1958). Nonlinear functional equations and the continuous analogies of the iterative methods, *Izv. Vyssh. Ucheb. Zaved. Mat.* **1958** (5). (In Russian.)
Gavurin, M.K. (1963). On the existence theorems in nonlinear functional equations. *In* "Metodi Vychislenii," Vol. 2. Leningrad State University Press, Leningrad. (In Russian.)
Gelfand, I.M., and Fomin, S.V. (1963). "Calculus of Variations." Prentice-Hall, Englewood Cliffs, N.J.
Gelfand, I.M., and Tsetlin, M.L. (1962). On certain methods of control in complex systems, *Usp. Mat. Nauk.* **17** (1). (In Russian.)
Germansky, R. (1934). Notiz über die Lösung von Extremaufgaber mittels Iteration, *Z. Angew. Math. Mech.* **14** (3).
Glazman, I.M. (1964). On the gradient relaxation for non-quadratic functionals, *Dokl. Akad. Nauk. SSSR* **154** (5). (In Russian.)
Glazman, I.M., and Senchuk, Yu.F. (1966). On a method of minimizing certain functionals in the calculus of variations. *In* " Teoria Funkciy Funkcionalniy Analiz i ikh Prilozhenia, Vol. 2. Harkov. (In Russian.)
Kalman, R.E., and Bertram, J.E. (1960). Control system analysis and design via the "Second Method" of Lyapunov, *Trans. ASME*, Series D **82** (2).

Kantorovich, L.V. (1956). Approximate solution of functional equations, *Usp. Mat. Nauk* **11** (6). (In Russian.)

Krasnoselskii, M.A. (1956). "Topological Methods in the Theory of Nonlinear Integral Equations." Gostehizdat, Moscow. (In Russian.)

Krasovskii, A.A. (1963). "Dynamics of Continuous Self-Organizing Systems." Fizmatgiz, Moscow. (In Russian.)

Künci, H.P., and Krelle, W. (1965). "Nonlinear Programming." Sovyetskoe Radio, Moscow. (Translated from German into Russian.)

Ledley, R.S. (1962). "Programming and Utilizing Digital Computers." McGraw-Hill, New York.

Lee, R.C.K. (1964). "Optimal Estimation, Identification and Control." MIT Press, Cambridge, Mass.

Levitin, E.S., and Polyak B.T. (1966). The methods of minimization under constraints, *Zh. Vychisl. Mat. Mat. Fiz.* **6** (5). (In Russian.)

Lurie, A.I. (1957). "Some Nonlinear Problems in the Theory of Automatic Control." Her Majesty's Stationery Office, London.

Lyubich, Yu.I. (1966). On the rate of convergence of a stationary gradient relaxation, *Zh. Vychisl. Mat. Mat. Fiz.* **6** (2). (In Russian.)

Lyusternik, L.A., and Sobolev, V.I.(1961). "Elements of Functional Analysis." Frederick Ungar, New York.

Matiash, I. (1965). Random optimization, *Automat. Remote Contr. (USSR)* **26** (2).

Mikhlin, S.G. (1966). "Numerical Realization of the Variational Methods." Nauka, Moscow. (In Russian.)

Mishkis, A.D. (1965). On the holes, *Zh. Vychisl. Mat. Mat. Fiz.* **5** (3). (In Russian.)

Mitter, S., Lasdon, L.S., and Waren, A.D. (1966). The method of conjugate gradients for optimal control problems, *Proc. IEEE* **54** (6).

Munson, I.K., and Rubin, A.I. (1959). Optimization by random search, *IRE Trans. Electron. Comput.* **EC-8** (2).

Ostrowski, A.M. (1960). "Solution of Equations and Systems of Equations." Academic Press, New York.

Polyak, B.T. (1963). Gradient methods for minimization of functionals, *Zh. Vychisl. Mat. Mat. Fiz.* **3** (4). (In Russian.)

Rastrigin, L.A. (1965). "Random Search in the Optimization Problems of Multivariable Systems." Zinatne, Riga. (In Russian.)

Ribashov, M.V. (1965a). Gradient method of solving the problems of nonlinear programming on an analog model, *Automat. Remote Contr. (USSR)* **26** (11).

Ribashov, M.V. (1965b). Gradient method for solving the problems of linear and nonlinear programming, *Automat. Remote Contr. (USSR)* **26** (12).

Ribashov, M.V., and Dudnikov, E.E. (1966). Application of the direct Liapunov method to the nonlinear programming problems, *Izv. Akad. Nauk SSSR Tekh. Kibern.* **1966** (6). (Engl. trans.: *Eng. Cybern. (USSR).*)

Shimanskii, V.E. (1966). "The Methods of Numerical Solutions of Boundary Value Problems on Digital Computers," Part II. Naukova Dumka, Kiev. (In Russian.)

Tompkins, C.B. (1956). Methods of steepest descent. *In* "Modern Mathematics for the Engineer" (E.B. Beckenbach, ed.) Chap. 18. McGraw-Hill, New York.

Traub, J.T. (1965). "Iterative Methods for the Solution of Equations." Prentice-Hall, Englewood Cliffs, N.J.

Tsypkin, Ya.Z. (1963). "Theory of Linear Impulsive Systems." Fizmatgiz, Moscow. (In Russian.)

Wegstein, J.H. (1958). Accelerating convergence of iterative processes, *Commun. ACM* **1** (6).

Wolff, A.C. (1966). The estimation of the optimum design function with a sequential random method, *IEEE Trans. Inform. Theory* **IT-12** (3).
Yakovliev, M.N. (1965). On some methods of solving nonlinear equations, *Tr. Mat. Inst. Steklova, Nauka* **84**. (In Russian.)
Yudin, D.B. (1965). The methods of quantitative analysis of complex systems, *Izv. Akad. Nauk SSSR Tekh. Kibern.* **1965** (1). (Engl. transl.: *Eng. Cybern. (USSR)*.)
Zoutendijk, G. (1960). "Methods of Feasible Directions, A Study of Linear and Nonlinear Programming," Elsevier, Amsterdam.

3

ADAPTATION AND LEARNING

3.1 Introduction

It is difficult to find more fashionable and attractive terms in the modern theory of automatic control than the terms of adaptation and learning. At the same time, it is not simple to find any other concepts which are less complex and more vague.

We shall explore this vast and important area of automatic control theory which only fifteen years ago was not even on the list of future problems in control theory. Regardless of whether we like it or not, we must give some working definitions of the terms which could be used in the formulation of the problems of learning (training), self-learning and adaptation. As it was frequently emphasized by academician A.A. Andronov, to have success in the solution of new problems, "it is necessary to reconstruct the present mathematical apparatus and discover a new mathematical apparatus which would be adequate for the processes under consideration, and at the same time be sufficiently efficient." It was possible and convenient to select the probabilistic iterative methods and, in particular, the stochastic approximation method. As we shall see, a relatively small development of these methods, and a change in the ideas for applications, give us a convenient mathematical apparatus for the theory of adaptation and learning. Formally, this apparatus is similar to the apparatus of the iterative method which was discussed in the preceding chapter. This formal analogy serves its purpose; it permits us to explain the generality and the differences between the regular and the adaptive approaches in the solution of optimization problems. The purpose of this chapter is not to examine the possibilities of solving various specific problems, but to develop a general approach for solving different problems (or classes of problems) which at this stage of development should be related to the area of adaptation and learning. The following chapters are devoted to the studies and the solutions of these problems.

3.2 Concepts of Learning, Self-Learning and Adaptation

There are many definitions of what should be considered under learning (training), self-learning and adaptation. Unfortunately, even in the case of automatic systems, such definitions are very contradictory. We shall not become involved with a comparative analysis and a critique of such definitions. This would take us slightly from our main course, and it is doubtful whether such a route would lead us to success. Instead, we shall attempt to give the definitions which are convenient for our purpose. Such definitions may also be criticized.

Under the term learning in a system, we shall consider a process of forcing the system to have a particular response to a specific input signal (action) by repeating the input signals and then correcting the system externally. Of course, it is assumed that the system is potentially "capable" of learning. An external correction (often called "reward" or "punishment") is performed by a "teacher" who knows the desired response on a particular external action (input signal). Therefore, during learning, a "teacher" provides the system with sufficient information regarding the correctness or incorrectness of its response.

Self-learning is learning without external corrections, i.e., without punishments or rewards. Any additional information regarding the correctness or incorrectness of the system's reaction is not given.

Adaptation is considered to be a process of modifying the parameters or the structure of the system and the control actions. The current information is used to obtain a definite (usually optimal) state of the system when the operating conditions are uncertain and time-varying.

Learning is sometimes identified with adaptation. There are many justifications for this, especially if a certain index of success in learning is introduced. However, we find it convenient to consider that learning is used in adaptation only to collect information about the state and the characteristics of the system which are necessary for optimal control under the conditions of uncertainty. When an initial uncertainty exists, it seems reasonable to try to reduce it by learning or self-learning during the process of control, and at the same time to use the collected information for an improvement of the performance index of control. Therefore, the most characteristic feature of adaptation is an accumulation and a slow usage of the current information to eliminate the uncertainty due to insufficient a priori information and for the purpose of optimizing a certain selected performance index.

The reader will perhaps notice that we identify adaptation with optimization under the conditions of insufficient a priori information.

3.3 Formulation of the Problem

Although the applications of the adaptive approach in the solution of optimization problems were already discussed in Chapter 2, it is useful to formulate the problem of adaptation from another viewpoint based on the definition given above. Let the criterion of optimality (performance index) be a functional of the vector **c**.

$$J(\mathbf{c}) = M_\mathbf{x}\{Q(\mathbf{x}, \mathbf{c})\} \tag{3.1}$$

It is assumed that the explicit form of $J(\mathbf{c})$ is unknown. This means that the probability density function $p(\mathbf{x})$ is unknown, and that only the realization $Q(\mathbf{x}, \mathbf{c})$, which depends upon the stationary random processes or sequences $\mathbf{x}$ and vector $\mathbf{c}$, is known. We also assume here that the constraints of the first kind are included in the functional. The constraints of the second kind will be considered later.

Many possibilities are hidden under the innocent words "a functional is not explicitly known," or "an unknown probability density function." These possibilities include first, the deterministic processes which are not completely understood, and second, those random processes with unknown or partially known probability distributions. For instance, the type of distribution may be given, but some of the parameters are unknown. Finally, we may not know whether the processes are deterministic or stochastic. In all these cases, when sufficient a priori information does not exist, there is a need for adaptation.

Our problem is to determine the optimal vector **c** which defines the extremum (and in order to be specific, let it be a minimum) of the functional (3.1). Obviously, the only possible way to solve this problem is related to the observations of the values **x** and their processing. It is clear that the regular iterative algorithms are not convenient here. However, the ideas of iterative algorithms are adequate to solve the problems of learning and adaptation. They will also show us a way to solve the basic problems of modern automation.

3.4 Probabilistic Iterative Methods

Probabilistic iterative methods are closely related to stochastic approximation, which, regardless of the broad popularity in the literature on statistics, did not find applicable solutions to engineering problems for a long time. In order to present the idea of probabilistic iterative methods, we consider again the condition of optimality (2.2), which can be written for (3.1) as

$$\nabla J(\mathbf{c}) = M_\mathbf{x}\{\nabla_\mathbf{c} Q(\mathbf{x}, \mathbf{c})\} = 0 \tag{3.2}$$

where

$$\nabla_c Q(\mathbf{x}, \mathbf{c}) = \left(\frac{\partial Q(\mathbf{x}, \mathbf{c})}{\partial c_1}, \frac{\partial Q(\mathbf{x}, \mathbf{c})}{\partial c_2}, \ldots, \frac{\partial Q(\mathbf{x}, \mathbf{c})}{\partial c_N} \right) \quad (3.3)$$

is the gradient of $Q(\mathbf{x}, \mathbf{c})$ with respect to $\mathbf{c}$.

In (3.2), we do not know the gradient of the functional, i.e., $\nabla_c J(\mathbf{x})$, but we do know the realization $\nabla_c Q(\mathbf{x}, \mathbf{c})$. It appears that for a suitable choice of the matrix $\Gamma[n]$, we can use many different forms of the regular methods if we substitute the gradient of the functional $\nabla J(\mathbf{c})$ by the sample values $\nabla_c Q(\mathbf{x}, \mathbf{c})$. In this lies the central idea of the probabilistic iterative methods. Therefore, the probabilistic algorithms of optimization, or briefly, the algorithms of adaptation, can be written in the recursive form:

$$\mathbf{c}[n] = \mathbf{c}[n-1] - \Gamma[n] \nabla_c Q(\mathbf{x}[n], \mathbf{c}[n-1]) \quad (3.4)$$

The algorithm of adaptation can also be written in the difference form

$$\Delta \mathbf{c}[n-1] = -\Gamma[n] \nabla_c Q(\mathbf{x}[n], \mathbf{c}[n-1]) \quad (3.5)$$

or, in the sum form

$$\mathbf{c}[n] = \mathbf{c}[0] - \sum_{m=1}^{n} \Gamma(m) \nabla_c Q(\mathbf{x}[m], \mathbf{c}[m-1]) \quad (3.6)$$

It is simple to see an analogy between the regular algorithms (2.4), (2.7), (2.8) and the probabilistic algorithms (3.4), (3.5), (3.6). But at the same time, they differ considerably, since for $\mathbf{c} = \mathbf{c}^*$,

$$\nabla_c Q(\mathbf{x}, \mathbf{c}) \neq 0 \quad (3.7)$$

Due to this characteristic, we have to impose certain specific constraints on the character of $\Gamma[n]$ in order to ensure convergence. These constraints will be discussed below.

If we consider the algorithms of adaptation (3.4) and (3.5) as the equations of a certain discrete feedback system, we can obtain a block diagram (Fig. 3.1). This block diagram differs from the one which corresponds to the regular algorithm of optimization (Fig. 2.1), since the external action $\mathbf{x}[n]$ exists at the

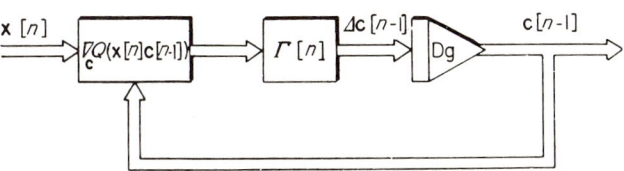

Fig. 3.1

input of the functional transformer in addition to the action $c[n]$. In adaptation, we have a nonautonomous system which is receiving the information about the signal $x[n]$ from the external world. The processing of this current information can help to compensate for insufficient a priori information.

3.5 Algorithms of Adaptation

The simplest algorithms of adaptation correspond to a particular choice of $\Gamma[n]$ in the form of a diagonal matrix

$$\Gamma[n] = \mathscr{A}\gamma[n] = \begin{pmatrix} \gamma_1[n] & 0 & \cdots & 0 \\ 0 & \gamma_2[n] & \cdots & 0 \\ \vdots & \vdots & & \vdots \\ 0 & 0 & \cdots & \gamma_N[n] \end{pmatrix} \tag{3.8}$$

where $\mathscr{A}$ is an operator which transforms a vector into a diagonal matrix. In a special case of equal components of the vector $\gamma[n]$, the operation of $\mathscr{A}$ is identical to multiplication with the identity matrix, i.e., $\mathscr{A}\gamma[n] = I\gamma[n]$. In this case, the algorithm can be written as

$$\mathbf{c}[n] = \mathbf{c}[n-1] - \gamma[n] \, \nabla_\mathbf{c} \, Q(\mathbf{x}[n], \mathbf{c}[n-1]) \tag{3.9}$$

or

$$\Delta\mathbf{c}[n-1] = -\gamma[n] \, \nabla_\mathbf{c} \, Q(\mathbf{x}[n], \mathbf{c}[n-1]) \tag{3.10}$$

or, finally,

$$\mathbf{c}[n] = \mathbf{c}[0] - \sum_{m=1}^{n} \gamma[m] \, \nabla_\mathbf{c} \, Q(\mathbf{x}[n], \mathbf{c}[m-1]) \tag{3.11}$$

It should be remembered that a choice of $\Gamma[n]$ in the form of a diagonal matrix very often permits an improvement in the properties of the algorithms of adaptation. A similar situation will be discussed in Section 3.15.

3.6 Search Algorithms of Adaptation

In the cases when for some reason it is not possible to obtain the gradient of the realization $\nabla_\mathbf{c} Q(\mathbf{x}, \mathbf{c})$, but the sample values of $Q(\mathbf{x}, \mathbf{c})$ can be measured, the search algorithms of adaptation can provide help. Let us introduce the following notation, which is similar to (2.18):

$$\begin{aligned} Q_+(\mathbf{x}, \mathbf{c}, a) &= (Q(\mathbf{x}, \mathbf{c} + a\mathbf{e}_1), \ldots, Q(\mathbf{x}, \mathbf{c} + a\mathbf{e}_N)) \\ Q_-(\mathbf{x}, \mathbf{c}, a) &= (Q(\mathbf{x}, \mathbf{c} - a\mathbf{e}_1), \ldots, Q(\mathbf{x}, \mathbf{c} - a\mathbf{e}_N)) \\ Q_0(\mathbf{x}, \mathbf{c}) &= (Q(\mathbf{x}, \mathbf{c}), \ldots, Q(\mathbf{x}, \mathbf{c})) \end{aligned} \tag{3.12}$$

3.6 Search Algorithms of Adaptation

where a is a scalar, and $\mathbf{e}_i$ ($i = 1, \ldots, N$) are the basis vectors (2.19). As before, we shall estimate the gradient by the ratios

$$\frac{Q_+(\mathbf{x}, \mathbf{c}, a) - Q_-(\mathbf{x}, \mathbf{c}, a)}{2a} = \tilde{\nabla}_{\mathbf{c}\pm} Q(\mathbf{x}, \mathbf{c}, a) \tag{3.13}$$

or

$$\frac{Q_+(\mathbf{x}, \mathbf{c}, a) - Q_0(\mathbf{x}, \mathbf{c})}{a} = \tilde{\nabla}_{\mathbf{c}+} Q(\mathbf{x}, \mathbf{c}, a)$$

$$\frac{Q_0(\mathbf{x}, \mathbf{c}) - Q_-(\mathbf{x}, \mathbf{c}, a)}{a} = \tilde{\nabla}_{\mathbf{c}-} Q(\mathbf{x}, \mathbf{c}, a) \tag{3.14}$$

which depends on the random vector $\mathbf{x}$. The search algorithm of adaptation can then be given in the recursive form

$$\mathbf{c}[n] = \mathbf{c}[n-1] - \gamma[n] \, \tilde{\nabla}_{\mathbf{c}\pm} Q(\mathbf{x}[n], \mathbf{c}[n-1], a[n]) \tag{3.15}$$

An estimate of the gradient can be obtained by using a synchronous detector. The block diagram of the corresponding extremal system is shown in Fig. 3.2. Here, we cannot select $a[n] = \text{const}$, since the additional generator of

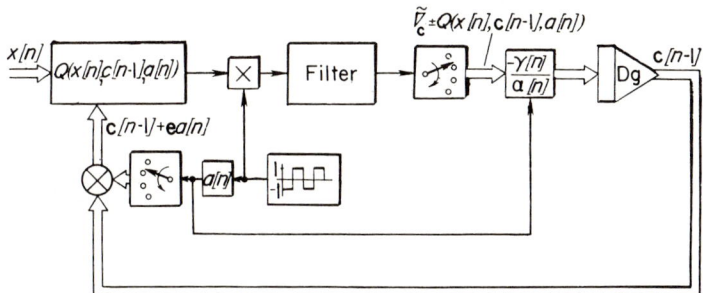

Fig. 3.2

square search pulses becomes more complicated. The amplitudes of the pulses must vary according to a specific law. The role of the commutators remains the same. They serve for a sequential selection of the arguments $\mathbf{c} \pm a\mathbf{e}_k$ and the components $\tilde{\nabla}_{\mathbf{c}\pm} Q(\mathbf{x}, \mathbf{c}, a)$.

In the realization of the algorithms of adaptation, as it was mentioned in Chapter 2, it is useful to consider various algorithms of search which were developed in the theory of extremal systems.

3.7 Constraints I

We shall assume that the constraints of the first kind are given as the equalities

$$g(c) = M_x\{h(x, c)\} = 0 \tag{3.16}$$

where the form of the vector function $g(c) = (g_1(c), \ldots, g_M(c))$ is unknown, and only the realization of the vector function $h(x, c)$ is known. Although the problem of minimizing (3.1) under the constraints (3.16) was not studied in the literature, its solution is relatively simple.

As in the regular case (see Section 2.8), we form a new functional

$$J(c, \lambda) = J(c) + \lambda^T g(c) \tag{3.17}$$

which is transformed into

$$J(c, \lambda) = M_x\{Q(x, c) + \lambda^T h(x, c)\} \tag{3.18}$$

by considering (3.1) and (3.16).

The problem now is to find the stationary point of the function $J(c, \lambda)$ by using its sample values $Q(x, c) + \lambda^T h(x, c)$. In order to do this, we apply the algorithm of adaptation (3.9) to (3.18). The algorithm of adaptation for such a case is then

$$c[n] = c[n-1] - \gamma[n](\nabla_c Q(x[n], c[n-1]) + H_c(x[n], c[n-1])\lambda[n-1])$$
$$\lambda[n] = \lambda[n-1] + \gamma_1[n]h(x[n], c[n-1]) \tag{3.19}$$

Here,

$$H_c(x, c) = \left\| \frac{\partial h_v(x, c)}{\partial c_\mu} \right\| \quad (v = 1, \ldots, M; \mu = 1, \ldots, N) \tag{3.20}$$

is an $N \times M$ matrix. Obviously, these algorithms also include the case when the constraints (3.16) are given in the explicit form, since we can always

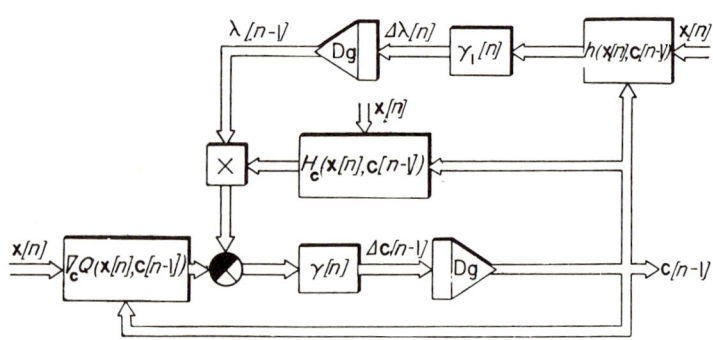

Fig 3.3

consider that $\mathbf{g}(\mathbf{c}) = M\{\mathbf{g}(\mathbf{c})\}$ with $\mathbf{h}$ and H substituted by $\mathbf{g}$ and G, respectively (see (2.18)).

The block diagram of the discrete system which corresponds to the algorithm of adaptation (3.19) is shown in Fig. 3.3. The scheme differs from the one shown in Fig. 3.1 by the presence of additional loops for the constraints.

3.8 Constraints II

For the constraints of the second kind, given as

$$\mathbf{g}(\mathbf{c}) = M_\mathbf{x}\{\mathbf{h}(\mathbf{x}, \mathbf{c})\} \leq 0 \quad (3.21)$$

where the form of the vector function $\mathbf{g}(\mathbf{c}) = (g_1(\mathbf{c}), \ldots, g_M(\mathbf{c}))$ is unknown, the algorithms of adaptation require a generalization of the Kuhn-Tucker theorem. We shall first consider that $J(\mathbf{c})$ and $\mathbf{g}(\mathbf{c})$ are known, and that $\mathbf{g}(\mathbf{c})$ satisfies Slater's condition (2.30). In this case, from the Kuhn-Tucker theorem (see Section 2.9),

$$\begin{aligned}
\nabla J(\mathbf{c}) + \lambda^T G(\mathbf{c}) &= 0 \\
\mathbf{g}(\mathbf{c}) + \delta &= 0 \\
\lambda^T \delta &= 0 \\
\lambda &\geq 0 \\
\delta &\geq 0
\end{aligned} \quad (3.22)$$

By using (3.1) and (3.21), we obtain

$$\begin{aligned}
M_\mathbf{x}\{\nabla_\mathbf{c} Q(\mathbf{x}, \mathbf{c}) + H_\mathbf{c}(\mathbf{x}, \mathbf{c})\lambda\} &= 0 \\
M_\mathbf{x}\{\mathbf{h}(\mathbf{x}, \mathbf{c}) + \delta\} &= 0 \\
\lambda^T \delta &= 0 \\
\lambda &\geq 0 \\
\delta &\geq 0
\end{aligned} \quad (3.23)$$

where $H_\mathbf{c}(\mathbf{x}, \mathbf{c})$ is defined in (3.20). Therefore, we have obtained the necessary conditions of optimality for the problem under consideration. Now by analogy to (3.19), we can write the following algorithm for finding $\mathbf{c}^*$ and λ^*:

$$\begin{aligned}
\mathbf{c}[n] &= \mathbf{c}[n-1] - \gamma[n][\nabla_\mathbf{c} Q(\mathbf{x}[n], \mathbf{c}[n-1]) + H_\mathbf{c}(\mathbf{x}[n], \mathbf{c}[n-1])\lambda[n-1]] \\
\lambda[n] &= \max\{0; \lambda[n-1] + \gamma_1[n]\mathbf{h}(\mathbf{x}[n], \mathbf{c}[n-1])\} \\
\lambda(0) &\geq 0
\end{aligned} \quad (3.24)$$

The block diagram of the discrete system which corresponds to the algorithm of adaptation (3.24) is shown in Fig. 3.4.

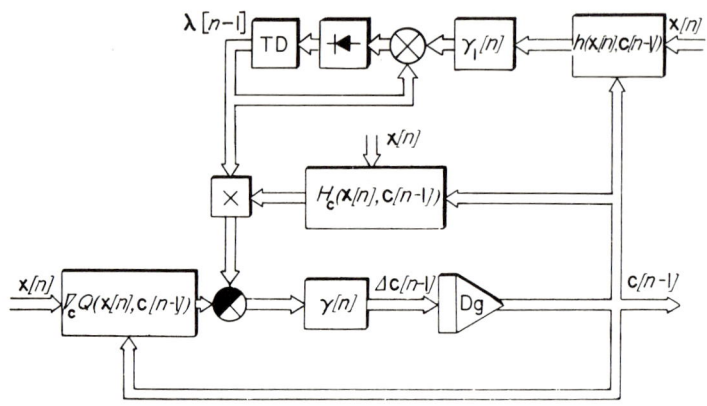

Fig. 3.4

3.9 A Generalization

It could happen that the constraints given in the form of equalities are obtained in the process of averaging the function **h**(**y**, **c**) of the process **x**, which does not depend on the random process **x**. It appears that this is not essential for obtaining the algorithms of adaptation. Actually, in this case

$$J(\mathbf{c}, \lambda) = M_\mathbf{x}\{Q(\mathbf{x}, \mathbf{c})\} + \lambda^T M_\mathbf{y}\{\mathbf{h}(\mathbf{y}, \mathbf{c})\} \tag{3.25}$$

can be written as

$$J(\mathbf{c}, \lambda) = M_{\mathbf{xy}}\{Q(\mathbf{x}, \mathbf{c}) + \lambda^T \mathbf{h}(\mathbf{y}, \mathbf{c})\} \tag{3.26}$$

where $Q(\mathbf{x}, \mathbf{c}) + \lambda^T \mathbf{h}(\mathbf{y}, \mathbf{c})$ corresponds to a random process with probability density function $p_1(\mathbf{x})p_2(\mathbf{y})$. This implies that the algorithm of adaptation for finding the maximum of the functional $J(\mathbf{x}) = M_\mathbf{x}\{Q(\mathbf{x}, \mathbf{c})\}$ under the constraints $M_\mathbf{y}\{\mathbf{h}(\mathbf{y}, \mathbf{c})\} = 0$ does not differ from the previously considered case (3.19).

It should be clear that the algorithms of adaptations are applicable in the case of inequality constraints.

3.10 Multistage Algorithms of Adaptation

As in the regular case, we can now construct multistage algorithms of adaptation which can serve as the means for finding the global minima. This class of algorithms can be written as

3.10 Multistage Algorithms of Adaptation

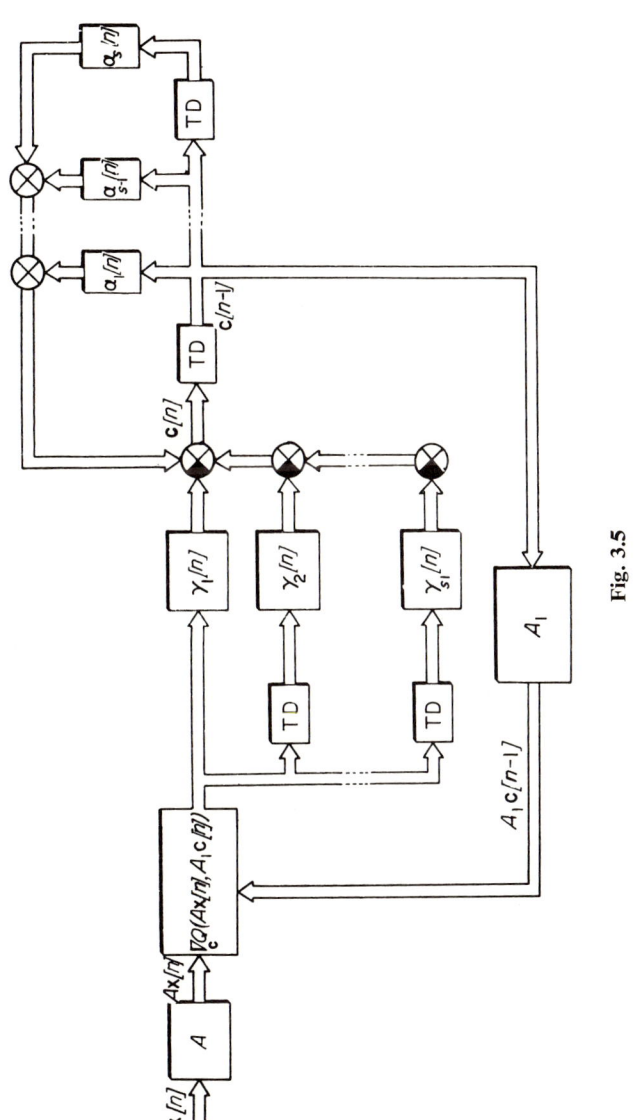

Fig. 3.5

$$\mathbf{c}[n] = \sum_{m=1}^{s} \alpha_m[n]\mathbf{c}[n-m] - \sum_{m=1}^{s_1} \gamma_m[n]\nabla_\mathbf{c} Q(A\mathbf{x}[n-m-1], A_1\mathbf{c}[n-m-1])$$
(3.27)

They are different from the regular algorithms, since the random process can be transformed by a device which is characterized by an operator A. Moreover, even the variable can be transformed by an operator A_1, such that $A_1\mathbf{c}^* = \mathbf{c}^*$. The block diagram of a discrete system which corresponds to the multistage algorithm of adaptation is presented in Fig. 3.5. The discrete filters and the transformers of information are now present in the nonautonomous system. The search algorithms of adaptation use an estimate $\tilde{\nabla}_{\mathbf{c}\pm} Q(\cdot)$ instead of the gradient $\nabla_\mathbf{c} Q(\cdot)$.

We expect a definite improvement from the multistage algorithms, since we believe that knowledge of the past makes us more certain of the future.

3.11 Continuous Algorithms

Continuous algorithms of adaptation can be obtained by a limiting process from the discrete algorithms of adaptation considered thus far. For instance, the continuous algorithm

$$\frac{d\mathbf{c}(t)}{dt} = -\gamma(t)\,\nabla_\mathbf{c} Q(\mathbf{x}(t), \mathbf{c}(t)) \tag{3.28}$$

corresponds to the discrete algorithm (3.10).

The continuous algorithms of adaptation represent the stochastic differential equations, since the right-hand side of (3.28) depends upon the random process $\mathbf{x}(t)$.

In addition to the algorithms of adaptation of the first order, we can construct the algorithms of a higher order. Finally, it is not difficult to imagine the continuous search algorithms.

The continuous algorithms of adaptation can be easily realized by analog computers. The block diagrams of the corresponding continuous algorithms differ from the block diagrams of the discrete systems only by using the continuous integrators instead of the discrete integrators.

3.12 Probabilistic Convergence and Stability

The values of the gradient $\nabla_\mathbf{c} Q(\mathbf{x}, \mathbf{c})$ and their estimates $\tilde{\nabla}_{\mathbf{c}\pm} Q(\mathbf{x}, \mathbf{c}, a)$, $\tilde{\nabla}_{\mathbf{c}+} Q(\mathbf{x}, \mathbf{c}, a)$, $\tilde{\nabla}_{\mathbf{c}-} Q(\mathbf{x}, \mathbf{c}, a)$, which depend on the random process $\mathbf{x}$, appear in the algorithms of adaptation. Therefore, the vectors $\mathbf{c}[n]$ are also random, and we cannot use the ordinary concepts of convergence which are well-known in mathematical analysis and which were used in Section 2.15. We must introduce new concepts of convergence, not in an ordinary sense, but in a stochastic one.

There are three basic types of convergence: convergence in probability, mean-square convergence, and convergence with probability one. A random

3.12 Probabilistic Convergence and Stability

vector $c[n]$ converges in probability to c^* as $n \to \infty$ if for any $\varepsilon > 0$, the probability that the norm $\|c[n] - c^*\|$ does not exceed ε converges to zero, or briefly, if

$$\lim_{n \to \infty} P\{\|c[n] - c^*\| > \varepsilon\} = 0 \qquad (3.29)$$

Convergence in probability does not imply that every sequence $c[n]$ of random vectors converges to c^* in the ordinary sense Moreover, we cannot guarantee this for any vector.

A random vector $c[n]$ converges to c^* in the mean-square sense as $n \to \infty$ if the mathematical expectation of the square of the norm $\|c[n] - c^*\|$ converges to zero, i.e., if

$$\lim_{n \to \infty} M\{\|c[n] - c^*\|^2\} = 0 \qquad (3.30)$$

Convergence in the mean-square sense implies the convergence in probability, but it does not imply the ordinary convergence for any random vector $c[n]$. Convergence in the mean-square sense is related to the investigations of the moments of the second order, which are easily computed and which also have a clear meaning in terms of energy. This can explain why such convergence is so extensively used in physics. However, they may not be satisfactory, since in both types of convergence, the probability that a given vector $c[n]$ converges to c^* in an ordinary sense is zero. We always operate with the sample gradient $\nabla_c Q(x[n], c[n-1])$ and the corresponding random vector $c[n]$, and it is desirable that the limit exists for that particular sequence of random vectors $c[n]$ ($n = 0, 1, 2, \ldots$) which is actually observed, and not for a family of sequences which may never be observed.

This can be accomplished if we introduce the concept of almost sure convergence. Since $c[n]$ is a random vector, we can consider the convergence of a sequence $c[n]$ to c^* as a random event. The sequence of random vectors $c[n]$ converges to c^* as $n \to \infty$ almost certainly or with probability one, if the probability of ordinary convergence of $c[n]$ to c^* is equal to one, i.e., if

$$P\left\{\lim_{n \to \infty} \|c[n] - c^*\| = 0\right\} = 1 \qquad (3.31)$$

Therefore, by neglecting the set of sequences of random vectors with total probability equal to one, we have an ordinary convergence. Of course, the rate of convergence then depends on the particular sample sequence and it has a random character.

The convergence of the algorithms of adaptation is equivalent to the stability of the systems described by stochastic difference or differential equations. The stability of these systems must be considered in a probabilistic sense; in probability, in the mean-square sense and almost surely (or with probability one). Stochastic stability is a relatively new subject in the theory of stability, and it is now being intensively developed.

3.13 Conditions of Convergence

The convergence of the algorithms of adaptation or, equivalently, the stability of nonautonomous stochastic feedback systems is a basic problem in the realization of the algorithms of adaptation. At present we can state certain necessary and sufficient conditions for convergence. Let us consider an algorithm of adaptation given by (3.10):

$$\Delta \mathbf{c}[n-1] = -\gamma[n]\, \nabla_{\mathbf{c}}\, Q(\mathbf{x}[n], \mathbf{c}[n-1]) \qquad (3.32)$$

If $\mathbf{c}[n]$ is to converge to $\mathbf{c}^*$ almost surely, we must at least request that the right-hand side of (3.32) converges to zero as $n \to \infty$, i.e.,

$$\lim_{n \to \infty} \gamma[n]\nabla_{\mathbf{c}}\, Q(\mathbf{x}[n], \mathbf{c}[n-1]) = 0 \qquad (3.33)$$

for practically any $\mathbf{x}[n]$. In the general case, the gradient $\nabla_{\mathbf{c}}\, Q(\mathbf{x}, \mathbf{c})$, as we have already mentioned earlier (see condition (3.7)), is different from zero and thus it is necessary that $\gamma[n]$ tends toward zero as n increases.

The sufficient conditions for convergence of the algorithms of adaptation can also be formulated. The algorithms of adaptation (3.9)–(3.11) converge almost surely if the following conditions are satisfied:

(a) $\gamma[n] > 0 \qquad \sum_{n=1}^{\infty} \gamma[n] = \infty \qquad \sum_{n=1}^{\infty} \gamma^2[n] < \infty$

(b) $\displaystyle\inf_{\varepsilon < \|\mathbf{c}-\mathbf{c}^*\| < 1/\varepsilon} M_{\mathbf{x}}[(\mathbf{c} - \mathbf{c}^*)^T \nabla_{\mathbf{c}}\, Q(\mathbf{x}, \mathbf{c})] > 0 \qquad (\varepsilon > 0) \qquad (3.34)$

(c) $M_{\mathbf{x}}[\nabla_{\mathbf{c}}^{T} Q(\mathbf{x}, \mathbf{c})\nabla_{\mathbf{c}}\, Q(\mathbf{x}, \mathbf{c})] \leq \alpha(1 + \mathbf{c}^T \mathbf{c}) \qquad (\alpha > 0)$

These conditions have a very simple physical and geometric meaning. Condition (a) requires that the rate of decrease in $\gamma[n]$ is such that the variance of the estimate of $J(\mathbf{c})$ is reduced to zero. A sufficiently large number of data must be observed during the variations of $\gamma[n]$ so that the law of large numbers is valid. Condition (b) defines the behavior of the surface $M_{\mathbf{x}}\{\nabla_{\mathbf{c}}\, Q(\mathbf{x}, \mathbf{c})\}$ near zero, and thus the sign of the increments in $\mathbf{c}[n]$. Finally, condition (c) means that the mathematical expectation of the quadratic form $M_{\mathbf{x}}\{\nabla_{\mathbf{c}}^{T} Q(\mathbf{x}, \mathbf{c})\nabla_{\mathbf{c}}\, Q(\mathbf{x}, \mathbf{c})\}$ has a rate of increase with $\mathbf{c}$ smaller than in a parabola of the second degree.

Naturally, for search algorithms of adaptation, there are some definite conditions imposed on $a[n]$, and the form of the conditions is modified, since the gradient is not available. The search algorithms of adaptation (3.15) converge almost surely, and in the mean-square sense, when the following conditions are satisfied:

(a) $\displaystyle\sum_{n=1}^{\infty} \gamma[n] = \infty \qquad \sum_{n=1}^{\infty} \gamma[n]a[n] < \infty \qquad \sum_{n=1}^{\infty} \left(\frac{\gamma[n]}{a[n]}\right)^2 < \infty$

(b) $(c - c^*)^T(Q_+(x, c, \varepsilon) - Q_-(x, c, \varepsilon))$
$$\geq K \|c - c^*\| \|Q_+(x, c, \varepsilon) - Q_-(x, c, \varepsilon)\| \tag{3.35}$$

where $\varepsilon > 0$, $K > 1/\sqrt{2}$

(c) $\|Q_+(x, c, a) - Q_-(x, c, a)\| \leq A \|c - c^*\| + B$

These conditions have almost the same meaning as those considered above, and thus we shall not discuss them here.

The stochastic algorithms are very stable when noise is present. Random additive noise with a mean value equal to zero is averaged and it does not influence the result, i.e., the optimal vector $c = c^*$.

If this noise does not exist, i.e., if its variance is equal to zero and for any x

$$\nabla_c Q(x, c^*) = 0 \tag{3.36}$$

condition (3.33) is satisfied even for $\gamma =$ const or for a γ which converges to a positive constant. In this case, in order to establish the maximal value γ_{max}, we can use the same approach that was employed in the regular algorithms of optimization. For $\gamma < \gamma_{max}$ we can almost surely obtain the convergence which is very close to the ordinary convergence.

The algorithms obtained in this case from (3.9) and (3.11) for $\gamma[n] = \gamma_0 =$ const are not very stable under noisy conditions. When noise has a variance σ^2, the convergence does not exist in the sense discussed previously, although

$$\lim_{n \to \infty} M_x[\|c[n] - c^*\|^2] \leq \mu(\gamma_0, \sigma) \tag{3.37}$$

where $\mu(\gamma_0, \sigma) \to 0$ when $\sigma^2 \to 0$. In practice, we may accept a milder condition which, in a certain sense, corresponds to the nonasymptotic Lyapunov's stability. Analogous conditions of convergence are known also for continuous algorithms. However, we shall not state them here. Considering the great enthusiasm with which they are studied in the theory of stochastic stability, we may expect more general and simpler conditions of stability. Thus far, Dvoretzky's theorem and the theory of martingales are fundamental and play a basic role in the proofs of convergence of stochastic iterative methods.

3.14 Stopping Rules

In practical applications of the algorithms, it is important to know the number of steps, n, after which the optimal vector c^* is determined with sufficient accuracy. There are special stopping rules for regular iterative methods which are based on the comparison of two last values $c[n - 1]$ and $c[n]$. Such stopping rules can be applied for sufficiently large n even in the case of probabilistic iterative algorithms if the convergence with probability one is guaran-

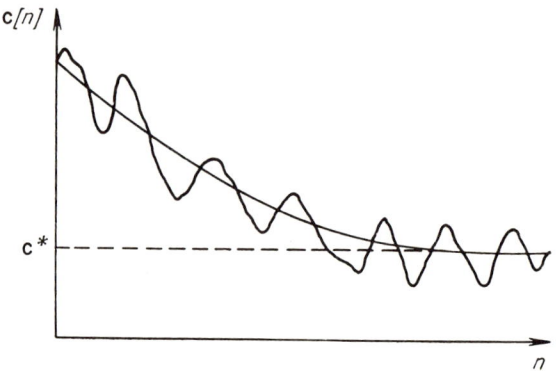

Fig. 3.6

teed, but since the sequence $\{c[n]\} = (c[0], c[1], \ldots)$ is random, such a rule needs an extremely large number of steps. Could we reduce this number of steps? The sequence $\{c[n]\}$ is qualitatively depicted by a continuous curve in Fig. 3.6. It could be considered that the stopping rule defines the index n_0, after which the sequence $\{c[n]\}$ has a stationary character.

In order to obtain a reliable n_0, we have to "smooth" the sequence in a certain way. One of many possible ways to do this is to use the running average

$$\mathbf{m}_N[n] = \frac{1}{N} \sum_{k=n}^{n+N} \mathbf{c}[k] \quad (n = 0, 1, 2, \ldots) \tag{3.38}$$

If, starting from a certain index k_0, for all $k \geq k_0$,

$$\|\mathbf{m}_N[kN] - \mathbf{m}_N[(k+1)N]\| < \varepsilon \tag{3.39}$$

where $\varepsilon > 0$ is a sufficiently small quantity, the index $n_0 = k_0 N$ defines that instant of time for which

$$M\{\mathbf{c}[n_0]\} \approx \mathbf{c}^* \tag{3.40}$$

The smoothing of $\mathbf{c}[n]$ can be accomplished by a modified algorithm which is a special case of (3.27):

$$\mathbf{c}[n] = \mathbf{c}[n-1] - \gamma[n] \, \nabla_\mathbf{c} \, Q(A\mathbf{x}[n], \mathbf{c}[n-1]) \tag{3.41}$$

where, for instance

$$A\mathbf{x}[n] = \frac{1}{n} \sum_{m=0}^{n} \mathbf{x}[m] \tag{3.42}$$

or

$$A\mathbf{x}[n] = \frac{1}{N} \sum_{m=n}^{n+N} \mathbf{x}[m] \tag{3.43}$$

Here, the smoothing of $\mathbf{c}[n]$ is accomplished by a preprocessing of $\mathbf{x}[m]$.

3.15 Acceleration of Convergence

The effectiveness of the algorithms of adaptation first depends on their rate of convergence. By using the methods of functional analysis for estimating the convergence rate of iterative methods, we can establish that for $\gamma[n] = \gamma_0 = \text{const}$ the rate of convergence is described by an exponential law, and for $\gamma[n] = d_0/n$, which satisfies (3.34), the rate of convergence is described by a polynomial law. Therefore, the rate of convergence for $\gamma[n] = d_0/n$ is slower than for $\gamma[n] = \gamma_0$. This fact has a physical explanation: A constant feedback provides a faster response than a feedback diminishing in time. This implies that it is more convenient to use $\gamma[n] = \gamma_0$ when there is no noise.

We can also increase the rate of convergence in the case when noise exists if we only vary $\gamma[n] = (\gamma_1[n], \ldots, \gamma_N[n])$ in a suitable fashion. This can be explained in the following way: The difference $\Delta c_\nu[n-1]$ will have the same sign far from the optimal value $\mathbf{c}^*$ regardless of noise. Close to the optimal value, the sign of this difference considerably depends upon noise. Therefore, the closeness to the optimum can be characterized by a number of changes in the sign of $\Delta c_\nu[n-1]$ per unit of time. The value of $\gamma_\nu[n]$ should be changed only when this sign begins to vary frequently. In order to determine the difference $\Delta c_\nu[n-1]$, it is necessary to perform two measurements. Therefore, $\gamma_\nu[0]$, $\gamma_\nu[1]$ $(\nu = 1, \ldots, N)$ are chosen arbitrarily, but usually equal to one. The future variations are specified according to

$$\gamma_\nu[n] = \gamma_\nu[s[n]] \tag{3.44}$$

where $s[n]$ is an integer-valued function defined by

$$s[n] = 2 + n - \sum_{m=1}^{n} \text{sgn}\,[\Delta c_\nu[m-1]\,\Delta c_\nu[m-2]] \tag{3.45}$$

with

$$\text{sgn}\,z = \begin{cases} 1 & \text{if } z \geq 0 \\ 0 & \text{if } z < 0 \end{cases} \tag{3.46}$$

The rate of convergence can also be improved by another approach. The successive values $\gamma_\nu[n]$ are chosen as in the relaxation method, i.e., by minimizing the gradient of the functional at each step. After each $c[n]$ is determined, we observe

$$v_r = Q(\mathbf{x}[n], \mathbf{c}[n-1] - ra\nabla_{\mathbf{c}} Q(\mathbf{x}[n], \mathbf{c}[n-1])) \tag{3.47}$$

for a certain fixed a until for $r = r_k$ the inequality $v_{r_k+1} > v_{r_k}$ is detected. Then,

$$\gamma[n] = r_k a \tag{3.48}$$

is selected. There are many other ways to increase the rate of convergence, but they are all very specific.

3.16 Measure of Quality for the Algorithms

The choice of $\gamma[n]$ considerably depends on the properties of the algorithms. There is a strong desire to select $\gamma[n]$ in such a way that not only the conditions of an almost certain convergence are satisfied, but also that the conditions under which our algorithm can be considered to be optimal are fulfilled.

Unfortunately, general methods of optimal control theory are not directly applicable to this type of problem. Although this narrows the possibility of discovering the theory of optimal algorithms, we should not stop trying. What kind of measure of quality can be used to evaluate whether a found algorithm is the best? Obviously, a measure of the quality of the algorithms has to depend on n, and it has to express the closeness of $\mathbf{c}[n]$ to the optimal value $\mathbf{c}^*$. The algorithm is then considered to be the best if it optimizes a measure of quality at each step n. Usually such an extremum corresponds to the minimum, since the proximity of $\mathbf{c}[n]$ to $\mathbf{c}^*$ is characterized by a certain generalized distance. Although a measure of quality of the algorithms is analogous to the performance indices of optimization, we emphasize its specific features by using a terminology which makes a distinction between the problems of optimizations solved by the algorithms, and the problems of selecting the best algorithms.

A very natural and traditional measure of the quality of the algorithms is the mean-square deviation of the current vector $\mathbf{c}[n]$ from the unknown optimal vector $\mathbf{c}^*$:

$$\bar{V}^2[n] = M\{\|\mathbf{c}[n] - \mathbf{c}^*\|^2\} \tag{3.49}$$

This measure of quality lies in the foundation of the classical Bayesian approach. A linear combination of the mean-square deviation of the current vector from the optimal vector and the mean-square value of the first difference of the current vector can serve as a generalization of the measure of quality (3.49):

$$\bar{V}_g^2[n] = M\{\|\mathbf{c}[n] - \mathbf{c}^*\|^2\} + \alpha[n]M\{\|\mathbf{c}[n] - \mathbf{c}[n-1]\|^2\} \tag{3.50}$$

This measure of quality requires a definite "smoothness" in the variations of the current vector $\mathbf{c}[n]$.

A slightly different measure of the quality of the algorithms is the performance index for the current values of the vector $\mathbf{c}[n]$ ($n = 1, 2, \ldots$), i.e.,

$$J(\mathbf{c}[n]) = M\{Q(\mathbf{x}, \mathbf{c}[n])\} \tag{3.51}$$

We shall say that the functionals (3.49)–(3.51) are defined for the algorithms if the current values $\mathbf{c}[n]$ in the measures of qualities are defined by the algorithms. Therefore, the problem of finding the best algorithms, for instance of the type (3.9), is reduced to one of finding such $\gamma[n] = \gamma_{\text{best}}[n]$ for which the corresponding functionals (3.49)–(3.51), defined over such algorithms, are minimized.

The functionals (3.49)–(3.51) give either exact or approximate expressions for $\gamma[n] = \gamma_{\text{best}}[n]$ which always contain mathematical expectations of certain functions of $\mathbf{x}[n]$ and of the estimates $\mathbf{c}[n-1]$. Unfortunately, the distributions of $\mathbf{x}[n]$ and $\mathbf{c}[n-1]$ are unknown to us. Such a result is expected, since for the chosen functionals (3.49)–(3.51), we obtain Bayes estimates in the recursive form. In order to compute such estimates, we need sufficiently complete a priori information. This difficulty can be overcome if instead of the functionals based on mathematical expectation, we use empirical or sample mean values

$$J_e(\mathbf{c}[n]) = \frac{1}{n} \sum_{m=1}^{n} Q(\mathbf{x}[m], \mathbf{c}[n]) \qquad (3.52)$$

which corresponds to a substitution of a true probability density function by an empirical one. Such functionals, as we shall see, lead to the expressions $\gamma[n] = \gamma_{\text{best}}[n]$, which depend on the quantities that can be determined from $\mathbf{x}[n]$ and the estimates $\mathbf{c}[n]$. The substitution of the functional (3.51) by (3.52) corresponds to the proverb, "Лучше синицу в руки, чем журавля в небе."* But the situation is actually better in our case.

3.17 The Best Algorithms

Let us consider the algorithm

$$\mathbf{c}[n] = \mathbf{c}[n-1] - \Gamma[n]\nabla_\mathbf{c} Q(\mathbf{x}[n], \mathbf{c}[n-1]) \qquad (3.53)$$

where

$$\Gamma[n] = \mathscr{A}\gamma[n] = \begin{pmatrix} \gamma_1[n] & 0 & \cdots & 0 \\ 0 & \gamma^2[n] & \cdots & 0 \\ \multicolumn{4}{c}{\dotfill} \\ 0 & 0 & \cdots & \gamma_N[n] \end{pmatrix} \qquad (3.54)$$

is a diagonal matrix, and $\mathscr{A}$ is an operator which transforms the vector $\gamma[n]$ into a diagonal matrix. Let us define the functional (3.52) over the algorithm (3.53). Then,

$$J_e(\mathbf{c}[n]) = \frac{1}{n} \sum_{m=1}^{n} Q(\mathbf{x}[m], \mathbf{c}[n-1] - \mathscr{A}\gamma[n]\nabla_\mathbf{c} Q(\mathbf{x}[n], \mathbf{c}[n-1])) \qquad (3.55)$$

The condition of the minimum of this functional is written as

$$\nabla_\gamma J_e(\mathbf{c}[n]) = -\mathscr{A}\nabla_\mathbf{c} Q(\mathbf{x}[n], \mathbf{c}[n-1]\} \frac{1}{n} \sum_{m=1}^{n} \nabla_\mathbf{c} Q(\mathbf{x}[m], \mathbf{c}[n-1]$$

$$-\mathscr{A}\gamma[n]\nabla_\mathbf{c} Q(\mathbf{x}[n], \mathbf{c}[n-1])) = 0 \qquad (n = 1, 2, \ldots) \qquad (3.56)$$

In a general case, the dependence of $\nabla_\mathbf{c} Q(\cdot)$ on $\mathbf{c}[n]$, and thus on $\gamma[n]$, is nonlinear. Therefore, the equations (3.56) cannot be solved explicitly with re-

---

* "A feather in hand is better than a bird in air."

spect to $\gamma[n]$. The only way to find $\gamma[n]$ is to use the regular iterative algorithm

$$\gamma[n, k] = \gamma[n, k-1] - \gamma_0 \nabla_\gamma J_e(\mathbf{c}[n-1] - \mathscr{A}\gamma[n, k-1] \nabla_\mathbf{c} Q(\mathbf{x}[n], \mathbf{c}[n-1])) \quad (n = 1, 2, \ldots) \quad (3.57)$$

where $\gamma_0 = \text{const}$ and $n = \text{const}$ in the time interval between the $(n-1)$st and the nth step. This algorithm then determines $\gamma_{\text{best}}[n]$ as a value of $\gamma[n, k]$ for a large k. Of course the iterative procedure (3.57) has to be carried at a faster rate.

If we assume that the norm $\|\gamma[n]\|$ is small (according to (3.53), this is equivalent to the assumption that $\|\mathbf{c}[n] - \mathbf{c}[n-1]\|$ is small), we can define $\gamma[n]$ approximately in an explicit form. Let the diagonal matrix $\mathscr{A}\nabla_\mathbf{c} Q(\cdot)$ be nonsingular; then, considering only linear approximations, we write condition (3.56) approximately as

$$\sum_{m=1}^{n} \nabla_\mathbf{c} Q(\mathbf{x}[m], \mathbf{c}[n-1]) - \sum_{m=1}^{n} \nabla_\mathbf{c}^2 Q(\mathbf{x}[m], \mathbf{c}[n-1])$$
$$\times \mathscr{A}\nabla_\mathbf{c} Q(\mathbf{x}[n], \mathbf{c}[n-1])\gamma[n] \approx 0 \quad (3.58)$$

Here,

$$\nabla_\mathbf{c}^2 Q(\mathbf{x}, \mathbf{c}) = \left\| \frac{\partial^2 Q(\mathbf{x}, \mathbf{c})}{\partial c_v \, \partial c_\mu} \right\| \quad (v, \mu = 1, \ldots, N)$$

is the matrix of second derivatives.

Considering now that for any n we want to have

$$\sum_{m=1}^{n-1} \nabla_\mathbf{c} Q(\mathbf{x}[m], \mathbf{c}[n-1]) = 0 \quad (3.59)$$

we obtain from (3.58)

$$\gamma_{\text{best}}[n] \approx \left[\sum_{m=1}^{n} \nabla_\mathbf{c}^2 Q(\mathbf{x}[m], \mathbf{c}[n-1]) \mathscr{A}\nabla_\mathbf{c} Q(\mathbf{x}[n], \mathbf{c}[n-1]) \right]^{-1}$$
$$\times \nabla_\mathbf{c} Q(\mathbf{x}[n], \mathbf{c}[n-1]) \quad (3.60)$$

Using the inverse of the product of two matrices, we write (3.60) as

$$\gamma_{\text{best}}[n] \approx [\mathscr{A}\nabla_\mathbf{c} Q(\mathbf{x}[n], \mathbf{c}[n-1])]^{-1} \left[\sum_{m=1}^{n} \nabla_\mathbf{c}^2 Q(\mathbf{x}[m], \mathbf{c}[n-1]) \right]^{-1}$$
$$\times \nabla_\mathbf{c} Q(\mathbf{x}[n], \mathbf{c}[n-1]) \quad (3.61)$$

We can now determine $\Gamma_{\text{best}}[n] = \mathscr{A}\gamma_{\text{best}}[n]$ according to (3.54), and then the algorithm (3.53) which corresponds to this matrix has the form

$$\mathbf{c}[n] = \mathbf{c}[n-1] - \left[\sum_{m=1}^{n} \nabla_\mathbf{c}^2 Q(\mathbf{x}[m], \mathbf{c}[n-1]) \right]^{-1} \nabla_\mathbf{c} Q(\mathbf{x}[n], \mathbf{c}[n-1]) \quad (3.62)$$

Since this algorithm is only correct for small $\|\mathbf{c}[n] - \mathbf{c}[n-1]\|$, it is appropriate to call it only an approximately optimal algorithm. We can define (in a similar manner) the best continuous algorithms. It is interesting to note that the constraints (small γ) which offered us only approximately best algorithms do not exist in the case of the best continuous algorithms.

3.18 Simplified Best Algorithms

Frequently, the expression $\Gamma_{\text{best}}[n] = \mathscr{A}\gamma_{\text{best}}[n]$ is very complex, and this can cause certain difficulties. We can then pose the problem of finding the best algorithms in which the diagonal matrix $\Gamma[n]$ is replaced by a single scalar quantity $\gamma[n]$. In this case, $\gamma_{\text{best}}[n]$ will be found by the steepest descent method discussed in Section 2.19. Instead of (3.56), we obtain

$$\frac{dJ_c(\mathbf{c}[n])}{d\gamma[n]} = -\nabla_c^T Q(\mathbf{x}[n], \mathbf{c}[n-1]) \frac{1}{n} \sum_{m=1}^{n} \nabla_c Q(\mathbf{x}[m], \mathbf{c}[n-1])$$

$$- \gamma[n])\nabla_c Q(\mathbf{x}[n], \mathbf{c}[n-1]) = 0 \quad (3.63)$$

and for small $\|\mathbf{c}[n] - \mathbf{c}[n-1]\|$,

$$\gamma_{\text{best}}[n]$$
$$\approx \frac{\nabla_c^T Q(\mathbf{x}[n], \mathbf{c}[n-1]) \sum_{m=1}^{n} \nabla_c Q(\mathbf{x}[m], \mathbf{c}[n-1])}{\nabla_c^T Q(\mathbf{x}[n], \mathbf{c}[n-1]) \sum_{m=1}^{n} \nabla_c^2 Q(\mathbf{x}[m], \mathbf{c}[n-1])\nabla_c Q(\mathbf{x}[n], \mathbf{c}[n-1])}$$

$$(3.64)$$

We shall now consider one more method of obtaining a simplified approximately best algorithm for which it is not necessary that $\|\Delta \mathbf{c}[n-1]\| = \|\mathbf{c}[n] - \mathbf{c}[n-1]\|$ is small. The measure of quality of the algorithms is chosen in the form of the functional (3.49). Using $\boldsymbol{\eta}[n] = \mathbf{c}[n] - \mathbf{c}^*$, we write the algorithm (3.53) in which the diagonal matrix is replaced by a scalar:

$$\boldsymbol{\eta}[n] = \boldsymbol{\eta}[n-1] - \gamma[n]\,\nabla_c Q(\mathbf{x}[n], \mathbf{c}^* + \boldsymbol{\eta}[n-1]) \quad (3.65)$$

Let us now find the mathematical expectation of the euclidean norm

$$M\{\|\boldsymbol{\eta}[n]\|^2 \,|\, \boldsymbol{\eta}[n-1]\} = \|\boldsymbol{\eta}[n-1]\|^2$$
$$- 2\gamma[n]\boldsymbol{\eta}^T[n-1]M_x\{\nabla_c Q(\mathbf{x}[n], \mathbf{c}^* + \boldsymbol{\eta}[n-1])\}$$
$$+ \gamma^2[n]M\{\|\nabla_c Q(\mathbf{x}[n], \mathbf{c}^* + \boldsymbol{\eta}[n-1])\|^2\} \quad (3.66)$$

Let us assume now that for all

$$\|M\{\nabla_c Q(\mathbf{x}[n], \mathbf{c}^* + \boldsymbol{\eta}[n-1])\}\| \leq k_1 \|\boldsymbol{\eta}[n-1]\|$$
$$k_0 \|\boldsymbol{\eta}[n-1]^2\| \leq \boldsymbol{\eta}^T[n-1]M\{\nabla_c Q(\mathbf{x}[n], \mathbf{c}^* + \boldsymbol{\eta}[n-1])\}$$
$$(3.67)$$

Then, by considering the condition

$$M\{\|\nabla_c Q(\mathbf{x}[n], \mathbf{c}^* + \boldsymbol{\eta}[n-1])\|^2\} = \|M\{\nabla_c Q(\mathbf{x}[n], \mathbf{c}^* + \boldsymbol{\eta}[n-1])\}\|^2 + \sigma^2$$

$$(3.68)$$

and substituting the estimates (3.67) in (3.66), we obtain

$$M\{\|\boldsymbol{\eta}[n]\|^2 \,|\, \boldsymbol{\eta}[n-1]\} \leq \|\boldsymbol{\eta}[n-1]\|^2 \{1 - 2k_0 \gamma[n] + k_1^2 \gamma^2[n]\} + \gamma^2[n]\sigma^2 \tag{3.69}$$

In order to find the unconditional mathematical expectation of (3.49), we average (3.69) for given $\boldsymbol{\eta}[n-1]$. Using notation (3.49), we obtain

$$\overline{V}^2[n] \leq \overline{V}^2[n-1](1 - 2k_0 \gamma[n] + k_1^2 \gamma^2[n]) + \gamma^2[n]\sigma^2 \tag{3.70}$$

We can now find the optimal value $\gamma[n]$. By differentiating the right-hand side of (3.70) with respect to $\gamma[n]$, and then setting the result equal to zero, we obtain

$$\gamma_{\text{best}}[n] = \frac{k_0 \overline{V}^2[n-1]}{k_1^2 \overline{V}^2[n-1] + \sigma^2} \tag{3.71}$$

Obviously, the obtained value, γ_{best}, indicates what can be achieved in a general case for an arbitrary norm $\|\mathbf{c}[n] - \mathbf{c}[n-1]\|$ and when there is no information about probability density functions. By replacing $\gamma_{\text{best}}[n]$ in (3.70), we have

$$\overline{V}^2[n] \leq \overline{V}^2[n-1]\left(1 - \frac{k_0^2 \overline{V}^2[n-1]}{k_1^2 \overline{V}^2[n-1] + \sigma^2}\right) \tag{3.72}$$

From here, we can determine the number of steps after which $\overline{V}^2[n]/\overline{V}^2[0]$ reaches a certain sufficiently small value,

$$n_0 \approx \frac{\sigma^2 + k_0^2 \overline{V}^2[0]}{k_1^2 \overline{V}^2[n]} \tag{3.73}$$

3.19 A Special Case

In the special case when $k_0 = k = 1$, which corresponds to the linear dependence of $\nabla_{\mathbf{c}} Q(\mathbf{x}[n], \mathbf{c}^* + \boldsymbol{\eta}[n-1])$ on $\boldsymbol{\eta}[n-1]$, and when the inequalities (3.67) are replaced by equalities, we have, from (3.71) and (3.72),

$$\gamma_{\text{best}}[n] = \frac{\overline{V}^2[n-1]}{\overline{V}^2[n-1] + \sigma^2}$$
$$\overline{V}^2[n] = \frac{\overline{V}^2[n-1]\sigma^2}{\overline{V}^2[n-1] + \sigma^2} \tag{3.74}$$

We can easily determine from (3.74) that

$$\gamma_{\text{best}}[n] = \frac{1}{n + \frac{\sigma^2}{\overline{V}^2[0]}} \qquad \overline{V}^2[n] = \frac{\sigma^2}{n + \frac{\sigma^2}{\overline{V}^2[0]}} \tag{3.75}$$

If there is no a priori information about the initial mean-square deviation, we can assume that $\overline{V}^2[0]$ is equal to infinity, and then

$$\gamma_{\text{best}}[n] = \frac{1}{n} \qquad \overline{V}^2[n] = \frac{\sigma^2}{n} \qquad (3.76)$$

This is a well-known result for the linear estimates of $\mathbf{c}^*$.

3.20 Relationship to the Least-Square Method

If $Q(\mathbf{x}, \mathbf{c})$ is a quadratic function of $\mathbf{c}$, for instance,

$$Q(\mathbf{x}, \mathbf{c}) = (y - \mathbf{c}^T \boldsymbol{\phi}(\mathbf{x}))^2$$

the substitution of (3.56) by (3.58) will not be approximate but exact. This means that the expression for $\gamma_{\text{best}}[n]$ in (3.61) is exact, and in this case it becomes

$$\gamma_{\text{best}}[n] = [\mathscr{A}\boldsymbol{\phi}(\mathbf{x}[n])]^{-1} \left[\sum_{m=1}^{n} \boldsymbol{\phi}(\mathbf{x}[m])\boldsymbol{\phi}^T(\mathbf{x}[m]) \right]^{-1} \boldsymbol{\phi}(\mathbf{x}[n]) \qquad (3.77)$$

Algorithm (3.53) then has the form

$$\mathbf{c}[n] = \mathbf{c}[n-1] + K[n](y[n] - \mathbf{c}^T[n-1]\boldsymbol{\phi}(\mathbf{x}[n]))\boldsymbol{\phi}(\mathbf{x}[n]) \qquad (3.78)$$

where

$$K[n] = \left[\sum_{m=1}^{n} \boldsymbol{\phi}(\mathbf{x}[m])\boldsymbol{\phi}^T(\mathbf{x}[m]) \right]^{-1} \qquad (3.79)$$

is Kalman's matrix, which can be computed by a recursive formula of the least-square method. At each stage n we obtain the best estimate of $\mathbf{c}^*$ in the least-square sense. This is accomplished for the cost of tedious computations in (3.77). If we assume that $\phi_v(\mathbf{x}[m])$ ($v = 1, \ldots, N$) are independent, (3.77) is simplified, and

$$\gamma_{\text{best}}[n] = \left(\frac{1}{\sum_{m=1}^{n} \phi_1^2(\mathbf{x}[m])}, \frac{1}{\sum_{m=1}^{n} \phi_2^2(\mathbf{x}[m])}, \ldots, \frac{1}{\sum_{m=1}^{n} \phi_N^2(\mathbf{x}[m])} \right)$$
$$(3.80)$$

In this case, we naturally have a simpler algorithm.

The best or the approximately best algorithms are suitable for those cases when there is finite number of data.

3.21 Relationship to the Bayesian Approach

For a finite number of observations, the Bayesian method provides the best estimate of the optimal vector $\mathbf{c}^*$ by minimizing a certain loss function. This is accomplished by using the complete a priori information about the probability density functions, and unfortunately by very tedious computations.

For a particular class of probability distributions, the criterion of optimality can be written in the form of the sample mean

$$J(\mathbf{c}[n]) = M\{Q(\mathbf{x}, \mathbf{c}[n]) \mid \mathbf{x}[1], \ldots, \mathbf{x}[n]\} = \frac{1}{n} \sum_{m=1}^{n} Q(\mathbf{x}[m], \mathbf{c}[n]) \quad (3.81)$$

This equality is correct for exponential distributions in particular.

Therefore, the algorithm of adaptation (3.62) simultaneously minimizes the conditional expectation (3.81), and thus it is approximately the best algorithm. If $Q(\mathbf{x}, \mathbf{c})$ is a quadratic function of $\mathbf{c}$, the algorithm is the best without any approximations. It could then be concluded that for a special choice of $\gamma[n]$ the probabilistic iterative algorithms give the same results as the Bayesian algorithm in the case of the quadratic loss functions. The value of $\gamma_{\text{best}}[n]$ depends upon the available vectors $\mathbf{x}[1], \mathbf{x}[2], \ldots, \mathbf{x}[n]$. It should be noted that for a vector of large dimensions the computational difficulties are such that even modern digital computers cannot handle them.

However, the problem can be solved by using the probabilistic iterative algorithms between two successive samples with $\gamma[n, k]$ of the type a/k. The infinite number of samples necessary for such algorithms is substituted by a periodic repetition of the finite number of already observed samples. In such a case the iterative probabilistic algorithms give the same best estimate for $\gamma[n] = \gamma_{\text{best}}[n]$. Of course, this periodization has to be carried at a faster rate so that we can define an estimate $\mathbf{c}[n]$ within every interval between the $(n-1)$st and the nth observation.

3.22 Relationship to the Maximum Likelihood Method

The maximum likelihood method is widely used in statistics. This method is based on a belief that the best estimate has to give the highest probability to the samples actually observed during the experiment.

Let $\mathbf{x}[n]$ be a sequence of independent random variables with identical probability density function $p(\mathbf{x}[n], \mathbf{c})$, where $\mathbf{c}$ is a vector parameter which has to be estimated. The likelihood function is defined as a function of the vector $\mathbf{c}$ which is obtained from the joint distribution of the samples $\mathbf{x}[m]$ ($m = 1, \ldots, n$):

$$L(\mathbf{x}, \mathbf{c}) = \prod_{m=1}^{n} p(\mathbf{x}[m], \mathbf{c}) \quad (3.82)$$

The estimate $\mathbf{c}^*$ is found from the condition satisfied by the maximum of the likelihood function with respect to the parameter $\mathbf{c}$:

$$\nabla_{\mathbf{c}} L(\mathbf{x}, \mathbf{c}) = 0 \quad (3.83)$$

3.22 Relationship to the Maximum Likelihood Method

Frequently, instead of the likelihood function $L(\mathbf{x}, \mathbf{c})$, we use $\log L(\mathbf{x}, \mathbf{c})$. Condition (3.83) is then replaced by

$$\nabla_\mathbf{c} \log L(\mathbf{x}, \mathbf{c}) = \sum_{m=1}^{n} \nabla_\mathbf{c} \log p(\mathbf{x}[m], \mathbf{c}) = 0 \qquad (3.84)$$

and the problem is to determine the real roots of equation (3.83) or (3.84), which are the sought estimates. This can be achieved with the help of the regular iterative methods. But if the likelihood function is varying with the number of observations, it appears that it is more convenient to use the probabilistic iterative methods.

Let us write (3.84) as

$$\frac{1}{n} \sum_{m=1}^{n} \nabla_\mathbf{c} \log p(\mathbf{x}[m], \mathbf{c}) = 0 \qquad (3.85)$$

It is assumed that all the conditions under which (3.81) can be used are satisfied. Equation (3.85) can then be written as

$$M\{\nabla_\mathbf{c} \log p(\mathbf{x}, \mathbf{c}) \mid \mathbf{x}[1], \ldots, \mathbf{x}[n]\} = 0 \qquad (3.86)$$

The algorithm

$$\mathbf{c}[n] = \mathbf{c}[n-1] + \gamma[n] \nabla_\mathbf{c} \log p(\mathbf{x}[n], \mathbf{c}[n-1]) \qquad (3.87)$$

can now be used. This algorithm can also be obtained by slightly different considerations. For instance, let us observe the mathematical expectation

$$M\{\nabla_\mathbf{c} \log p(\mathbf{x}, \mathbf{c})\} = \int_X \nabla_\mathbf{c} p(\mathbf{x}, \mathbf{c}) \frac{1}{p(\mathbf{x}, \mathbf{c})} p(\mathbf{x}, \mathbf{c}) \, d\mathbf{x} \qquad (3.88)$$

It is assumed here that the observed samples belong to the population with the distribution $p(\mathbf{x}, \mathbf{c}^*)$. After obvious transformations from (3.88), we obtain for $\mathbf{c} = \mathbf{c}^*$,

$$M\{\nabla \log p(\mathbf{x}, \mathbf{c})\} = \nabla \int_X p(\mathbf{x}, \mathbf{c}) \, d\mathbf{x} = 0 \qquad (3.89)$$

since

$$\int_X p(\mathbf{x}, \mathbf{c}) \, d\mathbf{x} = 1 \qquad (3.90)$$

Therefore, the mathematical expectation of $\nabla \log p(\mathbf{x}, \mathbf{c})$ is equal to zero for $\mathbf{c} = \mathbf{c}^*$. We can now conclude that the algorithm for estimating $\mathbf{c}^*$ has the form (3.87). For a suitable choice of $\gamma[n]$, we can obtain an unbiased, asymptotically unbiased and asymptotically efficient estimate.

3.23 Discussion

The problem of selecting the best algorithms is related to the solution of optimization problems, but it also has some special characteristics. We are not interested in the intermediate values of the vector **c**[n] defined by the algorithm, as long as it lies within the acceptable limits. We would also be completely satisfied if the algorithm defined the optimal vector with a prespecified accuracy in the smallest number of steps, i.e., in the shortest time. But what can we do with such a measure of the quality of the algorithms? It appears that the minimum-time algorithms cannot be realized. Therefore, we have to be satisfied with the best algorithms which minimize certain empirical functionals. If the gradient of realization $\nabla_\mathbf{c} Q(\mathbf{x}, \mathbf{c})$ is a linear function of **c**, the best algorithms are found without any difficulties (see, for instance, Section 3.9). In any other case, we can find only the approximately best algorithms on the basis of linear approximations (see, for instance, Section 3.18). In obtaining the best algorithms, it is profitable to use the results of statistical decision theory, conditional Markov chains, etc. It should also be emphasized that a substitution of a nonlinear problem by a linear approximation leads to the results which are accurate only for small deviations $\|\mathbf{c}[n] - \mathbf{c}[n-1]\|$. Such algorithms do not necessarily converge. It is then natural to ask whether it is worthwhile to consider the optimization of linearized processes, and if it is, then in what cases?

But even if we could find the best algorithms in the general form, their realization would be so complex that we would reject it. Therefore, the role of the optimal algorithms should not be overemphasized. A similar conclusion was reached in the theory of optimal systems where the optimal algorithms are only used to evaluate simplified but realizable algorithms. From this point of view, the studies of different simplified optimal algorithms which can be used in the time interval between two successive observed samples are of interest. In the following chapters we shall not emphasize the optimal algorithms. Using the results presented above, the reader can determine the best or the approximately best algorithms when they are needed.

3.24 Certain Problems

An important problem in the development of the algorithms of adaptation is the problem of formulating and constructing the best algorithms of adaptation. Since time is important in adaptation, the main efforts should be directed to the algorithms of adaptation with the fastest rate of convergence.

The adaptation time can be reduced if we do not ask for an almost certain convergence "into a point," but rather, "into a region."

$$P\left\{\lim_{n \to \infty} \|\mathbf{c}[n] - \mathbf{c}^*\| \leq \varepsilon\right\} = 1 \qquad (3.91)$$

We can ask for the minimum of the expected value of the number of steps after which the estimate of c^* falls into a desired region.

An explanation of the conditions of almost certain convergence into a region would be of great interest, especially in the case of the multistage algorithms. Finally, a further reduction in the adaptation time can probably be achieved on the basis of the rules which define the moment when the algorithm becomes stationary. It would be important to establish similar rules.

In order to broaden the class of the problems under consideration, it would be very desirable to develop and apply the apparatus of linear and nonlinear programming in the cases when the functionals and the constraints are not given in explicit form.

It would be useful to develop the algorithms of adaptation on the basis of the methods of feasible directions and random search.

3.25 Conclusion

We have become familiar with the probabilistic iterative methods, and in particular with the stochastic approximation method, which lie in the foundation of the developed concepts of adaptation and learning in automatic systems. These methods are employed in the construction of various algorithms of adaptation which use the observed samples to minimize the functionals which are not explicitly defined. The fact that we need time to determine the minimum of the functional or the optimal vector c^* is sad, but unavoidable —it is a cost that we have to pay in order to solve a complex problem in the presence of uncertainty. At this point, one should remember the regular algorithms for minimization of known functionals which also require time for convergence.

From the presented ideas, it could be concluded that adaptation and learning are characterized by a sequential gathering of information and the usage of current information to eliminate the uncertainty created by insufficient a priori information.

It is now time to show that the presented concepts of adaptation and learning allow us to consider different problems from a certain general viewpoint which not only unifies them, but also provides an effective method for their solution. This is the purpose of the remaining chapters of the book.

COMMENTS

3.2   At the present time, there is no shortage of definitions describing the terms learning and adaptation (Bush and Mosteller, 1955; Pask, 1962, 1963; Zadeh, 1963; Gibson, 1962; Sklansky, 1964, 1965, 1966; Jacobs, 1964; Feldbaum, 1964; etc.). The author found it much easier to create one more definition than to try to unify and apply the definitions that are widely scattered in the literature.

3 Adaptation and Learning

3.4 The basic idea of the stochastic approximation method was apparently born long ago, but it was first formulated in a clear and understandable way by Robbins and Monro (1951). They have indicated an iterative procedure (which uses sample realizations) for finding the roots of the regression equation. This procedure was actually a stochastic variant of the iterative procedure of von Mises and Pollaczek-Geiringer (1929). The work by Robbins and Monro has caused an avalanche of studies developing the idea.

Wolfowitz (1952) has relaxed the conditions of convergence. Callianpur (1954) and Blum (1954a) have independently shown, under even less strict conditions than those by Wolfowitz (1952), that the Robbins-Monro procedure converges not only in probability, but also with probability one. The Robbins-Monro procedure was generalized by Dvoretzky (1956). A simpler proof of this generalized procedure was given by Wolfowitz (1952) and Derman and Sacks (1959). A modification of this procedure, which consists in the selection of $\gamma[n]$, was proposed by Friedman (1963). The iterative procedure was generalized to the multidimensional case by Blum (1954a). A simple and clear proof of convergence for the multidimensional procedure was given by Gladishev (1965). For the linear regression functions, the modified procedure described by Dupač (1958) seems to be convenient.

Asymptotic properties of the iterative procedure are explained by Schmetterer (1953), Chung (1954) and Gladishev (1965), and, in the case of the linear regression function, by Hodges and Lehman (1956). Krasulina (1962a, b) discovered a relationship between the iterative procedure of Blum (1954a) and the generalized iterative procedure of Dvoretzky (1956), and presented a brief survey of the stochastic approximation methods.

The Robbins-Monro procedure was mentioned in the book by van der Waerden (1957), but a systematic presentation can be found in the book by Wetherill (1966). The papers by Gutman and Gutman (1959) in biometrics and Cochran and Davis (1965) in biology are a few of the publications about various applications of the Robbins-Monro procedure which fall outside the area of our interest.

A comparison of various methods of optimization in the presence of noise was performed by Gurin (1966a, b) and Movshovich (1966). See also the paper by Hasminskii (1965). Comprehensive surveys of the basic results of the stochastic approximation method were given by Derman (1956a), Schmetterer (1954b) and Loginov (1966).

3.6 In 1952, the idea of stochastic approximation was extended by Kiefer and Wolfowitz (1952) to the case of finding an extremum of the regression function. This was the first probabilistic search algorithm which was a variation on the search iterative procedure proposed by Germansky (1934). The convergence of the Kiefer-Wolfowitz search procedure with probability one was proved by Blum (1954a) and Burkholder (1956). There are many different generalizations for the multidimensional case (see the publications by Blum (1954b) and Gray (1964)). The asymptotic properties of the Kiefer-Wolfowitz procedure were studied by Derman (1956a), Dupač (1957) and Sacks (1958). The surveys mentioned above (Derman, 1956b; Schmetterer, 1953; and Loginov, 1966) also devote much attention to the Kiefer-Wolfowitz procedure.

Dvoretzky (1956) has shown that the Robbins-Monro and the Kiefer-Wolfowitz procedures are the special cases of a general procedure proposed by him. A generalization of Dvoretzky's procedure for the case of finite euclidean spaces is given by Derman and Sacks (1959) and for the case of Hilbert spaces by Schmetterer (1953). Recently, all these results were generalized by Venter (1966).

3.8 The questions regarding the convergence of the algorithms of adaptation under the inequality constraints are not yet completely answered.

3.9 Ya.I. Hurgin has pointed out to us the possibility of such a generalization.

3.11 The continuous algorithms have been thoroughly studied by Spaček and his students. The results of the studies in this direction for the algorithms without search are presented in the publications by Driml and Nedoma (1959), Driml and Hanš (1959b) and Hanš and Spaček (1959). Sakrison (1964) has described the search algorithms.

3.12 Rigorous formulations of various types of convergence are presented in the book by Loèv (1955); see also the book by Middleton (1960).

3.13 The conditions of convergence, which can be obtained on the basis of the stochastic approximation results listed above, are presented here. It would be very interesting to obtain the conditions of convergence as a special case of the conditions of stability for the stochastic systems. This problem was treated in the publications by Katz and Krasovskii (1960), Hasminskii (1962) and Kushner (1965).

Inequality (3.37) was obtained by Comer (1964). An explicit expression for η is given in his paper.

3.15 The described methods of accelerating the convergence were proposed by Kesten (1958) and Fabian (1960). See also the book by Wilde (1964).

3.16 The measure of quality of the algorithms similar to (3.49) was used by Dvoretzky (1956) and by Nikolic and Fu (1966) for obtaining the best algorithms for estimating the mean value.

3.17 The best value $\gamma[n]$ for the scalar case was obtained on the basis of different considerations by Stratonovich (1968).

3.18 The conditions (3.70)–(3.72) for a special case of $k_0 = k_1 = 1$ were obtained by Kirvaitis and Fu (1966).

3.19 The considerations presented here are the developments of an idea due to Dvoretzky (1956). See also the paper by Block (1957). These ideas were applied by Pervozvanskii (1965) in the selection of the optimal step for the simplest impulsive extremal systems.

3.20 A detailed presentation of the recursive least-square method was given in the publications by Albert and Sittler (1965). The recursive formula for computing $\Gamma[n]$ from the preceding values $\Gamma[n-1]$ can be obtained on the basis of a very popular recursive formula for matrix inversion proposed by R. Penrose (see the book by Lee (1964)).

The interesting method proposed by Kaczmarz is described in Tompkin's paper (Beckenbach, 1956). One can become familiar with the method by reading the original paper by Kaczmarz (1937).

3.21 Formula (3.81) for fixed n was obtained by Stratonovich (1968). The proof of convergence of recursive procedures with the repetition of a finite number of data was given by Litvakov (1966).

3.22 We use certain relationships presented in the book by Cramér (1954), and the ideas developed in the works by Sakrison (1964, 1965). In these works, the reader can find a detailed presentation of the recursive algorithms and their application in the solution of certain radar problems.

3.23 The questions related to the best algorithms were discussed from different points of view in the papers by Stratonovich (1968) and the author (Tsypkin, 1968).

BIBLIOGRAPHY

Albert, A., and Sittler, R.W. (1965). A method of computing least squares estimators that keep up with the data, *J. SIAM Contr.* Series A **3** (3).

Beckenbach, E.B. (ed.) (1956). "Modern Mathematics for the Engineer." McGraw-Hill, New York.

Block, H.D. (1957). Estimates of error for two modifications of the Robbins-Monro stochastic approximation process, *Ann. Math. Stat.* **28** (4).

Blum, J.R. (1954a). Approximation methods which converge with probability one, *Ann. Math. Stat.* **25** (2).

Blum, J.R. (1954b). Multidimensional stochastic approximation methods, *Ann. Math. Stat.* **25** (4).

Blum, J.R. (1958). A note on stochastic approximation, *Proc. Am. Math. Soc.* **9** (3).

Burkholder, D.L. (1956). On a class of stochastic approximation processes, *Ann. Math. Stat.* **27** (4).

Bush, R.R., and Mosteller, F.L. (1955). "Stochastic Models for Learning." Wiley, New York.

Callianpur, G. (1954). A note on the Robbins-Monro stochastic approximation method, *Ann. Math. Stat.* **25** (2).

Chung, K.L. (1954). On a stochastic approximation method, *Ann. Math. Stat.* **25** (3).

Cochran, W.G., and Davis, M. (1965). The Robbins-Monro method for estimating the median lethal dose, *J. Roy. Stat. Soc.* **27** (1).

Comer, J.P. (1964). Some stochastic approximation procedures for use in process control, *Ann. Math. Stat.* **35** (3).

Cramér, H. (1954). "Mathematical Methods of Statistics." Princeton University Press, Princeton, N.J.

Derman, C.T. (1956a). An application of Chung's lemma to Kiefer-Wolfowitz stochastic approximation procedure, *Ann. Math. Stat.* **27** (2).

Derman, C.T. (1956b). Stochastic approximation, *Ann. Math. Stat.* **27** (4).

Derman, C.T., and Sacks, J. (1959). On Dvoretzky's stochastic approximation theorem, *Ann. Math. Stat.* **30** (2).

Driml, M., and Hanš, O. (1959a). On experience theory problems. *In* "Transactions of the Second Prague Conference on Information Theory, Statistical Decision Functions Random Processes, Prague, 1959."

Driml, M., and Hanš, O. (1959b). Continuous stochastic approximations. *In* "Transactions of the Second Prague Conference on Information Theory, Statistical Decision Functions, Random Processes, Prague, 1959."

Driml, M. and Nedoma, T. (1959). Stochastic approximation for continuous random processes. *In* "Transactions of the Second Prague Conference on Information Theory, Statistical Decision Functions, Random Processes, Prague, 1959."

Dupač, V. (1957). O. Kiefer-Wolfowitzvé approximaćni methodé, *Cas. Pestováni Mat.* **82** (1).

Dupač, V. (1958). Notes on stochastic approximation methods, *Czech. Math. J.* **8** (1), 83.

Dupač, V. (1966a). A dynamic stochastic approximation method, *Ann. Math. Stat.* **36** (6).

Dupač, V. (1966b). Stochastic approximation in the presence of trend, *Czech. Math. J.* **16** (3), 91.

Dvoretzky, A. (1956). On stochastic approximation. *In* "Proceedings of the Third Berkeley Symposium on Mathematical Statistics and Probability," Vol. 1. University of California Press, Berkeley.

Fabian, V. (1960). Stochastic approximation methods, *Czech. Math. J.* **10** (1), 85.

Fabian, V. (1964). A new one-dimensional stochastic approximation method for finding a local minimum of a function. *In* "Transactions of the Third Prague Conference on Information Theory, Statistical Decision Functions, Random Processes, Prague, 1964."

Fabian, V. (1965). Přehled deterministických a stochastickych approximačnich metod pro minimalizaci funkci, *Kibern. Dok.* **1** (6).

Fabius, T. (1959). Stochastic approximation, *Stat. Nederlandica* **13** (2).

Feldbaum, A.A. (1964). The process of learning in men and automata. *In* "Kibernetika, Mishlenie, Zhizn," Misl. (In Russian.)

Friedman, A.A. (1963). On stochastic approximations, *Ann. Math. Stat.* **34** (1).

Germansky, R. (1934). Notiz über die Lösing von Extremaufgaber mittels Iteration, *Z. Angew. Math. Mech.* **14** (3).

Gibson, J.E. (1962). Adaptive learning systems, *Proc. Nat. Electron. Conf.* **18**.

Gladishev, E.G. (1965). On stochastic approximation, *Teor. Veroyat. Ee Primen.* **10** (2). (In Russian.)

Gray, K. (1964). Application of stochastic approximation to the optimization of random circuits, *Proc. Symp. Appl. Math.* **16**.

Gurin, L.S. (1964). On a question of optimization in stochastic models, *Zh. Vychisl. Mat. Mat. Fiz.* **4** (6). (In Russian.)

Gurin, L.S. (1966a). On a question of relative evaluation of different methods of optimization, *Avtomatika Vychisl. Tekh. Riga* (10). (In Russian.)

Gurin, L.S. (1966b). Random search under disturbances, *Izv. Akad. Nauk SSSR Tekh. Kibern.* **1966** (3). (Engl. trans.: *Eng. Cybern. (USSR).*)

Gutman, T.L., and Gutman, R. (1959). An illustration of the use of stochastic approximation, *Biometrics* **15** (4).

Hanš, O., and Spaček, A. (1959). Random fixed point approximation by differentiable trajectories. *In* "Transactions of the Second Prague Conference on Information Theory, Statistical Decision Function, Random Process, Prague, 1959."

Hasminskii, R.Z. (1962). On stability of the trajectories of Markov processes, *Prikl. Mat. Mekh.* **26** (6). (In Russian.)

Hasminskii, R.Z. (1965). Application of random noise in the problems of optimization and pattern recognition, *Probl. Peredachi Inform.* **1** (3). (In Russian.)

Hawkins, J.K. (1961). Self-organizing systems—review and a commentary, *Proc. IRE* **49** (1), 31–48.

Hodges, J.L., and Lehman, E.L. (1956). Two approximations to the Robbins-Monro process. *In* "Proceedings of the Third Berkeley Symposium on Mathematical Statistics and Probability," Vol. 1. University of California Press, Berkeley.

Ivanov, A.Z., Krug, G.K., Kushelev, Yu.N., Letskii, E.K., and Svetchinskii, V.V. (1962). Learning control systems, *Automat. Telemekhan. Tr. MEI* **44**. (In Russian.)

Jacobs, O.L.R. (1964). Two uses of the term "adaptive" in automatic control, *IEEE Trans. Automat. Contr.* **AC-9** (4).

Kaczmarz, S. (1937). Angenäherte Auflösung von Systemen linearer Gleichungen, *Bull. Inter. Acad. Pol. Sci. Lett. Cl. Sci. Math. Natur.* **1937**.

Katz, I.Ya., and Krasovskii, N.N. (1960). On stability of systems with random parameters, *Prikl. Mat. Mekh.* **24** (5). (In Russian.)

Kesten, H. (1958). Accelerated stochastic approximation, *Ann. Math. Stat.* **29** (1).

Kiefer, E., and Wolfowitz, J. (1952). Stochastic estimation of the maximum of a regression function, *Ann. Math. Stat.* **23** (3).

Kirvaitis, K., and Fu, K.S. (1966). Identification of nonlinear systems by stochastic approximation. *In* "Joint Automatic Control Conference, Seattle, Washington, 1966."

Kitagawa, T. (1959). Successive processes of statistics control, *Mem. Fac. Sci. Kyushu Univ.* Series A **13** (1).

Krasulina, T.P. (1962a). On stochastic approximation, *Tr. IV Vses. Sovershch. Teor. Veroyat. Mat. Stat. Vilnus* **1962**.

Krasulina, T.P. (1962b). Some comments on certain stochastic approximation procedures, *Teor. Veroyat. Ee Primen.* **6** (1). (In Russian.)

Kushner, H.T. (1963). Hill climbing methods for the optimization on multiparameter noise distributed system, *Trans. ASME J. Basic Eng.* Series D **85** (2).

Kushner, H.T. (1965). On the stability of stochastic dynamic systems, *Proc. Nat. Acad. Sci.* **53** (1).

Lee, R.C.K. (1964). "Optimal Estimation, Identification and Control." MIT Press, Cambridge, Mass.

Leonov, Yu.P., Raevskii, S.Ya., and Raibman, N.S. (1961). "A Handbook in Automatic Control (Statistical Dynamics in Automation)." Academy of Sciences, USSR, Moscow.

Litvakov, B.M. (1966). On an iterative method in the problem of approximation of a function, *Automat. Remote Contr.* (*USSR*) **27** (4).

Loèv, M. (1955). "Probability Theory." Van Nostrand, New York.

Loginov, N.V. (1966). The methods of stochastic approximation, *Automat. Remote Contr.* (*USSR*) **27** (4).

Middleton, D. (1960). "An Introduction to Statistical Communication Theory." McGraw-Hill, New York.

Movshovich, S.M. (1966). Random search and the gradient method in the problems of optimization, *Izv. Akad. Nauk SSSR Tekh. Kibern.* **1966** (6). (Engl. transl.: *Eng. Cybern.* (*USSR*).)

Nikolic, Z.J., and Fu, K.S. (1966). A mathematical model of learning in an unknown random environment, *Proc. Nat. Electron. Conf.* **22**.

Pask, G. (1962). Learning machines, *Theory Self-Adapt. Contr. Syst. Proc. IFAC Symp. 2nd.* **1962**.

Pask, G. (1963). Physical and linguistic evolution in self-organizing systems, *Theory Self-Adapt. Contr. Syst. Proc. IFAC Symp. 2nd* **1963**.

Pervozvanskii, A.A. (1965). "Random Processes in Nonlinear Control Systems." Academic Press, New York.

Robbins, H., and Monro, S. (1951). A stochastic approximation method, *Ann. Math. Stat.* **22** (1).

Sacks, J. (1958). Asymptotic distributions of stochastic approximations, *Ann. Math. Stat.* **29** (2).

Sakrison, D.T. (1964). A continuous Kiefer-Wolfowitz procedure for random processes, *Ann. Math. Stat.* **35** (2).

Sakrison, D.T. (1965). Efficient recursive estimation, application to estimating the parameters of a covariance function, *Inter. J. Eng. Sci.* **3** (4).

Schmetterer, L. (1953). Bemerkungen zum Verfahren der stochastischen Iteration, *Oesterr. Ing. Archiv.* **7** (2).

Schmetterer, L. (1954a). Sur l'approximation stochastique, *Bull. Inst. Inter. Stat.* **24** (1).

Schmetterer, L. (1954b). Zum Sequentialverfahren von Robbins und Monro, *Monatsh. Math.* **58** (1).

Schmetterer, L. (1958). Sur l'iteration stochastique. *Colloq. Inter. Cent. Nat. Rech. Sci.* **87** (1).

Schmetterer, L. (1961). Stochastic approximation. *In* "Proceedings of the Fourth Berkeley Symposium on Mathematical Statistics," Vol. 1. University of California Press, Berkeley.

Sklansky, J. (1964). Adaptation, learning, self-repair and feedback, *IEEE Spectrum* **1** (5).
Sklansky, J. (1965). "Adaptation Theory and Tutorial Introduction to Current Research." RCA Laboratories David Sarnoff Research Center, Princeton, N.J.
Sklansky, J. (1966). Learning systems for automatic control, *IEEE Trans. Automat. Contr.* **AC-11** (1).
Stratonovich, R.L. (1968). Is there a theory of synthesis of optimal adaptive, self-learning and self-organizing systems?, *Automat. Remote Contr. (USSR)* **29** (1).
Tsypkin, Ya.Z. (1968). Is there a theory of optimal adaptive systems?, *Automat. Remote Contr. (USSR)* **29** (1).
van der Waerden, B.L. (1957). "Mathematical Statistics." Springer, Berlin.
Venter, J.H. (1966). On Dvoretzky stochastic approximation theorems, *Ann. Math. Stat.* **37** (6).
von Mises, R., and Pollaczek-Geiringer, H. (1929). Praktische Verfahren der Gleichungsauflösung, *Z. Angew. Math. Mech.* **9**.
Wetherill, G.B. (1966). "Sequential Methods in Statistics." Methuen, London.
Wilde, D.J. (1964). "Optimum Seeking Methods." Prentice-Hall, Englewood Cliffs, N.J.
Wolfowitz, J. (1952). On the stochastic approximation method of Robbins and Monro, *Ann. Math. Stat.* **23** (3).
Yudin, D.B. (1966). The methods of quantitative analysis of complex systems, *Izv. Akad. Nauk SSSR Tekh. Kibern.* **1966** (6). (Engl. transl.: *Eng. Cybern. (USSR).*)
Yudin, D.B., and Hazen, E.M. (1966). Certain mathematical aspects of statistical methods of search, *Avtomat. Vychisl. Tekh. Riga* **1966** (13). (In Russian.)
Zadeh, L.A. (1963). On the definition of adaptivity, *Proc. IRE* **51** (3), 3.

4

PATTERN RECOGNITION

4.1 Introduction

The problem of pattern recognition is a very general one. It was created in the solution of certain specific problems such as recognition of visual patterns (numerals, letters, simple images), sounds (spoken words, noise), diagnoses of diseases or malfunctions in systems, etc. Pattern recognition is the first and most important step in the processing of information which is obtained through our senses and various devices. At first we recognize objects, and then the relationships between objects and between objects and ourselves. In other words, we recognize situations. Finally, we recognize changes in these situations; i.e., we recognize events. This actually causes discoveries of laws which control events. On the basis of such laws it is then possible to predict the course and development of events.

The capabilities for pattern recognition in human beings and in machines have already been the subject of broad and serious discussions that have attempted to determine whether men or machines are superior in this respect. Such discussions are not yet crowned by any reasonable agreement. We shall not do the same because of our conviction that to ask such a question is unfair. Machines have always been used to amplify human resources—first physical and then intellectual. Otherwise, they would be useless creations.

In this chapter, we shall formulate the problem of training, and show that the adaptive approach offers an effective solution to this problem. Technical literature is rife with the algorithms of training in pattern recognition, but we shall see that such algorithms, originally found by the heuristic method (the trial-and-error method), can be relatively easily obtained as special cases of the general algorithms of training.

4.2 Discussion of the Pattern Recognition Problem

Before the problem of pattern recognition is formulated, it is expedient to discuss some of its features without restricting ourselves to rigorous formulations and exact concepts. The basic problem of pattern recognition is to classify the presented object into one of the pattern classes which, generally speaking, are previously unknown. A pattern class consists of the objects with common, similar properties. That common property which unites the objects into a pattern class is called the pattern. In solving the problem of pattern recognition, it is first necessary to consider the problem of training on the presented sample patterns of known pattern classification.

By using the particular features to describe the patterns, we can reduce the problem of pattern recognition to a comparison of the features characterizing the presented object of unknown pattern classification and all the objects of previously known pattern classifications. For instance, the design of reading machines for standard texts is based on this principle. This is an important case, but it will not be discussed here. We shall only mention that this principle has long been applied in the automatic sorting systems for detecting defective parts. Although the objective existence of the common properties of the objects is never questioned, we shall consider only the case in which such common properties are either impossible or inconvenient to obtain. In this case, we want simply to classify the objects, but do not know how. Hopeful groping for the features sometimes leads to the solution of some important specific problems. However, another road is also possible: training by examples. It consists of a preliminary presentation of a sequence of objects of known pattern classification.

In order to state the problem of training in pattern recognition more rigorously, it is convenient to employ geometrical concepts. Each object can correspond to a point in a certain multidimensional space. It is natural then to expect that the points corresponding to similar objects are close to each other, and that the pattern classes are easily distinguished when the points corresponding to them are densely distributed.

This intuitively obvious fact can be defined as the "hypothesis of compactness." However this will not help us unless we can first learn whether the hypothesis is justified. To verify such a hypothesis is extremely difficult. We can therefore only talk about the best classifications under given conditions. The only time that we can obtain a perfect classification is when the pattern classes can be easily distinguished.

4.3 Formulation of the Problem

Let us now try to formulate the problem of training in pattern recognition more accurately. The geometrical problem of training to recognize the patterns is the problem of finding a surface which separates the multidimensional space into the regions corresponding to different pattern classes. Such a separating surface must be the best in a certain sense. The partitioning is performed on the basis of a certain number of observed objects (patterns) which correspond to the pattern classes under consideration. Pattern recognition performed after the end of the training period is then the process of examining a new and unknown object, and deciding to which pattern class it belongs.

The first part of the formulation of this problem of pattern recognition, called here the problem of training, can be cast into an "algebraic" formulation. Training in pattern recognition is then the process of "extrapolation" in which a discriminant function is obtained by using the observed patterns of known classification. We shall limit our discussion to only the case of two pattern classes: A and B, or 1 and 2. This case is frequently called a dichotomy. The general case, when the number of pattern classes is greater than 2, can be reduced sequentially into a dichotomy.

Let us designate the discriminant function by

$$y = f(\mathbf{x}) \tag{4.1}$$

where $\mathbf{x}$ is the one-dimensional vector characterizing the pattern, and y is the quantity defining the pattern class to which the pattern $\mathbf{x}$ belongs. We may ask that the discriminant function have the following properties:

$$\operatorname{sign} f(\mathbf{x}) = \begin{cases} 1 & \text{if } \mathbf{x} \in A \\ -1 & \text{if } \mathbf{x} \in B \end{cases} \tag{4.2}$$

In other words, the sign of $f(\mathbf{x})$ is used to decide whether $\mathbf{x}$ belongs to A or B. Other than a deterministic formulation, we can have a stochastic formulation of the problem. In this case $f(\mathbf{x})$ is considered to be the probability that the pattern corresponding to the vector $\mathbf{x}$ belongs to the pattern class A, and $1 - f(\mathbf{x})$ is the probability that such a pattern belongs to the pattern class B. Generally speaking, from (4.2) it follows that there exists a set of functions which define the separating surface. These functions are called discriminant functions. It is obvious that such a set exists at least when the pattern classes are easy to distinguish. However, if this is not the case, then there usually exists only one best discriminant function.

In formulating an extrapolation problem, or, if it is convenient, an approximation problem, we must first select a class of approximating functions and an error measure characterizing the accuracy of approximation. Let us designate the class of approximating functions by $\hat{f}(\mathbf{x}, \mathbf{c})$, where $\mathbf{c}$ is still an unknown vector of the coefficients, and the error measure is defined as a certain convex function of $y = f(\mathbf{x})$ and $\hat{f}(\mathbf{x}, \mathbf{c})$; for instance, $F(y; \hat{f}(\mathbf{x}, \mathbf{c}))$. If the observed vectors are random, the error measure is also random. Therefore, as a measure of approximation it is convenient to select the functional which is the mathematical expectation of the error measure:

$$J(\mathbf{c}) = M\{F(y; \hat{f}(\mathbf{x}, \mathbf{c}))\} \tag{4.3}$$

The best approximation corresponds to the vector $c = c^*$ for which $J(c)$ reaches its minimum.

In a number of cases, the error measure is defined as a convex function of the difference $y - \hat{f}(x, c)$. Then instead of the functional (4.3), we obtain

$$J(c) = M\{F(y - \hat{f}(x, c))\} \qquad (4.4)$$

From now on, we shall consider only such a functional. If we do not know the probability density function $p(x)$, and thus the mathematical expectation (4.4), the only way to obtain the solution $c = c^*$ is to use the vectors x describing the observed patterns, and then to apply the corresponding algorithms of adaptation or training. We shall now turn our attention to this problem.

4.4 General Algorithms of Training

Let us first specify the form of the approximating function. We must remember that the choice of an approximating function is not arbitrary, but is related to the constraints of the first kind. A sufficiently broad range of problems can be covered if it is assumed that $\hat{f}(x, c)$ is a finite sum

$$\hat{f}(x, c) = \sum_{v=1}^{N} c_v \phi_v(x) \qquad (4.5)$$

or

$$\hat{f}(x, c) = c^T \phi(x) \qquad (4.6)$$

where the coefficients c_v are the components of the N-dimensional vector c, and the linearly independent functions $\phi_v(x)$ are the components of the N-dimensional vector $\phi(x)$.

By substituting (4.6) into the functional (4.4), we obtain

$$J(c) = M\{F(y - c^T \phi(x))\} \qquad (4.7)$$

Since the functional (4.7) is not known explicitly, we shall use the measured gradients in order to find the minimum of $J(c)$.

In the case under consideration,

$$\nabla_c F(y - c^T \phi(x)) = -F'(y - c^T \phi(x))\phi(x) \qquad (4.8)$$

By applying the algorithm of adaptation (3.9), and assuming that $Q(x, c) = F(y - c^T \phi(x))$, we obtain

$$c[n] = c[n-1] + \gamma[n] F'(y[n] - c^T[n-1]\phi(x[n]))\phi(x[n]) \qquad (4.9)$$

This algorithm, approximately named an algorithm of training, indeed defines the optimal vector $c = c^*$ and the optimal discriminant function when $n \to \infty$.

The algorithm of training (4.9) can also be written in a slightly different form. Let us define

$$f_n(\mathbf{x}) = \boldsymbol{\phi}^T(\mathbf{x})\mathbf{c}[n] \tag{4.10}$$

and

$$K(\mathbf{x}, \mathbf{x}[n]) = \boldsymbol{\phi}^T(\mathbf{x})\boldsymbol{\phi}(\mathbf{x}[n]) \tag{4.11}$$

By scalar multiplication of both sides in (4.9) with $\boldsymbol{\phi}(\mathbf{x})$, and using definitions (4.10) and (4.11), we obtain an algorithm of training in the form of a functional recursive relationship,

$$f_n(\mathbf{x}) = f_{n-1}(\mathbf{x}) + \gamma[n]F'(y[n] - f_{n-1}(\mathbf{x}[n])K(\mathbf{x}, \mathbf{x}[n])) \tag{4.12}$$

Fundamentally different algorithms can be obtained on the basis of the search algorithm of adaptation (3.15). In such a case,

$$\mathbf{c}[n] = \mathbf{c}[n-1] - \gamma[n]\,\tilde{\nabla}_{\mathbf{c}\pm}F(\mathbf{x}[n], \mathbf{c}[n-1], a[n]) \tag{4.13}$$

where an estimate of the gradient is

$$\tilde{\nabla}_{\mathbf{c}\pm}F(\mathbf{x}[n], \mathbf{c}[n-1], a[n])$$
$$= \frac{\mathbf{F}_+(\mathbf{x}[n], \mathbf{c}[n-1], a[n]) - \mathbf{F}_-(\mathbf{x}[n], \mathbf{c}[n-1], a[n])}{2a[n]} \tag{4.14}$$

and the vectors $\mathbf{F}_+(\cdot)$ and $\mathbf{F}_-(\cdot)$ are defined, as we had already agreed, in a manner analogous to (3.12). Most probably it is not justifiable to use a search algorithm of training when the function $F(\cdot)$ is known and differentiable. However, when for some reasons it is difficult to define the gradient $\nabla_\mathbf{c} F(\cdot)$, such and only such algorithms can be applied.

In a number of cases it may be convenient to apply the continuous algorithms of training; for instance, the algorithms of the type

$$\frac{d\mathbf{c}(t)}{dt} = \gamma(t)F'(y(t) - \mathbf{c}^T(t)\boldsymbol{\phi}(x(t))\boldsymbol{\phi}(x(t)) \tag{4.15}$$

The meaning of this algorithm will be better explained slightly later in a specific example.

4.5 Convergence of the Algorithms

In considering algorithms of training (4.9) we shall distinguish two cases:

(1) The values of the function $f(\mathbf{x})$ for every fixed $\mathbf{x}$ are measured with an error which has a finite variance and a mean value equal to zero.

(2) The values of the function $f(\mathbf{x})$ for every fixed $\mathbf{x}$ are known exactly.

Let us start with the first case. If $\mathbf{c}[n]$ is to converge to $\mathbf{c}^*$ with probability one, then conditions (3.35) have to be satisfied. Conditions (3.35a) and (3.35b) are fulfilled by an appropriate selection of $\gamma[n]$ and $F(\cdot)$; for instance, $\gamma[n] = 1/n^\alpha$, $\tfrac{1}{2} + \varepsilon \leq \alpha \leq 1$, $\varepsilon > 0$, and $F(\cdot)$ includes all piecewise continuous functions with the rate of increase with $\mathbf{c}$ not greater than in a parabola of second degree. Regarding condition (3.35b), it is easy to show that such a condition is satisfied by all convex functions.

Let us now consider the second case. We assume that the discriminant function $y = f(\mathbf{x})$ satisfies the "hypothesis of representation." This implies that the function $f(\mathbf{x})$ is accurately represented by a finite sum

$$f(\mathbf{x}) = \sum_{v=1}^{N} c_v \phi_v(\mathbf{x}) = \mathbf{c}^T \boldsymbol{\phi}(\mathbf{x}) \tag{4.16}$$

In this case,

$$J(\mathbf{c}^*) = 0 \tag{4.17}$$

Now, the optimal vector $\mathbf{c}^*$ can be determined after a finite number of training samples; it is obvious that in this case the smallest number of training samples is equal to N. Any other training sample, described by an arbitrary $\mathbf{x}$ and its corresponding y, satisfies the equation

$$F(y - \mathbf{c}^{*T}\boldsymbol{\phi}(\mathbf{x}))\boldsymbol{\phi}(\mathbf{x}) = 0 \tag{4.18}$$

In this case the algorithm converges with probability one when the quantity γ is a constant, $\gamma = \gamma_0$. The limiting value γ_0 can be found by the stochastic principle of contraction mapping.

If the "hypothesis of representation" is not fulfilled, equation (4.18) cannot be written, since $\mathbf{c}^*$ will depend on the training sample $\mathbf{x}$. But even in such a case, the algorithm (4.9) will converge for a constant γ_0 which satisfies specific conditions. For instance, in the case of a quadratic criterion, the algorithm will converge with probability one when

$$\gamma_0 \leq 1/\max \sum_{v=1}^{N} \phi_v^2(\mathbf{x}) \tag{4.19}$$

4.6 Perceptrons

The algorithms presented above are written in the form of nonlinear difference equations which, generally speaking, have variable coefficients. They correspond, as we already know, to the multivariable sampled-data systems—simple (without search) and extremal (with search). These systems, which realize the algorithms of training, in fact define the discriminant function and thus perform the classification of the objects. A scheme of such a system is given in Fig. 4.1. This system consists of functional transformers $\boldsymbol{\phi}(\mathbf{x})$ and

82   4   Pattern Recognition

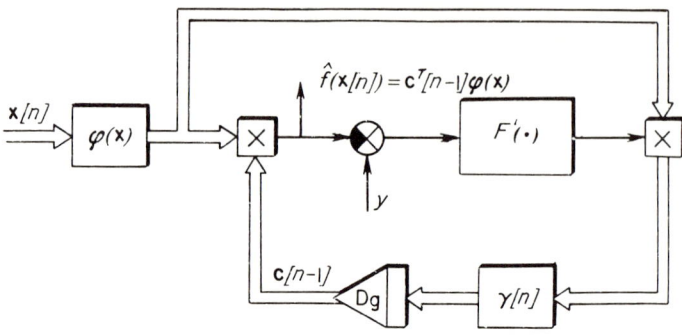

Fig. 4.1

$F'(\cdot)$, a multiplier which forms the scalar product $\mathbf{c}^T[n-1]\boldsymbol{\phi}(\mathbf{x}[n])$, a conventional multiplier, an amplifier with the variable gain coefficient $\gamma[n]$, and a digital integrator (digrator) Dg. We do not introduce special symbols for multipliers since they can be easily distinguished according to whether their input and output variables are the vectors or the scalars. The discriminant function is defined at the output of the device which performs scalar multiplication.

A detailed scheme of the system which realizes an algorithm of training is presented in Fig. 4.2. This scheme corresponds to a perceptron. In contrast to the original version proposed by Rosenblatt, arbitrary linearly independent functions $\phi_\nu(\mathbf{x})$ are used here instead of threshold functions. Perceptrons with threshold devices are usually considered as analog neural nets. There exists a definite relationship between the pattern recognition systems, neural nets, and

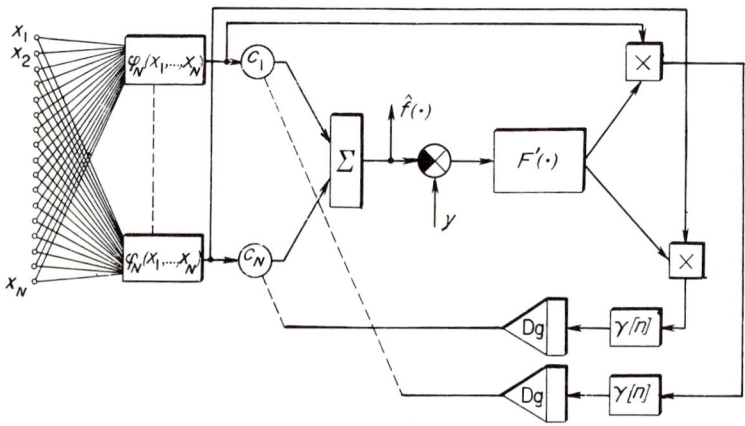

Fig. 4.2

the theory of finite automata and sequential machines. These questions will be examined in Chapter 10 of this book.

A new and original type of perceptron—a searching perceptron—can be synthesized from a conventional perceptron. Its scheme is given in Fig. 4.3.

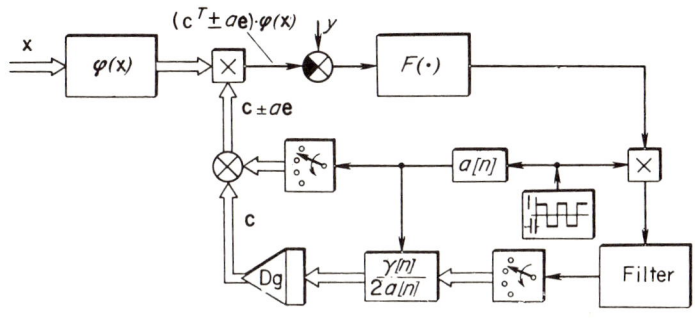

Fig. 4.3

The stability of perceptrons, and thus the convergence of the training process, depends considerably on the functions $\phi_\nu(\mathbf{x})$.

It is interesting to note that there exists a possibility of building the perceptrons not only from threshold elements but also from any other kind of linearly independent functional transformers.

4.7 Discrete Algorithms of Training

By selecting a specific form of the function $F(\cdot)$ and $\gamma[n]$ in the general algorithms of training (4.9) and (4.13), we obtain different particular algorithms which minimize corresponding functions. Typical algorithms of training are listed for convenience and comparison in Table 4.1. Some of these algorithms coincide with the algorithms which were obtained on the basis of heuristic considerations and have been reported in a number of papers on pattern recognition. The algorithms, the criteria of optimality (the functionals which are minimized by these algorithms), and the names of the authors who have proposed these algorithms, are shown in Table 4.1.

In the absence of noise, as we have already mentioned, $\gamma[n]$ can be a constant without affecting the convergence of the algorithms. Algorithms 1–3 of Table 4.1 correspond to deterministic problems of training, and algorithm 4 corresponds to a probabilistic problem of training.

Let us emphasize that the functional which generates such an algorithm is a random variable, since $R[D(\mathbf{x})]$ is an operator of a random experiment with two outcomes (a biased coin): 1 with probability $D(\mathbf{x})$ and -1 with probability $1 - D(\mathbf{x})$.

Table 4.1

TYPICAL ALGORITHMS OF TRAINING

Number	Functional	Algorithms	Comments	Authors
1	$J(\mathbf{c}) = M\{(\operatorname{sgn} y - \operatorname{sgn} \mathbf{c}^T \boldsymbol{\phi}(\mathbf{x}))\mathbf{c}^T\boldsymbol{\phi}(\mathbf{x})\}$	$\mathbf{c}[n] = \mathbf{c}[n-1] + \gamma[n][\operatorname{sign} y[n] - \operatorname{sign} \mathbf{c}^T[n-1] \times \boldsymbol{\phi}(\mathbf{x}[n])]\boldsymbol{\phi}(\mathbf{x}[n])$		Aizerman, Braverman, Rozonoer, Yakubovich
2	$J(\mathbf{c}) = M\{|y - \mathbf{c}^T\boldsymbol{\phi}(\mathbf{x})|\}$	$\mathbf{c}[n] = \mathbf{c}[n-1] + \gamma[n]\operatorname{sign}(y[n] - \mathbf{c}^T[n-1]) \times \boldsymbol{\phi}(\mathbf{x}[n])$		Aizerman, Braverman, Rozonoer
3	$J(\mathbf{c}) = M\{(y - \mathbf{c}^T\boldsymbol{\phi}(\mathbf{x}))^2\}$	$\mathbf{c}[n] = \mathbf{c}[n-1] + \gamma[n](y[n] - \mathbf{c}^T[n-1])\boldsymbol{\phi}(\mathbf{x}[n])$	L-optimality according to Yakubovich	Aizerman, Braverman, Rozonoer, Yakubovich
		$\mathbf{c}[n] = \mathbf{c}[n-1] + \Gamma[n](y[n] - \mathbf{c}^T[n-1])\boldsymbol{\phi}(\mathbf{x}[n])$	$y = \begin{cases} 1 & \text{if } \mathbf{x}[n] \in A \\ 0 & \text{if } \mathbf{x}[n] \notin A \end{cases}$ $\Gamma^{-1}[n] = \Gamma^{-1}[n-1] - \boldsymbol{\phi}\boldsymbol{\phi}^T$	Blaydon, Ho
4	$J(\mathbf{c}) = M\{R(D(\mathbf{x})) - R(\mathbf{c}^T\boldsymbol{\phi}(\mathbf{x}))\mathbf{c}^T\boldsymbol{\phi}(\mathbf{x})\}$	$\mathbf{c}[n] = \mathbf{c}[n-1] + \gamma[n]\left[R(D(\mathbf{x}[n])) - R(\mathbf{c}^T[n-1] \times \boldsymbol{\phi}(\mathbf{x}[n]))\right] \times \boldsymbol{\phi}(\mathbf{x}[n])$	$R(D(\mathbf{x}))$ is an operator of a random experiment with two outcomes: $+1$ with probability $D(\mathbf{x})$, and -1 with probability $1 - D(\mathbf{x})$	Aizerman, Braverman, Rozonoer
5	$J(\mathbf{c}) = \sum_{i=1}^{l} F^2(1 - \mathbf{c}^T\mathbf{k}^i) + \sum_{j=l+1}^{N} F_2^{\;2}(k - \mathbf{c}^T\mathbf{k}^j)$	$\dfrac{dc_i(t)}{dt} = \gamma(t)F_1(1 - \mathbf{c}^T(t)\mathbf{k}^i) \times F_1(1 - \mathbf{c}^T(t)\mathbf{k}^i),$ $(1 \leq i \leq l)$ $\dfrac{dc_j(t)}{dt} = \gamma(t)F_2(k - \mathbf{c}^T(t)\mathbf{k}^k) \times F_2(k - \mathbf{c}^T(t)\mathbf{k}^j),$ $(l+1 \leq j \leq N)$	$F_1(z) = z - |z|,$ $F_2(z) = |z| - z$	Vapnik, Lerner, Chervonenkis

When the teacher knows with certainty to which pattern class each observed sample belongs, then functional 4 coincides with functional 1, and the probabilistic problem becomes a deterministic one. However, if such a classification is known to the teacher only to a certain degree of certainty, the mean square error

$$J(\mathbf{c}) = \int_X [D(\mathbf{x}) - \overline{\mathbf{c}^T \boldsymbol{\phi}(\mathbf{x})}]^2 p(\mathbf{x})\, d\mathbf{x} \qquad (4.20)$$

can be taken as a performance index. In this case the algorithm of training coincides with the algorithm

$$\mathbf{c}[n] = \mathbf{c}[n-1] + \gamma[n][D(\mathbf{x}[n]) - \overline{\mathbf{c}^T[n-1]\boldsymbol{\phi}(\mathbf{x}[n])}]\boldsymbol{\phi}(\mathbf{x}[n]) \qquad (4.21)$$

Knowledge of functionals gives a possibility of comparing the algorithms with each other. Some algorithms, for instance, are not completely suitable for comparison. This is the case with the algorithms 1 and 4 for which the corresponding functionals are not convex in $\mathbf{c}$: they have the second minimum equal to zero at $\mathbf{c} = 0$. In order to avoid the convergence to such a trivial solution, it is necessary to choose initial conditions $\mathbf{c}(0)$ sufficiently far from the origin.

Let us mention in conclusion that the algorithms given in Table 4.1 are basically discrete algorithms; the exceptions are algorithms 5. These continuous algorithms will be discussed at greater length in Section 4.9.

4.8 Search Algorithms of Training

Although each discrete algorithm given in Table 4.1 can be related to a particular search algorithm of training, it is not necessary in such cases to give any preference to the search algorithms. We should recall that in the algorithms with which we became familiar in Section 4.7, the gradient was computed by using the values of the chosen function $F(\cdot)$ only. However, in a number of cases (and in a slightly different formulation of the pattern recognition problem), the search algorithms of training only are applicable. We shall now discuss such a formulation of the pattern recognition problem. Furthermore, it should be assumed that the pattern classes are separated in the feature space; i.e., it is necessary to demand that the hypothesis of representation (4.16) be fulfilled. If this is not the case, then such algorithms are not operational. However, any other algorithm from this table can produce, under the same conditions, a result which is best in the sense of the chosen functional.

Let us consider two pattern classes: A and B, or 1 and 2. If we classify a pattern from the pattern class ν into the pattern class μ, then this event is evaluated by a penalty $w_{\nu\mu}$ ($\nu, \mu = 1, 2$). The amounts of these penalties are specified by the matrix $\|w_{\nu\mu}\|$.

Average risk, equal to the mathematical expectation of the penalties, can be written in the form which is a special case of the functional (1.3):

$$R(d) = \int_\Lambda \{[d(\mathbf{x}) - 1][w_{12} P p_1(\mathbf{x}) + w_{22}(1 - P)p_2(\mathbf{x})] \\ + [2 - d(\mathbf{x})][w_{21}(1 - P)p_2(\mathbf{x}) + w_{22} P p_1(\mathbf{x})]\} \, d\mathbf{x} \quad (4.22)$$

where P is the probability of occurrence of the patterns from the pattern class 1, $1 - P$ is the probability of occurrence of the patterns from the pattern class 2, $p_1(\mathbf{x})$ and $p_2(\mathbf{x})$ are conditional probability density functions for the patterns of both pattern classes, and $d(\mathbf{x})$ is the decision rule: $d(\mathbf{x}) = 1$ if a pattern $\mathbf{x}$ is classified into the pattern class 1, and $d(\mathbf{x}) = 2$ if the pattern is classified into the pattern class 2. In the solution of the stated problem one can use the powerful results of the decision theory based on the Bayesian approach where the probability density functions $p_1(\mathbf{x})$ and $p_2(\mathbf{x})$ are known. But in our case these probability densities are unknown, and we cannot directly use the results of that elegant theory.

Certainly, we could apply such results if the probability density functions were first determined; this is done occasionally. However, it may be better to use the algorithms of training instead of spending time on a "needless" step. The risk $R(d)$, expressed by (4.22), is already a special case of the expected risk (1.5). Therefore, $R(d)$ can be written as

$$R(d) = M\{z \mid d\} \quad (4.23)$$

where $z = w_{\nu\mu}$, if for the specified decision rule we classify this pattern $\mathbf{x}$ into the pattern class μ when it actually belongs to the pattern class ν ($\nu, \mu = 1, 2$).

The decision rule $d(\mathbf{x})$ will be sought in the form of a known function $\hat{d}(\mathbf{x}, \mathbf{c})$ with an unknown parameter vector $\mathbf{c} = (c_1, \ldots, c_N)$. If z depends only implicitly on the vector $\mathbf{c}$, then it is convenient to determine the optimal vector by the search algorithm

$$\mathbf{c}[n] = \mathbf{c}[n-1] - \gamma[n] \, \tilde{\nabla}_{\mathbf{c}\pm} z[n] \quad (4.24)$$

with

$$\tilde{\nabla}_{\mathbf{c}\pm} z[n] = \frac{z_+[n] - z_-[n]}{2a[n]}$$

where

$$z_+[n] = w_{\nu\mu}$$

if $\mathbf{x}[2n - 1]$ belongs to the pattern class ν and

$$\hat{d}(\mathbf{x}[2n-1], \mathbf{c}[n-1] + a[n]\mathbf{e}) = \mu$$

and

$$z_-[n] = w_{\nu\mu}$$

if $\mathbf{x}[2n]$ belongs to the pattern class v and
$$\hat{a}(\mathbf{x}[2n], \mathbf{c}[n-1] - a[n]\mathbf{e}) = \mu$$

The search algorithm of adaptation (4.24) indeed solves the problem of training. This algorithm forms a bridge between the algorithmic approach and the statistical decision theory based on the Bayes criterion. A similar relationship will again be encountered in the filtering problems, or, more exactly, in the problems of detection and extraction of signals; this will be discussed in Chapter 6.

4.9 Continuous Algorithms of Training

Let us select the functional

$$J(\boldsymbol{\psi}) = \sum_{\mu=1}^{l} F_1^2(1 - \boldsymbol{\psi}^T \mathbf{x}_\mu) + \sum_{\eta=l+1}^{N} F_2^2(k - \boldsymbol{\psi}^T \mathbf{x}_\eta) \qquad (4.25)$$

where $0 < k < 1$, and $\mathbf{x}_v$ is the vector corresponding to the observed pattern; the vectors indexed from 1 to l correspond to A, and those indexed from $l+1$ to N correspond to B. The functions $F_1(\cdot)$ and $F_2(\cdot)$ are

$$F_1(\mathbf{x}) = \begin{cases} 0 & \text{for } \mathbf{x} < 0 \\ F(\mathbf{x}) & \text{for } \mathbf{x} \geq 0 \end{cases}$$
$$F_2(\mathbf{x}) = \begin{cases} F(\mathbf{x}) & \text{for } \mathbf{x} < 0 \\ 0 & \text{for } \mathbf{x} \geq 0 \end{cases} \qquad (4.26)$$

where $F(\mathbf{x})$ is any convex function. Vector $\boldsymbol{\psi}$, the so-called generalized portrait, defines the separating hyperplane. By the conventional gradient method we can obtain an algorithm for finding the optimal value of the vector $\boldsymbol{\psi}$:

$$\frac{d\boldsymbol{\psi}}{dt} = \gamma(t)\Bigg\{ \sum_{\mu=1}^{l} F_1(1 - \boldsymbol{\psi}^T\mathbf{x}_\mu) F_1'(1 - \boldsymbol{\psi}^T\mathbf{x}_\mu)\mathbf{x}_\mu $$
$$+ \sum_{\eta=l+1}^{N} F_2(k - \boldsymbol{\psi}^T\mathbf{x}_\eta) F_2'(k - \boldsymbol{\psi}^T\mathbf{x}_\eta)\mathbf{x}_\eta \Bigg\} \qquad (4.27)$$

It is customary to present the generalized portrait in the form

$$\boldsymbol{\psi} = \sum_{v=1}^{N} c_v \mathbf{x}_v \qquad (4.28)$$

If it is assumed that the vectors $\mathbf{x}_v$ of the observed patterns are linearly independent, and that the number of observed patterns is equal to the number of components in the vector $\mathbf{c}$, then the continuous algorithm (4.27) together with (4.28) yield the following equations:

$$\frac{dc_\mu(t)}{dt} = \gamma(t) F_1(1 - \mathbf{c}^T(t)\mathbf{k}^\mu) F_1'(1 - \mathbf{c}^T(t)\mathbf{k}^\mu) \qquad (1 \leq \mu \leq l)$$
$$\frac{dc_\mu(t)}{dt} = \gamma(t) F_2(k - \mathbf{c}^T(t)\mathbf{k}^\eta) F_2'(k - \mathbf{c}^T(t)\mathbf{k}^\eta) \qquad (l+1 \leq \mu \leq N) \qquad (4.29)$$

where k^s represents the sth column in the matrix $\|k_{\mu\eta}\|$; the elements of $\|k_{\mu\eta}\|$ are pairwise scalar products of the vectors which correspond to the observed patterns.

In a special case, when

$$F_1(\mathbf{x}) = \begin{cases} 0 & \text{for } \mathbf{x} < 0 \\ \mathbf{x} & \text{for } \mathbf{x} \geq 0 \end{cases}$$

$$F_2(\mathbf{x}) = \begin{cases} \mathbf{x} & \text{for } \mathbf{x} < 0 \\ 0 & \text{for } \mathbf{x} \geq 0 \end{cases}$$

the system of equations (4.29) can be rewritten in the form

$$\frac{dc_\mu(t)}{dt} = \gamma(t)(1 - \mathbf{c}^T(t)\mathbf{k}^\mu) \qquad (1 \leq \mu \leq l)$$

$$\frac{dc_\eta(t)}{dt} = \gamma(t)(k - \mathbf{c}^T(t)\mathbf{k}^\eta) \qquad (l+1 \leq \eta \leq N)$$

(4.30)

where $\gamma(t)$, in particular can be a constant. This algorithm was presented in Table 4.1 An analog computer can be used in the realization of this algorithm; one possible scheme is given in Fig. 4.4, where $\mathbf{c}_\mathrm{I} = (c_1, \ldots, c_b)$ and $\mathbf{c}_\mathrm{II} = (c_{b+1}, \ldots, c_N)$.

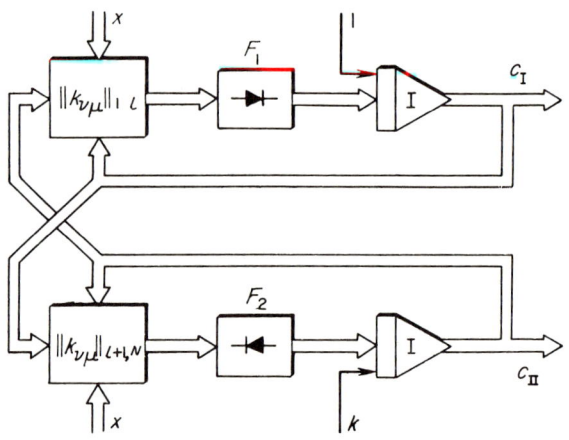

Fig. 4.4

4.10 Comments

All the algorithms discussed above are attempts to approximate teachers' decisions expressed in the form

$$\operatorname{sign} f(\mathbf{x}) = \begin{cases} +1 & \text{for } \mathbf{x} > 0 \\ -1 & \text{for } \mathbf{x} < 0 \end{cases}$$

or $y = +1$, i.e., a discontinuous function in $\mathbf{x}$, by a continuous function $\hat{f}(\mathbf{x}, \mathbf{c}) = \mathbf{c}^T \boldsymbol{\phi}(\mathbf{x})$. However, the desired discriminant function does not always have to be a natural one. It is obvious that in the problems of pattern recognition it is sometimes more reasonable to determine $\hat{f}(\mathbf{x}, \mathbf{c})$ so that its sign only coincides with the sign of y. It is possible that such considerations have led the authors of algorithm 1 (Table 4.1) to propose the corrections only when the signs of y and $\mathbf{c}^T \boldsymbol{\phi}(\mathbf{x})$ are different. Such an algorithm will be written here as

$$\mathbf{c}[n] = \mathbf{c}[n-1] + \gamma[n](\operatorname{sign} f(\mathbf{x}[n]) - \operatorname{sign} \mathbf{c}^T[n-1]\boldsymbol{\phi}(\mathbf{x}[n]))\boldsymbol{\phi}(\mathbf{x}[n]) \quad (4.31)$$

where $\gamma[n]$ can be a constant.

From the gradient

$$\nabla_c F(\cdot) = (\operatorname{sign} f(\mathbf{x}[n]) - \operatorname{sign} \mathbf{c}^T[n-1]\boldsymbol{\phi}(\mathbf{x}[n]))\boldsymbol{\phi}(\mathbf{x}[n]) \quad (4.32)$$

it is easy to restore the criterion of optimality

$$J(\mathbf{c}) = M\{(\operatorname{sign} f(\mathbf{x}) - \operatorname{sign} \mathbf{c}^T \boldsymbol{\phi}(\mathbf{x}))\mathbf{c}^T \boldsymbol{\phi}(\mathbf{x})\} \quad (4.33)$$

This criterion was listed in Table 4.1. As we have already stated in Section 4.7, it is not a suitable one since the function under the expectation is not convex. Other than the optimal solution $\mathbf{c} = \mathbf{c}^*$, the functional has a trivial solution $\mathbf{c} = 0$. The existence of this solution (after establishing this fact) is obvious even from algorithm (4.31). This property makes algorithm (4.31) unfit for the cases in which the pattern classes A and B are not completely separated.

4.11 More about Another General Algorithm of Training

Let us now consider another possibility of obtaining general algorithms. It is based on the coincidence of the signs of y and $\mathbf{c}^T \boldsymbol{\phi}(\mathbf{x})$. Such a condition is fulfilled if $\mathbf{c}$ is obtained from the system of inequalities

$$y[n](\mathbf{c}^T[n-1]\boldsymbol{\phi}(\mathbf{x}[n])) > 0 \quad (n = 1, 2, \ldots) \quad (4.34)$$

This system of inequalities can be replaced by the system of equations

$$y[n](\mathbf{c}^T[n-1]\boldsymbol{\phi}(\mathbf{x}[n])) = \alpha[n-1] \quad (n = 1, 2, \ldots) \quad (4.35)$$

if it is assumed that

$$\alpha[n-1] > 0 \quad (4.36)$$

Considering the random character of the input patterns, we introduce the criterion of optimality

$$J(\mathbf{c}, a) = M\{F(y(\mathbf{c}^T \boldsymbol{\phi}(\mathbf{x})) - \alpha)\} \quad (4.37)$$

with the constraint

$$\alpha > 0 \quad (4.38)$$

where $F(\cdot)$ is a strictly convex function.

Using an algorithm of adaptation similar to (3.24) and which takes into consideration the constraints of the type (4.38), we obtain, in this case,

$$\mathbf{c}[n] = \mathbf{c}[n-1] - \gamma_1[n]F'(y[n](\mathbf{c}^T[n-1]\phi(\mathbf{x}[n])) - \alpha[n-1])y[n]\phi(\mathbf{x}[n])$$
$$\alpha[n] = \alpha[n-1] + \gamma_2[n][F'(y[n](\mathbf{c}^T[n-1]\phi(\mathbf{x}[n])) - \alpha[n-1])$$
$$+ |F'(y[n](\mathbf{c}^T[n-1]\phi(\mathbf{x}[n])) - \alpha[n-1])|]$$
$$\alpha(0) > 0 \qquad (4.39)$$

Let us designate

$$z(\mathbf{c}, \alpha[n]) = y[n](\mathbf{c}^T\phi(\mathbf{x}[n]) - \alpha[n]) \qquad (4.40)$$

Here, $z(\mathbf{c}, \alpha[n])$ is a random variable which characterizes the inequalities (4.34). Considering these notations, algorithm (4.39) is rewritten as

$$\mathbf{c}[n] = \mathbf{c}[n-1] - \gamma_1[n]F'(z(\mathbf{c}[n-1], \alpha[n-1]))y[n]\phi(\mathbf{x}[n])$$
$$\alpha[n] = \alpha[n-1] + \gamma_2[n][F'(z(\mathbf{c}[n-1], \alpha[n-1]))$$
$$+ |F'(z(\mathbf{c}[n-1], \alpha[n-1]))|] \qquad (4.41)$$
$$\alpha[0] > 0$$

A perceptron (a discrete system) which realizes this algorithm is shown in Fig. 4.5. It differs from the classical perceptron by the existence of an addi-

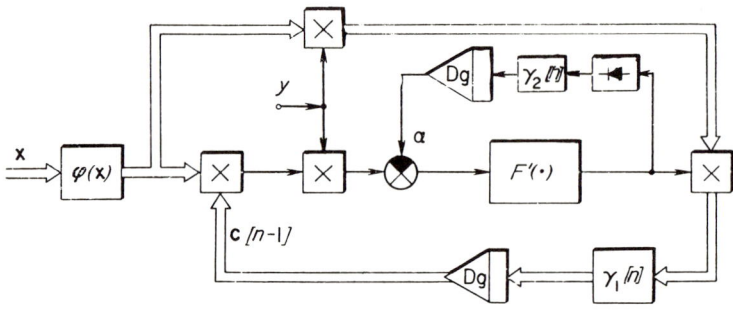

Fig. 4.5

tional loop for finding α^*. Such a modification of the basic perceptron structure can lead to an accelerated rate of convergence.

4.12 Special Cases

Special cases of the algorithm of training (4.41) are given in Table 4.2, and the particular forms of the function $F(z)$ appearing in the algorithms are shown in Fig. 4.6.

4.12 Special Cases

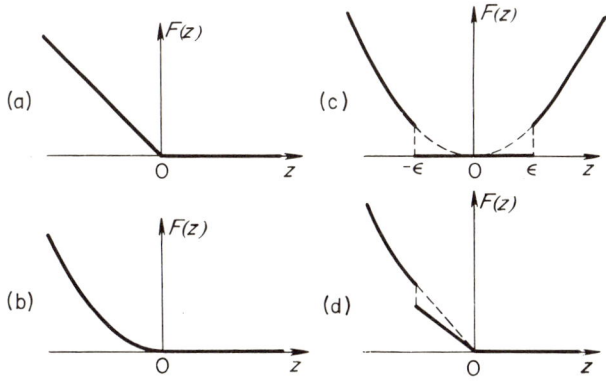

Fig. 4.6

In cases when not only the coincidence in signs but also the distance between the desired and approximating functions is required, $F(z)$ is different from zero even for positive z (algorithm 3 of Table 4.2, Fig. 4.6b). We should notice that algorithms 1a and 1b (Table 4.2) are actually identical; this can be easily verified by inspection. A special feature of algorithm 1b is that $y[n]$ is not only a function of n but also of the vectors $\mathbf{c}[n-1]$. This offers a possibility for improving the convergence.

In algorithms 1–4 it is assumed that $\alpha[n] = \text{const}$. The convergence of these algorithms can be considerably improved if $\alpha[n]$ is not considered to be a constant, but if it is chosen according to the expression (4.39). This property was actually used in algorithm 5. In the general case the top line of the algorithm has the form

$$\mathbf{c}[n] = \mathbf{c}[n-1] + \Gamma[n][z(\mathbf{c}[n-1], \alpha[n-1])]y[n]\phi(\mathbf{x}[n]) \quad (4.42)$$

where $\Gamma[n]$ is a certain matrix.

When $z(\mathbf{c}[0], \alpha[0]) < 0$, algorithm (4.41) coincides with algorithm 3 in Table 4.2. If, on the other hand, $z(\mathbf{c}[0], \alpha[0]) \geq 0$, then $\mathbf{c}[0]$ and $\alpha[0]$ are the desired solutions.

If the set of patterns from each pattern class is finite ($\nu = 1, \ldots, M$), i.e., if all the patterns can be observed, the problem of discrimination is a deterministic one.

Let us form a sequence of samples

$$\phi(\mathbf{x}_\mu[n]) \quad \text{where} \quad \mu = 1, \ldots, M \text{ for every } n = 1, 2, \ldots \quad (4.43)$$

Using (4.43) in algorithm (4.41), we obtain $\mathbf{c}[n] \to \mathbf{c}^0$ when $n \to \infty$, where $\mathbf{c}_0$ is the solution of a finite system of inequalities

$$y_\mu \mathbf{c}^T \phi(\mathbf{x}_\mu) > 0 \quad (\mu = 1, \ldots, M) \quad (4.44)$$

Table 4.2
Special Cases of the Algorithm of Training

Number	Functional	Algorithms	Comments	Authors
1	$J = M\{|z| - z\}$	$\mathbf{c}[n] = \mathbf{c}[n-1] + \gamma[n]$ $\times \{\text{sign}(z(\mathbf{c}[n-1],$ $\times \alpha[n])) - 1\}\gamma[n]\boldsymbol{\varphi}(\mathbf{x}[n])$	(a) $\gamma[n] = \frac{1}{2}\beta[n]$ $\times \left[\rho - \dfrac{\mathbf{c}^T[n-1]\boldsymbol{\varphi}(\mathbf{x}[n])}{\gamma[n]\boldsymbol{\varphi}^T(\mathbf{x}[n])\boldsymbol{\varphi}(\mathbf{x}[n])}\right];$ $0 < \beta[n] \leq 2;\ \rho > 0;$ (b) $\alpha[n] \equiv 0$ $0 < \gamma < 2;\ \alpha[n] \equiv 1$	Yakubovich
2	$J = M\{|z| - z\}^2$	$\mathbf{c}[n] = \mathbf{c}[n-1] + \gamma[n]$ $\times \{|z(\mathbf{c}[n-1],$ $\alpha[n])| - z(\mathbf{c}[n-1],$ $\alpha[n])\}\gamma[n]\boldsymbol{\varphi}(\mathbf{x}[n])$	$\alpha[n] \equiv 1$	Novikoff Agmon, Mays
3	$J = \begin{cases} M\{z^2\} & \text{for } |z| \geq \varepsilon \\ 0 & \text{for } |z| < \varepsilon \end{cases}$	$\mathbf{c}[n] = \begin{cases} \mathbf{c}[n-1] - \gamma[n] \\ \quad \times z(\mathbf{c}[n-1], \alpha[n]) \\ \quad \times y[n]\boldsymbol{\varphi}(\mathbf{x}[n]) \\ \quad \text{for } |z| \geq \varepsilon \\ \mathbf{c}[n-1] \quad \text{for } |z| < \varepsilon \end{cases}$	$\gamma[n] = \dfrac{1}{\gamma^2[n]\boldsymbol{\varphi}^T(\mathbf{x}[n])\boldsymbol{\varphi}(\mathbf{x}[n])}$	Yakubovich

4.12 Special Cases

4. $J = \begin{cases} M\{z^2\} & \text{for } z \leq a_n \\ M\{2a_n z\} & \text{for } a_n < z < 0 \\ 0 & \text{for } z \geq 0 \end{cases}$

    $\mathbf{c}[n] = \begin{cases} \mathbf{c}[n-1] - \gamma[n] \\ \quad \times z(\mathbf{c}[n-1], \alpha[n]) \\ \quad \times y[n]\boldsymbol{\phi}(\mathbf{x}[n]) \\ \quad \text{for } z_n \leq a_n \\ \mathbf{c}[n-1] - 2\gamma[n] \\ \quad \times a[n]y[n]\boldsymbol{\phi}(\mathbf{x}[n]) \\ \quad \text{for } a_n < z_n < 0 \\ \mathbf{c}[n-1] \quad \text{for } z_n \geq 0 \end{cases}$

    $a_n = y^2[n] \dfrac{\boldsymbol{\phi}^T(\mathbf{x}[n])\boldsymbol{\phi}(\mathbf{x}[n])}{\lambda[n]h[n]}$

    where $0 < \lambda[n] < 2$, $h[n]$ is the number of corrections of the vector $\mathbf{c}[n]$

    Fomin, Motzkin, Shoenberg

5. $J = \begin{cases} M\{z^2\} & \text{for } z < 0 \\ 0 & \text{for } z \geq 0 \end{cases}$

    $\mathbf{c}[n] = \mathbf{c}[n-1] - \Gamma[n]z(\mathbf{c}[n-1], \alpha[n-1])y[n]\boldsymbol{\phi}(\mathbf{x}[n])$
    $\alpha[n] = \alpha[n-1] - \gamma\{z(\mathbf{c}[n-1],\alpha[n-1]) + |z|(\mathbf{c}[n],\alpha[n-1])|\}$

    Depending on the choice of the matrix $\Gamma[n]$ we obtain different modified algorithms

    Ho, Kashyap

We should notice that in all algorithms considered above only one sample is shown on each iteration. If the number of samples is finite, then on each iteration we can use all the samples. In such a case we obtain the algorithm

$$\mathbf{c}[n] = \mathbf{c}[n-1] - \gamma_1[n] \sum_{\mu=1}^{M} F'(z_\mu(\mathbf{c}[n-1], \alpha[n-1]) y_\mu[n] \phi(\mathbf{x}_\mu[n]))$$

$$\alpha[n] = \alpha[n-1] + \gamma_2[n] \left[\sum_{\mu=1}^{M} F'(z_\mu(\mathbf{c}[n-1], \alpha[n-1])) \right.$$

$$\left. + \left| \sum_{\mu=1}^{M} F'(z_\mu(\mathbf{c}[n-1], \alpha[n-1])) \right| \right] \quad (4.45)$$

When $n \to \infty$, $\mathbf{c}$ converges to the vector $\mathbf{c}^*$, which is a solution of the system of inequalities (4.44), and at the same time minimizes the criterion of optimality,

$$J(\mathbf{c}, \alpha) = \sum_{\mu=1}^{M} F(y_\mu \mathbf{c}^T \phi(\mathbf{x}_\mu) - \alpha_\mu) \quad (4.46)$$

We have already discussed this in Section 3.21.

4.13 Discussion

In general, all the algorithms of training define the optimal vector $\mathbf{c}^*$ and the discriminant function $\hat{f}(\mathbf{x}, \mathbf{c}^*)$ with probability one theoretically only after an infinite number of steps. Practically, we always use a finite number of steps and this number of steps is determined by the accuracy with which we have to define $\mathbf{c}^*$ and $\hat{f}(\mathbf{x}, \mathbf{c}^*)$. There is a complete analogy between the training time and the transient time in automatic systems. In Section 4.5 we have indicated the conditions under which the number of steps (this also means the training time) is finite. These conditions imply that the noise is not present, and that the pattern classes A and B are such that the discriminant function $f(\mathbf{x})$ can be expressed by an approximating function $\hat{f}(\mathbf{x}, \mathbf{c}) = \mathbf{c}^{*T} \phi(\mathbf{x})$. In this case,

$$F(y - \mathbf{c}^{*T} \phi(\mathbf{x})) \phi(\mathbf{x}) = 0 \quad (4.47)$$

for every $\mathbf{x}$. When the noise does not exist, an N-dimensional optimal vector can be, in principle, completely determined after the first N observed samples.

Tables 4.1 and 4.2 can serve as guides in the complicated and confusing labyrinth of the algorithms. From the careful consideration of these tables, it can be concluded that all the paths of the labyrinth approach the same road which leads to the minimization of the corresponding functionals.

The following fact is also curious. Almost all the algorithms found by guesses do not escape the boundaries of the algorithms which minimize quadratic or simple piecewise linear functionals. What is the cause of this? Limited fantasy, computational simplicity, or a certain real advantage of these algorithms?

4.14 Self-Learning

Self-learning is learning without any external indications regarding the correctness or incorrectness of the reactions of a system on the presented samples. At first it seems that the self-learning of a system is in principle impossible. This feeling can even be transformed into a belief if, for instance, one accepts the following argument. Since the classified patterns possess different features, it seems impossible that a system designed to recognize the patterns can decide which of these features it has to take into consideration and which to discard. Therefore, the system cannot perform the classifications a priori envisioned by the designer. On the other hand, it is doubtful whether any classification performed by the system can satisfy anyone.

But such pessimism seems to be slightly hasty. By careful consideration of the problem, it is usually found that the designer solves many problems for the system even in the design stage. The features used for classification are first determined by the input devices of the system, i.e., by the selection of sensors. For instance, if the input device is a collection of photoelements, then the features can be the configuration and the dimensions, but not the density or the weight of the observed object. What can be the basis for self-learning of a system? We shall attempt to explain this in a general and perhaps slightly inexact way.

Let us assume that a set of patterns X consists of several disjoint subsets X_n, which correspond to different pattern classes. The patterns are characterized by the vectors $\mathbf{x}$. The occurrences of the patterns $\mathbf{x}$ from the subset X_k are random. Let us designate by P_k the probability of occurrence of a pattern $\mathbf{x}$ from the subset X_k, and by $p_k(\mathbf{x}) = p(\mathbf{x}|k)$ the conditional probability density function of the vectors $\mathbf{x}$ within the corresponding pattern class.

The conditional probability density functions $p_k(\mathbf{x})$ are such that their maxima occur over the "centers" of the pattern classes corresponding to the subsets X_k (Fig. 4.7). Unfortunately, when it is not known which pattern class

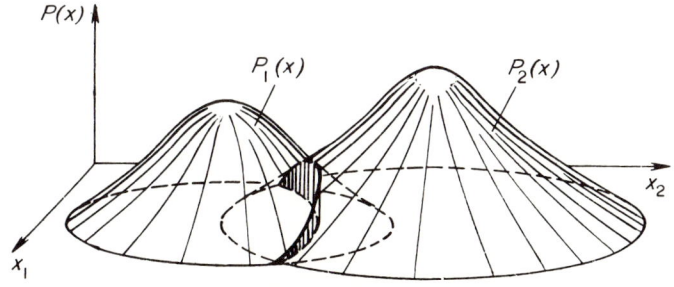

Fig. 4.7

a pattern **x** belongs to, these conditional probability density functions cannot be exactly determined.

The mixture probability density function

$$p(\mathbf{x}) = \sum_{k=1}^{M} P_k \, p_k(\mathbf{x}) \qquad (4.48)$$

contains sufficiently complete information about the subsets. In particular, it can be expected that the maxima of $p(\mathbf{x})$ will also correspond to the "centers" of the classes. Therefore, a problem of self-learning can frequently be reduced to a problem of estimating the mixture probability density function, its "centers" and the boundaries of the pattern classes.

In connection with this we shall first consider restoration (estimation) of the probability density function. Moreover, the solution of this type of problem is in itself an interesting one.

4.15 The Restoration of Probability Density Functions and Moments

In restoring (estimating) a mixture probability density function $p(\mathbf{x})$ by using the samples **x**, we assume that $p(\mathbf{x})$ can be approximated by a finite number of arbitrary, but now also orthonormal functions $\phi_v(\mathbf{x})$; i.e.,

$$\hat{p}(\mathbf{x}, \mathbf{c}) = \sum_{v=1}^{N} c_v \, \phi_v(\mathbf{x}) = \mathbf{c}^T \boldsymbol{\phi}(\mathbf{x}) \qquad (4.49)$$

The optimal value of the vector $\mathbf{c} = \mathbf{c}^*$ is considered to be one for which the quadratic measure of the approximation error

$$J(\mathbf{c}) = \int_X [p(\mathbf{x}) - \mathbf{c}^T \boldsymbol{\phi}(\mathbf{x})]^2 \, d\mathbf{x} \qquad (4.50)$$

is minimized. On the surface, this problem appears as an approximation problem which we have discussed so far. But there is an essential difference in the fact that now we do not have external training samples. Therefore, not only the vector **c** but also the function y, which in this case corresponds to the probability density function, are both unknown. However, this is not a serious difficulty. It is easy to see that due to the orthonormality of the functions $\phi_v(\mathbf{x})$, the performance index (4.50) has its minimum for

$$\mathbf{c}^* = \int_X \boldsymbol{\phi}(\mathbf{x}) p(\mathbf{x}) \, d\mathbf{x} = M[\boldsymbol{\phi}(\mathbf{x})] \qquad (4.51)$$

Therefore, the optimal value of the vector $\mathbf{c} = \mathbf{c}^*$ is equal to the mathematical expectation of the vector function $\boldsymbol{\phi}(\mathbf{x})$. In this case, the samples of $\boldsymbol{\phi}(\mathbf{x})$ are known to us.

From (4.51), in particular, it follows that if we choose the exponential functions, then the components of the vector c^* will define the moments of the corresponding order. Therefore, we have a possibility of solving the problems of moment estimation and the estimation of a probability density function by similar algorithms. However, it should always be remembered that if the exponential functions are not orthonormal, then the values of the vector c^*, found according to the formula (4.51), do not minimize the functional (4.50).

If the functions $\phi_1(x), \ldots, \phi_N(x)$ are considered to be only linearly independent (the condition of orthonormality is removed), then instead of a simple relation (4.51) we can obtain a more complicated one which can be realized by several interconnected systems. But is it worthwhile to employ a generalization which leads to a more complicated realization and which does not promise any obvious advantage?

4.16 Algorithms of Restoration

In order to define c^* we can rewrite (4.51) as

$$M[c - \phi(x)] = 0 \tag{4.52}$$

and we can now consider (4.52) as an equation in c. Let us apply to it an algorithm of training. We obtain

$$c[n] = c[n-1] - \gamma[n][c[n-1] - \phi(x[n])] \tag{4.53}$$

This algorithm, when the conditions of convergence (3.34) for a one-dimensional case are satisfied, defines the sought value $c = c^*$, and thus $\hat{p}(x, c^*)$, with probability one when $n \to \infty$. The algorithm (4.53), which can also serve in finding the moments when a special selection of the functions $\phi_\nu(x)$ is made, is realized in the form of simple linear sampled-data systems with variable coefficients. One such system is shown in Fig. 4.8 (for the νth com-

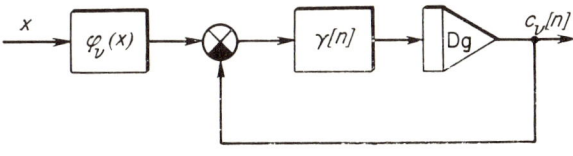

Fig. 4.8

ponent). A general scheme of the device for restoring (estimating) $p(x)$ only from the observed patterns is presented in Fig. 4.9.

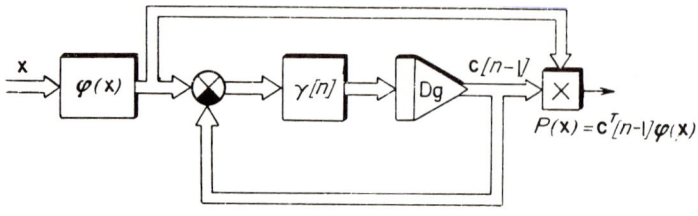

Fig. 4.9

If the observed patterns arrive continuously in time, then for the stationary case we can use continuous algorithms (3.28). Then, instead of (4.53), we obtain

$$\frac{d\mathbf{c}(t)}{dt} = -\gamma(t)[\mathbf{c}(t) - \boldsymbol{\phi}(\mathbf{x}(t))] \qquad (4.54)$$

The scheme which realizes this algorithm differs from the one shown in Fig. 4.9 only by the presence of continuous integrators in place of the discrete integrators Dg.

It can be shown by using the results of Sections 3.16 and 3.17 that the selection of $\gamma[n] = 1/n$ in an algorithm of the type (4.53) is optimal with respect to the minimization of the mean square error of the estimate of $\mathbf{c}$ for any fixed value of n. For a continuous algorithm of the type (4.54), the optimal selection is $\gamma(t) = 1/t$.

Instead of the algorithms (4.53) and (4.54) we can also use modified algorithms with an averaging operator of the type (3.42),

$$\mathbf{c}[n] = \mathbf{c}[n-1] - \gamma[n]\left(\mathbf{c}[n-1] - \frac{1}{n}\sum_{m=1}^{n} \boldsymbol{\phi}(\mathbf{x}[m])\right) \quad (n = 1, 2, \ldots) \quad (4.55)$$

and similarly,

$$\frac{d\mathbf{c}(t)}{dt} = -\gamma(t)\left(\mathbf{c}(t) - \frac{1}{t}\int_0^t \boldsymbol{\phi}(\mathbf{x}(\tau))\, d\tau\right) \qquad (4.56)$$

The scheme which realizes these algorithms is given in Fig. 4.10. It differs

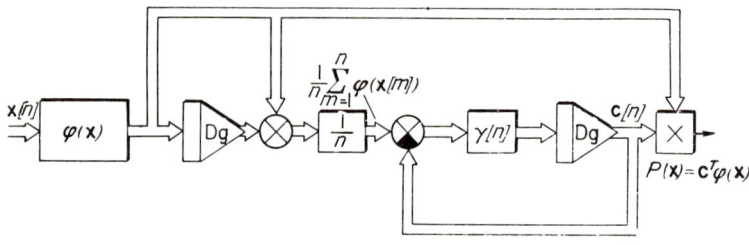

Fig. 4.10

from the schemes of the previous algorithms by the presence of an additional integrator. Another interpretation of the basic and modified algorithms will be given in Section 5.6.

In the modified algorithm (4.55), the choice of $\gamma[n] = 1/n$ is optimal with respect to the minimization of the sum of the mean square errors of the estimate and its weighted first difference, i.e., with respect to the minimization of a functional of the type (3.50),

$$\overline{V}_g^2[N] = M\{\|\mathbf{c}[n] - \mathbf{c}^*\|^2 + (n-1)\|\Delta\mathbf{c}[n-1]\|^2\} \tag{4.57}$$

4.17 Principle of Self-Learning

Self-learning in pattern recognition, or in the classification of a set of patterns, consists of finding the boundary between the pattern classes by using only the observed samples without any external supervision. In the favorable cases, this boundary is a ravine on a hilly surface formed by the mixture probability density function. Although we already know how to estimate a mixture probability density function, the problem of finding the boundary, i.e., the problem of generating a separating surface, has not become simpler. However, the problem is simplified if we have some a priori information; for instance, if the form of conditional probability density functions is given. When the conditional probability density functions are normal and differ only in their first moments, it is sufficient to find the weighted mean of the joint probability density function in order to obtain the boundary between two pattern classes. Assuming sufficiently complete a priori information, classification without external supervision can be performed in many ways, but we are not going to be concerned with them here. However, is it impossible to form a separating surface without an explicit estimation of the probability density function and any a priori information regarding these densities? It appears that such a possibility does exist.

One such possibility consists in using the answers of the perceptrons instead of external evaluations. It can be expected that the perceptron finds the boundaries between the two classes, but of course, it will not be capable of recognizing these pattern classes uniquely. The actions of such a perceptron are similar to the actions of a trustful pessimist or of a distrustful optimist, who accepts as actual anything desired or undesired.

Another possibility is more interesting. Let us imagine that all the elements **x** which correspond to the same pattern class are grouped in the neighborhood of the "center" of that class. Let us then introduce a function which in a general way characterizes the distance of **x** from still unknown centers α_k of these pattern classes.

Then the classification could be based on the requirement that each element of each pattern class is closer to the center of its pattern class than to the centers of other pattern classes. This requirement, as it is mentioned below, can be related to the minimization of a certain functional (average risk). Now it is not difficult to understand that a problem of self-learning can be considered solved if the centers of the sets X_k, and the boundaries of these sets, are found by minimizing the expected risk on the bases of the observed patterns $\mathbf{x} \in X$ only. In order to go on from this general consideration to a solution of the problem of self-learning, it is first of all necessary to define the expected risk and find its variation. The next two sections are devoted to these problems.

4.18 Average Risk

Let us now introduce the concept of distance ρ between the elements $\mathbf{x}$ of the set X and the "centers" α_k of the subsets X_k. We specifiy ρ by a certain convex function F of the difference $\phi(\mathbf{x}) - \phi(\alpha_k)$:

$$\rho(\mathbf{x}, \mathbf{u}_k) = F(\phi(\mathbf{x}) - \mathbf{u}_k) \tag{4.58}$$

where

$$\mathbf{u}_k = \phi(\alpha_k) \tag{4.59}$$

The function $\rho(\mathbf{x}, \mathbf{u}_k)$ can be considered as a loss function or a function of penalties for the kth pattern class. Average risk or average penalty for all pattern classes can be given by the expression

$$R = \sum_{k=1}^{M} P_k \int_{X_k} F(\phi(\mathbf{x}) - \mathbf{u}_k) p_k(\mathbf{x}) \, d\mathbf{x} \tag{4.60}$$

or, by introducing the mixture probability density function (4.48), in the form

$$R = \sum_{k=1}^{M} \int_{X_k} F(\phi(\mathbf{x}) - \mathbf{u}_k) p(\mathbf{x}) \, d\mathbf{x} \tag{4.61}$$

Let us recall that X_k ($k = 1, \ldots, M$) are disjoint regions.

If we introduce the characteristic function

$$\varepsilon_k(\mathbf{x}, \mathbf{u}_1, \ldots, \mathbf{u}_M) = \begin{cases} 1 & \text{when } \mathbf{x} \in X_k \\ 0 & \text{when } \mathbf{x} \notin X_k \end{cases} \tag{4.62}$$

we can then form a general loss function

$$S(\mathbf{x}, \mathbf{u}_1, \ldots, \mathbf{u}_M) = \sum_{k=1}^{M} \varepsilon_k(\mathbf{x}, \mathbf{u}_1, \ldots, \mathbf{u}_M) F(\phi(\mathbf{x}) - \mathbf{u}_k) \tag{4.63}$$

and the average risk (4.61) can be written in a slightly more convenient form:

$$R = M_{\mathbf{x}}\{S(\mathbf{x}, \mathbf{u}_1, \ldots, \mathbf{u}_M)\} \tag{4.64}$$

The quantities $\mathbf{u}_k$ and the sets X_k in the expressions for the average risk and thus the characteristic functions, are unknown.

4.19 Variation of the Average Risk

Minimization of the average risk requires computation of the variation of the average risk. Then, by setting the variation equal to zero, it is easy to find the conditions for the minimum of the average risk. Variation of the average risk, defined by expression (4.61), can be written in the form of a sum of two variations: the variation δR_1, related to the changes in the parameters $\mathbf{u}_1, \ldots, \mathbf{u}_M$, and the variation δR_2, related to the changes in the regions X_k:

$$\delta R = \delta R_1 + \delta R_2 \tag{4.65}$$

where

$$\delta R_1 = \sum_{k=1}^{M} \int_{X_k} \nabla_{\mathbf{u}_k} F(\phi(\mathbf{x}) - \mathbf{u}_k) p(\mathbf{x}) \, d\mathbf{x} \, \delta \mathbf{u}_k \tag{4.66}$$

and

$$\delta R_2 = \sum_{k=1}^{M} \int_{X} \sum_{m=1}^{N} \frac{\partial}{\partial x_m} [F(\phi(\mathbf{x}) - \mathbf{u}_k) p(\mathbf{x}) \delta x_m^{(k)}] \, d\mathbf{x} \tag{4.67}$$

Here, N is the dimension of the vector $\mathbf{x}$. Let us examine more carefully the expression for δR_2. By Green's formula, it can be transformed into

$$\delta R_2 = \sum_{k=1}^{M} \int_{\Lambda_k} F(\phi(\mathbf{x}) - \mathbf{u}_k) p(\mathbf{x}) \delta \Lambda_k \, d\mathbf{x} \tag{4.68}$$

where

$$\delta \Lambda_k = \sum_{\nu=1}^{N} (-1)^{\nu} \delta x_{\nu}^{(k)} \tag{4.69}$$

Since the regions X_k $(k = 1, \ldots, M)$ are disjoint, each border surface Λ_k can be decomposed into segments Λ_{km} with which the region X_k is bordered by the regions X_m $(m = 1, \ldots, M; m \ne k)$. Obviously,

$$\delta \Lambda_{km} = -\delta \Lambda_{mk} \tag{4.70}$$

Considering this, we obtain

$$\delta R_2 = \sum_{k,m=1}^{s} \int_{\Lambda_{km}} [F(\phi[\mathbf{x}] - \mathbf{u}_k) - F(\phi(\mathbf{x}) - \mathbf{u}_m)] p(\mathbf{x}) \delta \Lambda_{km} \, d\mathbf{x} \tag{4.71}$$

where s is the number of pairs of adjacent regions.

4.20 The Conditions for the Minimum of the Average Risk

By setting the total variation of the average risk equal to zero, we obtain the condition for the minimum of the average risk, which, according to (4.65), (4.66) and (4.71), is written as

$$\delta R = \sum_{k=1}^{M} \int_{X_k} \nabla_{\mathbf{u}_k} F(\phi(\mathbf{x}) - \mathbf{u}_k) p(\mathbf{x}) \, d\mathbf{x} \, \delta \mathbf{u}_k + \sum_{k,m=1}^{s} \int_{\Lambda_{km}} [F(\phi(\mathbf{x}) - \mathbf{u}_k)$$
$$- F(\phi(\mathbf{x}) - \mathbf{u}_m)] p(\mathbf{x}) \delta \Lambda_{km} \, d\mathbf{x} = 0 \tag{4.72}$$

Since the variations δu_k and $\delta \Lambda_{km}$ are arbitrary and independent, it follows from (4.72) that these conditions have to be satisfied:

$$\int_{X_k} \nabla_{u_k} F(\phi(x) - u_k) p(x) \, dx = 0 \qquad (4.73)$$

and

$$F(\phi(x) - u_k) - F(\phi(x) - u_m) = 0 \qquad (x \in \Lambda_{km}, m \neq k) \quad (4.74)$$

The conditions (4.73) define the optimal values $u_k = u_k^*$ which characterize the "centers" of the regions, and (4.74) is the equation of the surface which separates the regions X_k and X_m. This surface corresponds to the optimal decomposition of the region X into pattern classes (in the sense of minimizing average risk). Therefore, the problem of self-learning consists of solving the system of equations (4.73) with respect to u_k when the probability density function $p(x)$ is unknown, and when the equalities (4.74) are satisfied on the boundaries of the regions. For simplicity, we shall limit our discussions to the case of two pattern classes ($M = 2$). Extension of the results to a case $M > 0$ will not create any basic difficulties.

4.21 Algorithms of Self-Learning

If we use the characteristic function (4.62), then the conditions (4.73) for $M = 2$ (this corresponds to two pattern classes A and B) can be written as

$$\begin{aligned} \nabla_{u_1} R &= M\{\varepsilon_1(x, u_1, u_2) \nabla_{u_1} F(\phi(x) - u_1)\} = 0 \\ \nabla_{u_2} R &= M\{\varepsilon_2(x, u_1, u_2) \nabla_{u_2} F(\phi(x) - u_2)\} = 0 \end{aligned} \qquad (4.75)$$

Now, the problem is to find $u_1 = u_1^*$ and $u_2 = u_2^*$ which satisfy condition (4.75). Let us apply to (4.75) the algorithms of adaptation, or, as it is now appropriate to call them, algorithms of self-learning. Then we formally obtain

$$\begin{aligned} u_1[n] &= u_1[n-1] - \gamma_1[n] \varepsilon_1(x[n], u_1[n-1], u_2[n-1]) \\ &\quad \times \nabla_{u_1} F(\phi(x[n]) - u_1[n-1]) \\ u_2[n] &= u_2[n-1] - \gamma_2[n] \varepsilon_2(x[n], u_1[n-1], u_2[n-1]) \\ &\quad \times \nabla_{u_2} F(\phi(x[n]) - u_2[n-1]) \end{aligned} \qquad (4.76)$$

Of course, these algorithms cannot be immediately applied, as we do not know the characteristic functions ε_1 and ε_2 which appear in the algorithms. However, this difficulty is easily overcome by using condition (4.75). Let us designate

$$f(x, u_1, u_2) = F(\phi(x) - u_1) - F(\phi(x) - u_2) \qquad (4.77)$$

The function $f(x, u_1, u_2)$ is a discriminant function. As it follows from (4.74), this function is equal to zero on the boundary, and it has different signs in different regions. Its sign can always be determined by setting the specific values of $\mathbf{u}_1$ and $\mathbf{u}_2$ into (4.77). We can now write the algorithms of self-learning in the final form:

$$\begin{aligned}\mathbf{u}_1[n] &= \mathbf{u}_1[n-1] - \gamma_1[n]\nabla_{\mathbf{u}_1}F(\phi(\mathbf{x}[n]) - \mathbf{u}_1[n-1]) \\ \mathbf{u}_2[n] &= \mathbf{u}_2[n-1]\end{aligned} \quad (4.78)$$

if

$$f(\mathbf{x}[n], \mathbf{u}_1[n-1], \mathbf{u}_2[n-1]) \le 0 \quad (4.79)$$

and

$$\begin{aligned}\mathbf{u}_1[n] &= \mathbf{u}_1[n-1] \\ \mathbf{u}_2[n] &= \mathbf{u}_2[n-1] - \gamma_2[n]\nabla_{\mathbf{u}_2}F(\phi(\mathbf{x}[n]) - \mathbf{u}_2[n-1])\end{aligned} \quad (4.80)$$

if

$$f(\mathbf{x}[n], \mathbf{u}_1[n-1], \mathbf{u}_2[n-1]) > 0 \quad (4.81)$$

A structural scheme of such a self-learning perceptron is given in Fig. 4.11.

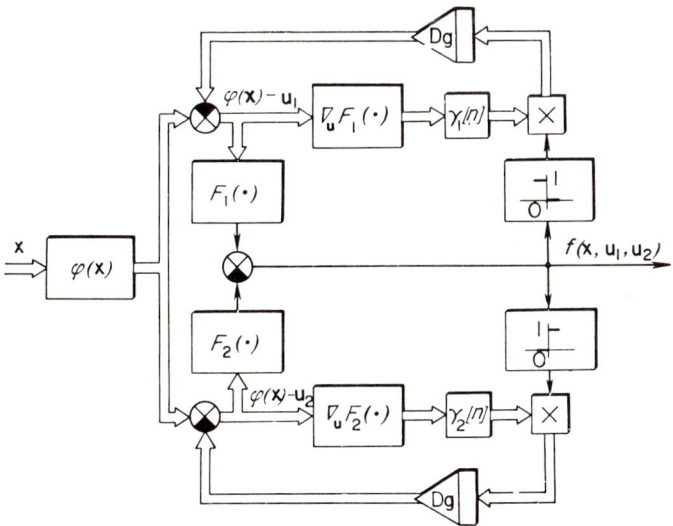

Fig. 4.11

4.22 A Generalization

Until now we have been assuming that the loss function for each pattern class depends only on one parameter vector $\mathbf{u}_k$ which defines the "center" of that pattern class. The algorithms obtained above can be easily generalized to a more complex case when the loss function of the kth pattern class depends

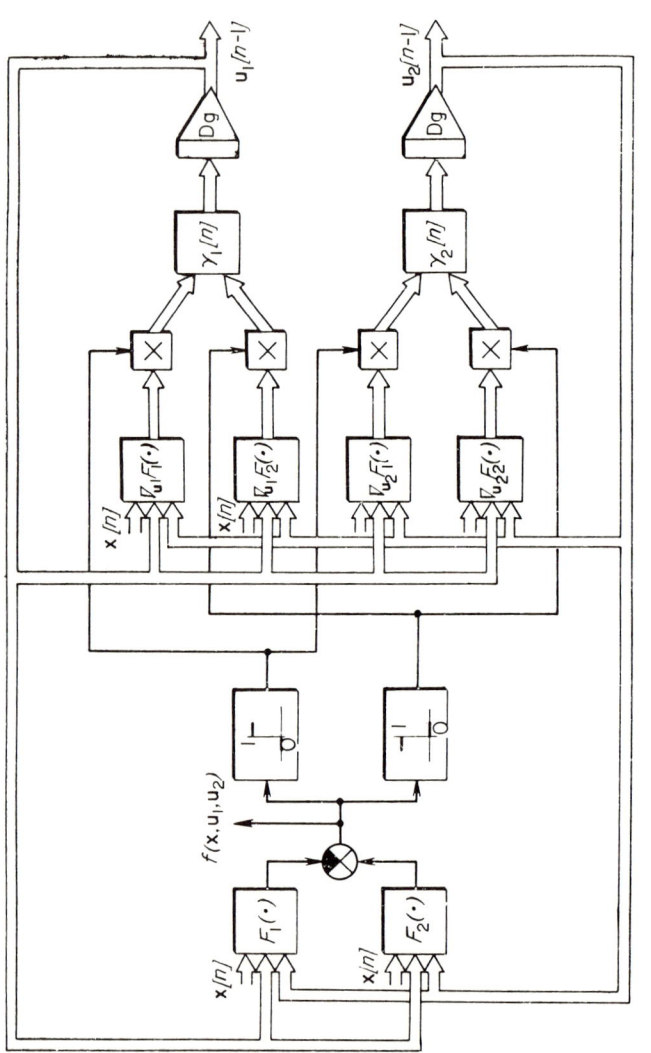

Fig. 4.12

on all the parameters $\mathbf{u}_1, \ldots, \mathbf{u}_N$, and it varies from one pattern class to another; for instance, when instead of the function $F(\boldsymbol{\phi}(\mathbf{x}) - \mathbf{u}_k)$, we choose the function $F_k(\mathbf{x}, \mathbf{u}_1, \ldots, \mathbf{u}_N)$.

If we limit ourselves, as we have been doing already, to the consideration of two pattern classes only, the general loss function can be written by analogy to (4.63) as

$$S(\mathbf{x}, \mathbf{u}_1, \mathbf{u}_2) = \varepsilon_1(\mathbf{x}, \mathbf{u}_1, \mathbf{u}_2) F_1(\mathbf{x}, \mathbf{u}_1, \mathbf{u}_2) + \varepsilon_2(\mathbf{x}, \mathbf{u}_1, \mathbf{u}_2) F_2(\mathbf{x}, \mathbf{u}_1, \mathbf{u}_2) \quad (4.82)$$

The conditions for the minimum of the average risk are now written as

$$\begin{aligned} M_\mathbf{x}\{\varepsilon_1(\mathbf{x}, \mathbf{u}_1, \mathbf{u}_2) \nabla_{\mathbf{u}_1} F_1(\mathbf{x}, \mathbf{u}_1, \mathbf{u}_2) + \varepsilon_2(\mathbf{x}, \mathbf{u}_1, \mathbf{u}_2) \nabla_{\mathbf{u}_1} F_2(\mathbf{x}, \mathbf{u}_1, \mathbf{u}_2)\} &= 0 \\ M_\mathbf{x}\{\varepsilon_1(\mathbf{x}, \mathbf{u}_1, \mathbf{u}_2) \nabla_{\mathbf{u}_2} F_1(\mathbf{x}, \mathbf{u}_1, \mathbf{u}_2) + \varepsilon_2(\mathbf{x}, \mathbf{u}_1, \mathbf{u}_2) \nabla_{\mathbf{u}_2} F_2(\mathbf{x}, \mathbf{u}_1, \mathbf{u}_2)\} &= 0 \end{aligned} \quad (4.83)$$

Hence the algorithms of self-learning can be written in the form

$$\begin{aligned} \mathbf{u}_1[n] &= \mathbf{u}_1[n-1] - \gamma_1[n] \nabla_{\mathbf{u}_1} F_1(\mathbf{x}[n], \mathbf{u}_1[n-1], \mathbf{u}_2[n-1]) \\ \mathbf{u}_2[n] &= \mathbf{u}_2[n-1] - \gamma_2[n] \nabla_{\mathbf{u}_2} F_1(\mathbf{x}[n], \mathbf{u}_1[n-1], \mathbf{u}_2[n-1]) \end{aligned} \quad (4.84)$$

when

$$\begin{aligned} f(\mathbf{x}[n], \mathbf{u}_1[n-1], \mathbf{u}_2[n-1]) &= F_1(\mathbf{x}[n], \mathbf{u}_1[n-1], \mathbf{u}_2[n-1]) \\ &\quad - F_2(\mathbf{x}[n], \mathbf{u}_1[n-1], \mathbf{u}_2[n-1]) \leq 0 \end{aligned} \quad (4.85)$$

and

$$\begin{aligned} \mathbf{u}_1[n] &= \mathbf{u}_1[n-1] - \gamma_1[n] \nabla_{\mathbf{u}_1} F_2(\mathbf{x}[n], \mathbf{u}_1[n-1], \mathbf{u}_2[n-1]) \\ \mathbf{u}_2[n] &= \mathbf{u}_2[n-1] - \gamma_2[n] \nabla_{\mathbf{u}_2} F(\mathbf{x}[n], \mathbf{u}_1[n-1], \mathbf{u}_2[n-1]) \end{aligned} \quad (4.86)$$

when

$$\begin{aligned} f(\mathbf{x}[n], \mathbf{u}_1[n-1], \mathbf{u}_2[n-1]) &= F_1(\mathbf{x}[n], \mathbf{u}_1[n-1], \mathbf{u}_2[n-1]) \\ &\quad - F_2(\mathbf{x}[n], \mathbf{u}_1[n-1], \mathbf{u}_2[n-1]) > 0 \end{aligned} \quad (4.87)$$

From these algorithms for $F_k(\mathbf{x}, \mathbf{u}_1, \mathbf{u}_2) = F(\boldsymbol{\phi}(\mathbf{x}) - \mathbf{u}_k)$, $k = 1, 2$, one can obtain the algorithms given in Section 4.21. A structural scheme of the general self-learning perceptron is shown in Fig. 4.12.

4.23 Specific Algorithms

We can easily obtain different specific algorithms of self-learning from the general ones by selecting different functions $\mathbf{F}_k(\cdot)$. Let us assume, for instance,

$$F_k(\mathbf{x}, \mathbf{u}_1, \mathbf{u}_2) = \|\boldsymbol{\phi}(\mathbf{x}) - \mathbf{u}_k\|^2 \qquad (k = 1, 2) \quad (4.88)$$

Then, it follows from (4.77) that the discriminant function will have the form

$$f(\mathbf{x}, \mathbf{u}_1, \mathbf{u}_2) = \|\boldsymbol{\phi}(\mathbf{x}) - \mathbf{u}_1\|^2 - \|\boldsymbol{\phi}(\mathbf{x}) - \mathbf{u}_2\|^2$$

Table 4.3

ALGORITHMS OF SELF-LEARNING

Number	Loss function	Algorithms	Authors
1	$F_1 = F_1(\mathbf{x}, \mathbf{u}_1, \mathbf{u}_2)$ $F_2 = F_2(\mathbf{x}, \mathbf{u}_1, \mathbf{u}_2)$	$\mathbf{u}_1[n] = \mathbf{u}_1[n-1] - \gamma_1[n]\nabla_{u_1}F_1(\mathbf{x}[n], \mathbf{u}_1[n-1], \mathbf{u}_2[n-1])$ $\mathbf{u}_2[n] = \mathbf{u}_2[n-1] - \gamma_2[n]\nabla_{u_2}F_1(\mathbf{x}[n], \mathbf{u}_1[n-1], \mathbf{u}_2[n-1])$ for $F_1(\mathbf{x}[n], \mathbf{u}_1[n-1], \mathbf{u}_2[n-1]) - F_2(\mathbf{x}[n], \mathbf{u}_1[n-1], \mathbf{u}_2[n-1]) \leq 0$ $\mathbf{u}_1[n] = \mathbf{u}_1[n-1] - \gamma_1[n]\nabla_{u_1}F_2(\mathbf{x}[n], \mathbf{u}_1[n-1], \mathbf{u}_2[n-1])$ $\mathbf{u}_2[n] = \mathbf{u}_2[n-1] - \gamma_2[n]\nabla_{u_2}F_2(\mathbf{x}[n], \mathbf{u}_1[n-1], \mathbf{u}_2[n-1])$ for $F_1(\mathbf{x}[n], \mathbf{u}_1[n-1], \mathbf{u}_2[n-1]) - F_2(\mathbf{x}[n], \mathbf{u}_1[n-1], \mathbf{u}_2[n-1]) > 0$	Braverman
2	$F_1 = \|\boldsymbol{\phi}(\mathbf{x}) - \mathbf{u}_1\|^2$ $F_2 = \|\boldsymbol{\phi}(\mathbf{x}) - \mathbf{u}_2\|^2$	$\mathbf{u}_1[n] = \mathbf{u}_1[n-1] + \gamma_1[n][\boldsymbol{\phi}(\mathbf{x}[n]) - \mathbf{u}_1[n-1]]$ $\mathbf{u}_2[n] = \mathbf{u}_2[n-1]$ for $2(\mathbf{u}_1^T[n-1] - \mathbf{u}_2^T[n-1])\boldsymbol{\phi}(\mathbf{x}[n])$ $- (\|\mathbf{u}_1[n-1]\|^2 - \|\mathbf{u}_2[n-1]\|^2) \geq 0$ $\mathbf{u}_1[n] = \mathbf{u}_1[n-1]$ $\mathbf{u}_2[n] = \mathbf{u}_2[n-1] + \gamma_2[n][\boldsymbol{\phi}(\mathbf{x}[n]) - \mathbf{u}_2[n-1]]$ for $2(\mathbf{u}_1^T[n-1] - \mathbf{u}_2^T[n-1])\boldsymbol{\phi}(\mathbf{x}[n])$ $- (\|\mathbf{u}_1[n-1]\|^2 - \|\mathbf{u}_2[n-1]\|^2) < 0$	
3	$F_1 = \|\boldsymbol{\phi}(\mathbf{x}) - \mathbf{u}_1\|^2 + \|\mathbf{u}_2\|^2$ $F_2 = \|\boldsymbol{\phi}(\mathbf{x}) - \mathbf{u}_2\|^2 + \|\mathbf{u}_1\|^2$	$\mathbf{u}_1[n] = \mathbf{u}_1[n-1] - \gamma_1[n](\mathbf{u}_1[n-1] - \boldsymbol{\phi}(\mathbf{x}[n]))$ $\mathbf{u}_2[n] = \mathbf{u}_2[n-1]$ for $(\mathbf{u}_1^T[n-1] - \mathbf{u}_2^T[n-1])\boldsymbol{\phi}(\mathbf{x}[n]) \geq 0$ i.e., for $\mathbf{x} \in X_1[n]$ $\mathbf{u}_1[n] = \mathbf{u}_1[n-1]$ $\mathbf{u}_2[n] = \mathbf{u}_2[n-1] - \gamma_2[n](\mathbf{u}_2[n-1] - \boldsymbol{\phi}(\mathbf{x}[n]))$ for $(\mathbf{u}_1^T[n-1] - \mathbf{u}_2^T[n-1])\boldsymbol{\phi}(\mathbf{x}[n]) < 0$ i.e., for $\mathbf{x} \in X_2[n]$ $\gamma_1[n] = \dfrac{1}{N[n]} \quad \gamma_2[n] = \dfrac{1}{n - N[n]}$	Dorofeyuk

where $N[n]$ is the number of samples which are from class 1

4.23 Specific Algorithms

or

$$f(\mathbf{x}, \mathbf{u}_1, \mathbf{u}_2) = -2(\mathbf{u}_1^T - \mathbf{u}_2^T)\boldsymbol{\phi}(\mathbf{x}) + (\|\mathbf{u}_1\|^2 - \|\mathbf{u}_2\|^2) \qquad (4.89)$$

In this special case,

$$\nabla_{\mathbf{u}} F(\boldsymbol{\phi}(\mathbf{x}) - \mathbf{u}) = -2(\boldsymbol{\phi}(\mathbf{x}) - \mathbf{u}) \qquad (4.90)$$

and the algorithms of self-learning (4.78)–(4.81) can be written in the following way:

$$\begin{aligned} \mathbf{u}_1[n] &= \mathbf{u}_1[n-1] + 2\gamma_1[n](\boldsymbol{\phi}(\mathbf{x}[n]) - \mathbf{u}_1[n-1]) \\ \mathbf{u}_2[n] &= \mathbf{u}_2[n-1] \end{aligned} \qquad (4.91)$$

if

$$-2(\mathbf{u}_1^T[n-1] - \mathbf{u}_2^T[n-1])\boldsymbol{\phi}(\mathbf{x}[n]) + (\|\mathbf{u}_1[n-1]\|^2 - \|\mathbf{u}_2[n-1]\|^2) \leq 0 \qquad (4.92)$$

and

$$\begin{aligned} \mathbf{u}_1[n] &= \mathbf{u}_1[n-1] \\ \mathbf{u}_2[n] &= \mathbf{u}_2[n-1] + 2\gamma_2[n](\boldsymbol{\phi}(\mathbf{x}[n]) - \mathbf{u}_2[n-1]) \end{aligned} \qquad (4.93)$$

if

$$-2(\mathbf{u}_1^T[n-1] - \mathbf{u}_2^T[n-1])\boldsymbol{\phi}(\mathbf{x}[n]) + (\|\mathbf{u}_1[n-1]\|^2 - \|\mathbf{u}_2[n-1]\|^2) > 0 \qquad (4.94)$$

A structural scheme of such a self-learning perceptron is shown in Fig. 4.13.

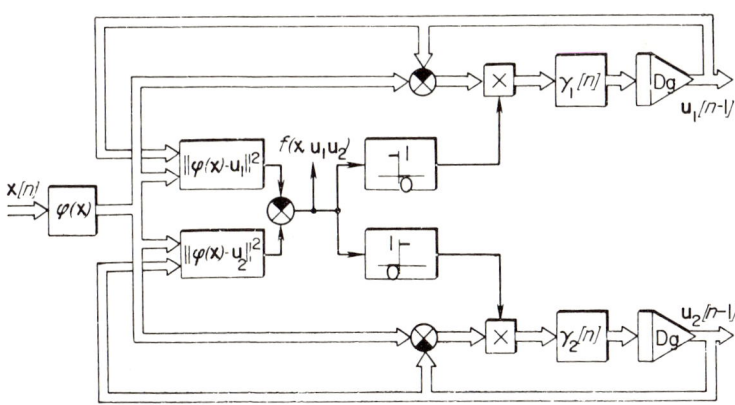

Fig. 4.13

The other specific algorithms of self-learning can be obtained in an analogous manner. Some of them are given in Table 4.3. The reader can complete this table without difficulty when it is needed.

4.24 Search Algorithms of Self-Learning

If for some reason it is not possible to compute $\nabla_{\mathbf{u}} F_k(\mathbf{x}, \mathbf{u}_1, \mathbf{u}_2)$ (for instance, due to discontinuity in $F_k(\mathbf{x}, \mathbf{u}_1, \mathbf{u}_2)$), then the search algorithms of self-learning can be used. Thus, for $M = 2$, from (4.64) and (4.63) we obtain

$$R = M\{S(\mathbf{x}, \mathbf{u}_1, \mathbf{u}_1)\} \tag{4.95}$$

where the general loss function is defined by the expression (4.82). Let us employ the search algorithm of adaptation (or self-learning) of the type (4.13) in finding the optimal values $\mathbf{u}_1^*$, and $\mathbf{u}_2^*$. In the case under consideration we obtain

$$\begin{aligned}\mathbf{u}_1[n] &= \mathbf{u}_1[n-1] - \gamma_1[n]\tilde{\nabla}_{\mathbf{u}_1+} S(\mathbf{x}[n], \mathbf{u}_1[n-1], \mathbf{u}_2[n-1]) \\ \mathbf{u}_2[n] &= \mathbf{u}_2[n-1] - \gamma_2[n]\tilde{\nabla}_{\mathbf{u}_2+} S(\mathbf{x}[n], \mathbf{u}_1[n-1], \mathbf{u}_2[n-1])\end{aligned} \tag{4.96}$$

These algorithms employ the estimates of the gradients

$$\tilde{\nabla}_{\mathbf{u}_1+} S(\mathbf{x}[n], \mathbf{u}_1[n-1], \mathbf{u}_2[n-1])$$

$$= \frac{1}{2a[n]} [S_+(\mathbf{x}[n], \mathbf{u}_1[n-1] + e a[n], \mathbf{u}_2[n-1])$$

$$- S_0(\mathbf{x}[n], \mathbf{u}_1[n-1], \mathbf{u}_2[n-1])] \tag{4.97}$$

and

$$\tilde{\nabla}_{\mathbf{u}_2+} S(\mathbf{x}[n], \mathbf{u}_1[n-1], \mathbf{u}_2[n-1])$$

$$= \frac{1}{2a[n]} [S_+(\mathbf{x}[n], \mathbf{u}_1[n-1], \mathbf{u}_2[n-1] + e a[n])$$

$$- S_0(\mathbf{x}[n], \mathbf{u}_1[n-1], \mathbf{u}_2[n-1])] \tag{4.98}$$

For obtaining the values of the characteristic functions $\varepsilon_1(\mathbf{x}, \mathbf{u}_1, \mathbf{u}_2)$ and $\varepsilon_2(\mathbf{x}, \mathbf{u}_1, \mathbf{u}_2)$ which appear in the estimates of the gradients (4.97) and (4.98) we use the same method we used in Section 4.21. The pairs of values $\mathbf{u}_1[n-1]$, $\mathbf{u}_2[n-1]$; $\mathbf{u}_1[n-1] + ea[n]$, $\mathbf{u}_2[n-1]$; and $\mathbf{u}_1[n-1]$, $\mathbf{u}_2[n-1] + ea[n]$ for given $\mathbf{x}[n]$ are used in the expression of the discriminant function (4.77). If the value of the discriminant function is negative, $\varepsilon_1 = 1$ and $\varepsilon_2 = 0$, and if it is positive, $\varepsilon_1 = 0$ and $\varepsilon_2 = 1$.

4.25 Discussion

The algorithms of self-learning differ from the algorithms of training by having two "vector" algorithms instead of one. These two algorithms replace each other according to the signs of the discriminant function $f(\mathbf{x}[n], \mathbf{u}_1[n-1], \mathbf{u}_2[n-1])$ in which the parameters $\mathbf{u}_1$ and $\mathbf{u}_2$ become more accurate after each observed pattern. This compensates for an apparent nonexistence of the evaluations by a teacher. However, a teacher actually exists in this case; his role is great even in the solution of a problem of self-learning. It is not difficult to discover that the role of a teacher appears in the selection of the penalty function made earlier. As it was shown above, the penalty function also defines the discriminant function uniquely.

Therefore, as we have discussed in Sections 4.14–4.24, self-learning is not, as it was sometimes considered to be, learning without a teacher. It is more similar to learning by correspondence where the student cannot continuously receive a qualified help or ask questions. Instead, he must use only methodical instructions (frequently very obsolete and sometimes of a questionable quality). Do not these methodical instructions remind us of the original definition of a penalty function?

4.26 Certain Problems

For practical applications of different algorithms of training, it is necessary to compare such algorithms not only according to the length of the "training" sequence but also according to the noise immunity. It would also be important to explain which functions $\phi_\nu(\mathbf{x})$ yield both a good approximation and a simple realization of the system. It would be interesting to discover how the accuracy of approximation depends upon the number of functions $\phi_\nu(\mathbf{x})$ ($\nu = 1, \ldots, N$), and also to estimate the "training" time. Of course, it is only reasonable to do it for specific functions $\phi_\nu(\mathbf{x})$.

4.27 Conclusion

We have established that different approaches to the solution of the problem of learning in pattern recognition differ from each other only by the selection of approximating functions, the form of the functional, and the manner in which such a functional is minimized. Therefore, it can be stated that an era of intuitive searches for the algorithms of training (at least of recursive algorithms) and of the inventive discoveries of such algorithms has passed. This is not extremely sad. Instead of a secret charm of artificial intelligence, which existed in many works on learning in pattern recognition, the problem of learning has now been reduced to a common problem of approximation. This is the reason for the present explanation of the sufficiently simple character of self-learning. It is true that we still cannot effectively use approximations but simple geometric representations allow us to penetrate even this seemingly mystic area. Therefore, the era of mystery and mysticism, even if it has not already passed, is definitely passing. But this should not cause despair. It happens thus often in science. As soon as a certain fact is clearly understood, its content begins to seem (perhaps only seem) trivial. We do not want to imply by this that the problem of training in pattern recognition, and thus of self-learning, is completely solved. Let us mention that the problem would actually be solved if we had limited ourselved to the formulation of the problem given above. By such a formulation, we have bypassed many difficulties in obtaining the complete solution of the problem.

The basic difficulty in the solution of this problem is to discover a universal method for obtaining useful features. However, we cannot rush into the search for such a universal method. This is true not only because we have to solve some immediate problems of adaptation, but because it is not yet known whether such a method exists at all.

COMMENTS

4.1 The fact that pattern recognition is the first stage in information processing was often emphasized by Harkevich (1959, 1965). One can spend much time reading about the discussions on the topic "man or a machine." The author has become familiar with such discussions by reading the books by Taube (1961) and Kobrinskii (1965), and he recommends them to the reader. The special problems in pattern recognition were treated in the books by Sebestyen (1962b) and Nilsson (1965).

4.2 The hypothesis of compactness was proposed by Braverman (1962). Its more rigorous form, obtained after a long period of time, is known as the hypothesis of representation, which is discussed in Section 4.5.

4.3 Similar functionals were introduced by Yakubovich (1965) for the quadratic loss functions, and by the author (Tsypkin, 1965b, 1966) for the general case.

4.4 Approximation of an arbitrary function using a system of linearly independent or orthogonal functions is broadly used in the solution of various engineering problems. The approach presented here is based on the correspondence by the author (Tsypkin, 1965a) and the paper by Devyaterikov et al. (1967).

The algorithms of the type (4.12) for certain particular cases ($F(\cdot)$ is either a linear or relay type of function) were described by Aizerman et al. (1964c) on the basis of their potential function method. Furthermore, they have also obtained the algorithms of the type (4.9) from such algorithms. According to the terminology of Aizerman et al. (1964a), the algorithms of the type (4.12) correspond to the "machine realization" and the treatment of the problem in the original space, while the algorithms of the type (4.9) correspond to the "perceptron realization" and the treatment of the problem in the "rectified" space. It follows from the results presented in Section 4.4 that these realizations are equivalent to each other.

4.5 The "hypothesis of compactness" was introduced by Aizerman et al. (1964a–c). We shall mention that they have proven the convergence of their algorithms only in probability when the system of function $\phi_\nu(\mathbf{x})$ ($\nu = 1, \ldots, N$) is assumed to be orthonormal. In particular, condition (4.19) was obtained in the second of the mentioned references. Braverman (1965) has removed one of the constraints and proved the convergence with probability one. The approach presented here does not require the hypothesis of representation, and the convergence with probability one, i.e., almost sure convergence, is proved.

4.6 Much has been written about Rosenblatt's perceptron. We shall mention here only those publications which are most interesting from the viewpoint of our approach. First of all, these are the works by Rosenblatt (1962a, b, 1964). Important facts about perceptrons can be found in the paper by Block (1962). The possibility of building a perceptron not only with the threshold elements was noticed by Yakubovich (1963, 1965) and Aizerman et al. (1964a).

4.7 The presented results are based on the work by Devyaterikov et al. (1967). The algorithms listed in Table 4.1, which represent the special cases of the general algorithm of learning correspond to the algorithms discovered by different authors for the past several years.

Algorithms 1–4 in Table 4.1 were obtained by Aizerman et al. (1964a–c). Yakubovich has obtained the algorithms of learning similar to algorithms 1 and 3 in Table 4.1. He has introduced the concepts of C- and L-optimality, and has given a comparison of these algorithms.

Shaikin (1966) has shown that algorithm 4 in Table 4.1 guarantees the convergence of $c[n]$ in probability to the vector c^* when the hypothesis of representation is satisfied. In that case, the mean square error is equal to zero. A simplification of the proof of convergence of this algorithm (with the help of a result by Dvoretzky regarding the stochastic approximations) was obtained relatively recently by Blaydon (1966).

4.8 The idea of obtaining the search algorithm presented here was first described by D.B. Cooper (1964) (see also Bialasiewicz (1965, 1966) and Leonov (1966)).

4.9 The algorithms of the type (4.30) were obtained by Vapnik et al. (1965) on the basis of their concept of generalized portrait (see also Vapnik and Chervonenkis (1964, 1966a, b)).

4.10 Algorithm (4.31) with $\gamma = $ const was obtained by Aizerman et al. (1964a). In a slightly different form, it was obtained earlier by Novikoff (1963) (see Section 4.12).

4.11 The substitution of the system of inequalities by the system of equalities was used slightly differently by Ho and Kashyap (1965, 1966).

4.12 See also the paper by Devyaterikov et al. (1967). Algorithm (4.42), when Γ is a matrix, was investigated by Novikoff (1963). Very interesting results regarding the convergence of the algorithms and the usage of the same samples periodically repeated were given by Litvakov (1966).

4.13 Certain results of learning in pattern recognition will now be considered. The perceptrons, and especially Widrow's (1963, 1964, 1966) perceptrons of the "adaline" type, were used for weather prediction, speech recognition, diagnosis, recognition of handwriting, and also as a learning device (Widrow, 1959). The results of such applications are also very interesting. In the first case, measurements of barometric pressure at different points were presented (after coding) at the input of the perceptron. The output of the perceptron corresponded to the answer—it will or it will not rain. A system of three perceptrons was used. The prediction of rain was made for three intervals of 12 hours:

I	Today	from 8.00 A.M. to 8.00 P.M.
II	Today	from 8.00 P.M. to 8.00 A.M.
III	Tomorrow	from 8.00 A.M. to 8.00 P.M.

The measurements were performed at 4 o'clock in the morning of the first day. In the experiment, the information was given in three different ways: A, the "today's" chart of the pressure at 4 o'clock in the morning; B, the "today's" and "yesterday's" charts obtained at 4 o'clock in the morning; C, the "today's" chart and the difference between the today's and yesterday's pressure. The experiments were conducted during 18 days. The results are given in the table below.

	Percentage of correct recognition		
Official forecast	I (78)	II (89)	III (67)
A	72	67	67
B	78	78	78
C	78	89	83

It is necessary to emphasize the success of this experiment, especially since only information about pressure was used.

In the second case, a word from a microphone was applied to eight bandpass filters distributed over the acoustic range of the spectrum. The output signal of these filters (proportional to the spectral energy) was then quantized, coded and transformed into impulses and as such applied to the input of the perceptron. In a typical experiment, the output of each filter was split on four levels. Ten impulses would then correspond to each word, and each level would be represented by a 3-bit code so that $8 \times 3 \times 10 = 240$ bits would correspond to each word. The perceptron was trained on the voice of one person. After a period of training (in which ten samples of each word of one person were used), the perceptron understood new samples of the same words (and of the same person) with an accuracy of 98% or better. When another person was speaking, the accuracy was of the order of 90%.

For diagnosis based on the electrocardiogram (EKG), three tracks were recorded simultaneously so that the phase information was also available. This information was given at the input of the perceptron in intervals of 10 msec. After 100 impulses, a medical doctor, who was also an EKG specialist, decided whether the conclusion "healthy" or "ill" was correct. The results of the experiment are presented in the table below.

	Answered correctly (%)	
Results	"Healthy" (27 cases)	"Ill" (30 cases)
Doctor's conclusion	95	54
Perceptron's conclusion	89	73

A perceptron based on the idea of a generalized portrait was used for investigation of oil wells. According to Vapnik et al. (1965), twelve parameters were used to describe each one of 104 layers of small thickness. From these 104 layers, 23 layers containing oil and 23 layers containing water, i.e., 46 layers, were used for learning. After a period of learning, there were no errors made in classifying the layers. In classifying 180 layers of greater thickness (45 of them were used for learning with the earlier found generalized portrait), a number of layers containing water were taken as layers containing oil. The details can be found by the reader in the mentioned paper, and also in the paper by Guberman et al. (1966), in which the algorithms proposed by M.M. Bongrad, A.G. Frantsuz and others were also used for the same purpose.

Interesting results were obtained by Kozinets (1966) in recognizing handwriting, and also by Yakubovich (1963) in recognizing profiles (contours). Many works have been devoted to the recognition of letters and numerals, printed and handwritten. We shall not distract the reader with these results. Surveys of the works describing pattern recognition devices have been written by Sochivko (1963, 1964).

Comments

4.15 The problem of restoring probability density function was treated by Frolov and Chentsov (1962), Chentsov (1962), Kirillov (1959) and Aizerman et al. (1964b). We, like the authors of these works, put aside the questions of normalizing $p(\mathbf{x})$.

4.16 The algorithms obtained by the author (Tsypkin, 1965a, 1967) are presented here. The modified algorithms (of the type (4.55)) were proposed by Nikolic and Fu (1966b). Their statement that the modified algorithms converge faster than the ordinary algorithms was incorrect.* This was clarified by V.S. Pugachev and the author in their evaluation of the nature of modified algorithms.

4.17 A survey of the principles of self-learning in pattern classification under sufficiently complete information can be found in the paper by Spragins (1966). This survey paper is recommended to the reader who is interested in different special approaches. The papers by D.B. Cooper (1964), P.W. Cooper (1962) and Patrick and Hancock (1965, 1966) follow such directions. The self-learning of the type of a "trustful optimist" or of an "untrusting pessimist," and also its combinations, were studied by Widrow (1966). A more general variational formulation of the problem of self-learning belongs to Shlezinger (1965).

4.18 The functionals of the type of the average risk (4.61) for the quadratic loss function were actually considered in the already mentioned paper by Shlezinger (1965), and also (in slightly different terms) in the paper by Braverman (1966).

4.19 In computing the variation of the average risk, we have used the results which can be found in the textbook of Gelfand and Fomin (1963), and in the paper by Elman (1966).

4.23 Special recursive algorithms of self-learning were discussed in a number of papers. In the papers by Dorofeyuk (1966) and Braverman and Dorofeyuk (1966), certain algorithms of self-learning were obtained on the basis of concepts related to the potential function method. Such a heuristic way of constructing algorithms leaves something to be desired. Special types of functionals were introduced by Shlezinger (1965) and Braverman (1966), and their minimization was supposed to lead to the algorithms of self-learning. By minimizing the average risk, Braverman has obtained the recursive algorithms listed in Table 4.3 after long and complex considerations.

Although the algorithms (4.91)–(4.94) differ externally from those obtained by Braverman (1966), all these algorithms are actually equivalent. In the algorithms presented here, a relationship which exists between the coefficients of the discriminant function and which, for some reason, was not taken into consideration by Braverman, was used here. We shall also mention that the constraints imposed by him on γ_1 and γ_2 in order to prove the convergence of the algorithm of self-learning can be relaxed. It appears that of five conditions, it is sufficient to satisfy only the first two, i.e., the ordinary conditions for the convergence of the probabilistic iterative methods (3.34a). Shlezinger has actually solved the problem in two steps. The probability density function of the mixture, $p(\mathbf{x})$, is first restored (estimated), and then it is assumed that an iterative method can be applied to the equation which defines the center of gravity of the sought regions. Shlezinger has not given this algorithm in an explicit form. However, he has written a recursive formula without any special justifications, which is similar to our algorithms if $F'(\cdot)$ is assumed to be linear and constant.

The solutions of the problem of self-learning presented in Sections 4.18–4.23 were given by the author and Kelmans (Tsypkin and Kelmans, 1967).

---

* The correction and the new interpretation of the modified algorithms can also be found in Y.T. Chien and K.S. Fu, "On Bayesian learning and stochastic approximation," *IEEE Trans. Syst. Sci. Cybern.* **SSC-3**, 28–38 (June 1967).—Redactor and Translator.

BIBLIOGRAPHY

Abramson, N., and Braverman, D. (1962). Learning to recognize patterns in a random environment, *IRE Trans. Inform. Theory* **8** (5).

Abramson, N., Braverman, D., and Sebastian, G. (1963). Pattern recognition and machine learning, *IEEE Trans. Inform. Theory* **1–9** (4).

Agmon, S. (1956). The relaxation method for linear inequalities, *Can. J. Math.* **6** (3).

Aizerman, M.A. (1962). Experiments with training a machine to recognize visual patterns. *In* "Biologicheskie Aspekty Kybernetiki." Soviet Academy of Sciences Publishing House, Moscow. (In Russian.)

Aizerman, M.A. (1963). The problem of training an automaton to perform classification of input situations (pattern recognition), *Theory Self-Adapt. Contr. Syst. Proc. IFAC Symp. 2nd* **1963**.

Aizerman, M.A., Braverman, E.M., and Rozonoer, L.I. (1964a). Theoretical foundations of the potential function method in the problem of training automata to classify input situations, *Automat. Remote Contr.* (*USSR*) **25** (6).

Aizerman, M.A., Braverman, E.M., and Rozonoer, L.I. (1964b) The probabilistic problem of training automata to recognize patterns and the potential function method, *Automat. Remote Contr.* (*USSR*) **25** (9).

Aizerman, M.A., Braverman, E.M., and Rozonoer, L.I. (1964c) The potential function method in the problem of estimating the characteristics of the functional transformer by randomly observed points, *Automat. Remote Contr.* (*USSR*) **25** (12).

Aizerman, M.A., Braverman, E.M., and Rozonoer, L.I. (1965). The Robbins-Monro process and the potential function method, *Automat. Remote Contr.* (*USSR*) **26**.

Albert, A. (1963). Mathematical theory of pattern recognition, *Ann. Math. Stat.* **34** (1).

Arkadiev, A.G., and Braverman, E.M. (1964). Experiments in teaching a machine to recognize patterns. *In* "Opiti po Obucheniyu Machin Raspoznavaniu Obrazov." Nauka, Moscow. (In Russian.)

Bashkirov, O.A., Braverman, E.M., and Muchnik, I.B. (1964). The algorithms of training a machine to recognize visual patterns, based on the potential function method, *Automat. Remote Contr.* (*USSR*) **25** (5).

Benenson, Z.M., and Hazen, E.M. (1966). The methods of sequential analysis in the problem of multi-hypothesis recognition, *Izv. Akad. Nauk SSSR Tekh. Kibern.* **1966** (4). (Engl. transl.: *Eng. Cybern.* (*USSR*).)

Bialasiewicz, J. (1965). Wielowymiarowy model procesu rozpoznawania obiektów z uczeniem opartym na metodzie aproksymacji stochastycznej, *Arch. Automat. Telemech.* **10** (2).

Bialasiewicz, J. (1966). Liniowe funkcje decyzyjne w zastosowaniu do rozpoznawania sytuacji technologicznych, *Arch. Automat. Telemech.* **11** (4).

Bishop, A.B. (1963). Adaptive pattern recognition, *WESCON Conv. Rec.* **7** (4).

Blaydon, C.C. (1966). On a pattern classification result of Aizerman, Braverman and Rozonoer, *IEEE Trans. Inform. Theory* **IT-12** (1).

Blaydon, C.C., and Ho, Y.H. (1966). On the abstraction problem in pattern classification, *Proc. Nat. Electron. Conf.* **22**.

Blazhkin, K.A., and Friedman, V.M. (1966). On an algorithm of training a linear perceptron, *Izv. Akad. Nauk SSSR Tekh. Kibern.* **1966** (6). (Engl. transl.: *Eng. Cybern.* (*USSR*).)

Block, H.D. (1962). The perceptron: a model for brain functioning, 1, *Rev. Mod. Phys.* **34** (1).

Bozhanov, E.S., and Krug, G.K. (1966). Classification of many objects using experimental data. *In* "Planirovanie Eksperimentov." Nauka, Moscow. (In Russian.)

Braverman, D. (1962). Learning filters for optimum pattern recognition, *IRE Trans. Inform. Theory* **IT-8** (4).

Braverman, E.M. (1962). The experiments with training a machine to recognize patterns, *Automat. Remote Contr.* (*USSR*) **23** (3).

Braverman, E.M. (1965). On the potential function method, *Automat. Remote Contr.* (*USSR*) **26** (12).

Braverman, E.M. (1966). The potential function method in the problem of learning without a teacher, *Automat. Remote Contr.* (*USSR*) **27** (10).

Braverman, E.M., and Dorofeyuk, A.A. (1966). The experiments in training a machine to recognize patterns without reinforcement. *In* "Samoobuchayushchiesia Automaticheskie Ustroistva." Nauka, Moscow. (In Russian.)

Braverman, E.M., and Pyatnitskii, E.S. (1966). Estimates of the rate of convergence of the algorithms based on the potential function method, *Automat. Remote Contr.* (*USSR*) **27** (1).

Chentsov, N.N. (1962). Estimation of an unknown probability density function using observations, *Dokl. Akad. Nauk SSSR* **147** (1). (In Russian.)

Chow, C.K. (1957). An optimum character recognition system using decision functions, *IRE Trans. Electron. Comp.* **EC-6** (4).

Chow, C.K. (1966). A class of nonlinear recognition procedures, *IEEE Int. Conv. Rec.* **1966**.

Cooper, D.B. (1964). Adaptive pattern recognition and signal detection using stochastic approximation, *IEEE Trans. Electron. Comp.* **EC-13** (3).

Cooper, D.B., and Cooper P.W. (1964). Non-supervized adaptive signal detection and pattern recognition, *Inform. Contr.* **7** (3).

Cooper, J.A. (1963). A decomposition resulting in linearly separable functions of transformed input variables, *WESCON Conv. Rec.* **7** (4).

Cooper, P.W. (1962). The hypersphere in pattern recognition, *Inform. Contr.* **5** (4).

Devyaterikov, I.P., Propoi, A.I., and Tsypkin, Ya.Z. (1967). On recursive algorithms of training in pattern recognition, *Automat. Remote Contr.* (*USSR*) **28** (1).

Dorofeyuk, A.A. (1966). Algorithms of training a machine to recognize patterns without a teacher, based on the potential function method, *Automat. Remote Contr.* (*USSR*) **27** (10).

Duda, R.O., and Fossum, H. (1966). Pattern classification by iteratively determined linear and piecewise linear discriminant functions, *IEEE Trans. Electron. Comp.* **EC-15** (2).

Elkin, V.N., and Zagoruiko, N.G. (1966a). Linear discriminant functions close to the optimal. *In* "Vichislitelnie systemi," No. 19. Novosibirsk.

Elkin, V.N., and Zagoruiko, N.G. (1966b). Classification problem of pattern recognition. *In* "Vichislitelnie systemi," No. 22. Novosibirsk. (In Russian.)

Elman, R.I. (1966). The questions of optimizing pattern recognition in noisy environment, *Izv. Akad. Nauk SSSR Tekh. Kibern.* **1966** (5). (Engl. transl.: *Eng. Cybern.* (*USSR*).)

Flores, I., and Gray, L. (1960). Optimization of reference signals for pattern recognition, *IRE Trans. Electron. Comp.* **EC-9** (1).

Fomin, V.N. (1965). On an algorithm of pattern recognition. *In* "Vichislitelnaia Tekhnika i Voprosi Programirovania," Vol. 4. Leningrad State University Press, Leningrad. (In Russian.)

Freeman, H. (1962). On the digital computer classification of geometric line patterns, *Proc. Nat. Electron. Conf.* **18**.

Freeman, H. (1965). A technique for the classification and recognition of geometric patterns. *In* "Third International Congress on Cybernetics, Namur, 1964." Acta Proceedings, Namur.

Frolov, A.S., and Chentsov, N.N. (1962). Application of the Monte-Carlo method in obtaining smooth curves. *In* "Tr. VI Vsesoyuznogo Soveshchania po Teorii Veroyatnostei i Matematicheskoi Statistiki." Vilnus. (In Russian.)

Fu, K.S. (1962). Sequential decision model of optimal pattern recognition, *Probl. Bionics* **1962**.

Fu, K.S. (1964). Learning control systems. *In* "Computer and Information Sciences" (J. Tou, ed.). Academic Press, New York.

Gelfand, I.M., and Fomin, S.V. (1963). "Calculus of Variations." Prentice-Hall, Englewood Cliffs, N.J.

Glushkov, V.M. (1966). "Introduction to Cybernetics." Academic Press, New York.

Goerner, T., and Gerhardt, L. (1964). Analysis of training algorithms for a class of self-organizing systems, *Proc. Nat. Electron. Conf.* **20**.

Guberman, S.A., Izvekova, M.L., and Hurgin, Ya.I. (1966). Application of pattern recognition to interpretation of geophysical data. *In* " Samoobuchayushchiesia Avtomaticheskie Systemi." Nauka, Moscow. (In Russian.)

Hansalik, W. (1966). Adaptation procedures, *J. SIAM* **14** (5).

Harkevich, A.A. (1959). Pattern recognition, *Radiotekh.* **14** (5). (In Russian.)

Harkevich, A.A. (1960). Design of reading machines, *Radiotekh.* **15** (2).

Harkevich, A.A. (1965). Certain methodological questions in the problem of pattern recognition, *Probl. Peredachi Inform.* **1** (3).

Hay, J.S., Martin, F.S., and Wightman, S.W. (1960). The Mark I perceptron-design and performance, *IRE Inter. Conv. Rec.* **8** (2).

Ho, Yu-Chi, and Kashyap, R.L. (1965). An algorithm for linear inequalities and its applications, *IEEE Trans. Electron. Comp.* **EC-14** (5).

Ho, Yu-Chi, and Kashyap, R.L. (1966). A class of iterative procedures for linear inequalities, *J. SIAM*, Series A **4** (1).

Johnson, D.D., Haugh, G.R., and Li, K.P. (1966). The application of few hyperplane design techniques to handprinted character recognition, *Proc. Nat. Electron. Conf.* **22**.

Kanal, L. (1962). Evaluation of a class of pattern recognition systems, *Bionics* **1962**.

Kanal, L., Slymaker, F., Smith, D., and Walker, W. (1962). Basic principles of some pattern recognition systems, *Proc. Nat. Electron. Conf.* **18**.

Kashyap, R.L., and Blaydon, C.C. (1966). Recovery of functions from noisy measurements taken at randomly selected points and its application to pattern recognition, *Proc. IEEE* **54** (8).

Kirillov, N.E. (1959). A method of measuring probability density functions based on their orthogonal polynomial expansion. *In* "Problemi Peredachi Informatsii." Akademii Nauk SSSR Press, Moscow. (In Russian.)

Kobrinskii, A.E. (1965). "Who Will Win?" Molodaia Gvardia, Moscow. (In Russian.)

Kozinets, B.N., Lantsman, R.M., Sokolov, B.M., and Yakubovich, V.A. (1966). Recognition and differentiation of pictures on digital computers. *In* "Samoobuchayushchiesia Avtomaticheskie Systemi." Nauka, Moscow. (In Russian.)

Laptev, V.A., and Milenkii, A.V. (1966). On pattern classification during learning period, *Izv. Akad. Nauk SSSR Tekh. Kibern.* **1966** (6). (Engl. transl.: *Eng. Cybern.* (*USSR*).)

Lawrence, G.R. (1960). Pattern recognition with an adaptive network, *IRE Inter. Conv. Rec.* **2**.

Leonov, Yu.P. (1966). Classification and statistical hypothesis testing, *Automat. Remote Contr.* (*USSR*) **27** (12).

Lewis, P.M. (1963). The characteristic selection problem in recognition systems, *IRE Trans. Inform. Theory* **IT-24** (6).

Litvakov, B.M. (1966). On an iterative method in the problem of approximation of a function, *Automat. Remote Contr.* (*USSR*) **27** (4).

McLaren, R.W. (1964). A Markov model for learning systems operating in unknown random environment, *Proc. Nat. Electron. Conf.* **20**.

Mays, C.H. (1964). Effects of adaptation parameters on convergence time and tolerance for adaptive threshold elements, *IEEE Trans. Electron. Comp.* **EC-13** (4).

Motzkin, T.S., and Schoenberg, I.J. (1954). The relaxation method for linear inequalities, *Can. J. Math.* **3** (3).

Mucciardi, A.N., and Gose, E.E. (1966). Evolutionary pattern recognition in incomplete nonlinear multithreshold networks, *IEEE Trans. Electron. Comp.* **EC-15** (2).

Nikolic, Z.J., and Fu, K.S. (1966). An algorithm for learning without external supervision and its application to learning control systems, *IEEE Trans. Automat. Contr.* **AC-11** (3).

Nikolic, Z.J., and Fu, K.S. (1966b). A mathematical model of learning in an unknown random environment, *Proc. Nat. Electron. Conf.* **22**.

Nilsson, N.J. (1965). "Learning Machines." McGraw-Hill, New York.

Novikoff, A. (1963). On convergence proofs for perceptrons. *In* "Proceedings of Symposium on Mathematical Theory of Automata," Vol. XII. Polytechnic Institute of Brooklyn, New York.

Patrick, E.A., and Hancock, J.C. (1965). The nonsupervised learning of probability spaces and recognition of patterns, *IEEE Inter. Conv. Rec.* **9**.

Patrick, E.A., and Hancock, J.C. (1966). Nonsupervised sequential classification and recognition of patterns, *IEEE Trans. Inform. Theory* **IT-12** (3).

Rosenblatt, F. (1962a). Perceptual generalization over transformation groups. *In* "Self Organizing Systems," (M. Yovitts and S. Cameron, eds.). Pergamon Press, New York.

Rosenblatt, F. (1962b). "Principles of Neurodynamics." Cornell Aeronautical Laboratory, Rep. 1196-G-8. Spartan, Washington, D.C.

Rosenblatt, F. (1964). Analytic techniques for the study of neural nets, *IEEE Trans. Appl. Ind.* **83** (74).

Rudowitz, D. (1966). Pattern recognition *J. Roy. Stat. Soc.* **129** (4).

Sears, R.W. (1965). Adaptive representation for pattern recognition, *IEEE Inter. Conv. Rec.* **13** (6).

Sebestyen, G.S. (1962a). Pattern recognition by an adaptive process of sample set construction, *IRE Inter. Symp. Inform. Theory* **IT-8** (2).

Sebestyen, G.S. (1962b) "Decision-Making Processes in Pattern Recognition." Macmillan, New York.

Shaikin, M.E. (1966). Proof of convergence for an algorithm of learning by stochastic approximation method, *Automat. Remote Contr.* (*USSR*) **27** (5).

Shlezinger, M.I. (1965). On arbitrary pattern recognition. *In* "Chitayushchie Automati." Kiev. (In Russian.)

Smith, F.W. (1965). Coding analog variables for threshold logic descrimination, *IEEE Trans. Electron. Comp.* **EC-14** (6).

Sochivko, V.P. (1963). "Pattern Recognition Devices." Sudpromgiz, Leningrad. (In Russian.)

Sochivko, V.P. (1964) "Electronic Pattern Recognition Devices." Energia, Moscow-Leningrad. (In Russian.)

Spragins, J. (1966). Learning without a teacher, *IEEE Trans. Inform. Theory*. **IT-12** (2).

Taube, M. (1961). "Computers and Common Sense, the Myth of Thinking Machines." Columbia, Washington, D.C.

Tsypkin, Ya.Z. (1965a). On restoration probability density functions using observed data, *Automat. Remote Contr.* (*USSR*) **26** (3).

Tsypkin, Ya.Z. (1965b). On estimation of the characteristic of a functional transformer using randomly observed points, *Automat. Remote Contr.* (*USSR*) **26** (11).

Tsypkin, Ya.Z. (1966). Adaptation, learning and self-learning in automatic systems, *Automat. Remote Contr.* (*USSR*) **27** (1).

Tsypkin, Ya.Z. (1967). On algorithms for restoration of probability density functions and moments using observations, *Automat. Remote Contr.* (*USSR*) **28** (7).

Tsypkin, Ya.Z., and Kelmans, G.K. (1967). On recursive algorithms of self-learning, *Izv. Akad. Nau kSSSR Tekh. Kibern.* **1967** (5). (Engl. transl.: *Eng. Cybern.* (*USSR*).)

Vapnik, V.N., and Chervonenkis, A.Ya. (1964). On a class of perceptrons, *Automat. Remote Contr.* (*USSR*) **25** (1).

Vapnik, V.N., and Chervonenkis, A.Ya. (1966a). Training of automata to perform optimization I, *Automat. Remote Contr.* **27** (5).

Vapnik, V.N., and Chervonenkis, A.Ya. (1966b). Training of automata to perform optimization II, *Automat. Remote Contr.* **27** (6).

Vapnik, V.N., Lerner, A.Ya., Chervonenkis, A.Ya. (1965). The systems of learning in pattern recognition based on generalized portraits, *Izv. Akad. Nauk SSSR Tekh. Kibern.* **1965** (1). (Engl. transl.: *Eng. Cybern.* (*USSR*).)

Widrow, B. (1959). Adaptive sampled-data systems. A statistical theory of adaptation, *WESCON Conv. Rec.* **3** (4).

Widrow, B. (1961). Self-adaptive discrete systems, *Theory Self-Adapt. Contr. Syst. Proc. IFAC Symp. 1st* **1961**.

Widrow, B. (1963). A statistical theory of adaptation. *In* "Adaptive Control Systems," (F.P. Caruthers and H. Levenstein, eds.). Pergamon Press, New York.

Widrow, B. (1964). Pattern recognition and adaptive control, *IEEE Trans. Appl. Ind.* **83** (74).

Widrow, B. (1966). "Bootstrap learning" in threshold logic. *In* "Automatic and Remote Control. Proceedings of the Third Congress of the International Federation of Automatic Control." Butterworths, London.

Wurtele, Z.S. (1965). A problem in pattern recognition, *J. SIAM* **13** (3).

Yakubovich, V.A. (1963). Machines that learn to recognize patterns. *In* "Metodi Vichisleniy," Vol. 2. Leningrad State University Press, Leningrad. (In Russian.)

Yakubovich, V.A. (1965). Certain general theoretical principles in the design of learning pattern recognition systems, Part I. *In* "Vichislitelnaia Tekhnika i Voprosi Programirovania," Vol. 4. Leningrad State University Press, Leningrad. (In Russian.)

Yakubovich, V.A. (1966). Recursive algorithms for solving a system of inequalities in a finite number of iterations, *Dokl. Akad. Nauk SSSR* **166** (6). (In Russian.)

5

IDENTIFICATION

5.1 Introduction

The problem of identification or, more appropriately, the problem of finding the characteristics of the controlled objects (plant characteristics) and of the control actions, is one of the basic problems in the design of automatic control systems. In deterministic problems, the control actions and the plant characteristics are usually found on the basis of theoretical investigations, definite hypothesis or experimental data. In statistical problems, probabilistic descriptions of external actions (probability density functions, correlation functions, spectral densities, etc.) are obtained on the basis of an ensemble of such actions, and the plant characteristics (equations, time characteristics, etc.) are found using well-known statistical methods, but again after processing a number of sample functions.

In the problems related to the applications of adaptation, these methods are not suitable since they need special actions (test signals), a long time for observation and processing, and, generally speaking, laboratory conditions.

For solving the problems of adaptation in the systems of automatic control, it is necessary to be able to determine the current characteristics of the control actions and the controlled plants. In other words, these characteristics have to be estimated during the process of normal operations, and in such a manner that the estimates can be directly used for an improvement of that normal operation.

The following two problems must be distinguished in an identification process:

(1) The problem of determining the structure and the parameters of the plant.

(2) The problem of determining the plant parameters for a given or an assumed structure.

If the first problem has to deal with a "black," nontransparent box, then the second is related to a "gray," semitransparent one. Any information about the possible plant structure, or a sufficiently general permissible structure can considerably accelerate the process of estimation. Therefore, as it is usually done in practice, the main attention will be given to the second problem.

In this chapter, it will be shown that the problems of identification can be observed from the same standpoint as the problems of pattern recognition (Chapter 4). Various examples of estimating the mean values, dispersions and correlation function, characteristics of nonlinear elements and plants with both lumped and distributed parameters, will serve as good illustrations of the solutions to the simplest identification problems.

5.2 Estimation of the Mean Value

In order to clarify both the physical meaning and a number of properties of the algorithms of adaptation, we begin with the simplest problem-estimation of the mean value of a random variable:

$$x = c^* + \xi \tag{5.1}$$

where c^* is an unknown constant, and ξ is noise with zero meaning and finite variance. This problem appears, for instance, in the processing of measurement results and in the extraction of a constant signal from a background noise. The observed quantity x is the only quantity which can be measured and processed.

If the error (sometimes called noise) is uniformly distributed, the best estimate after n observations is the arithmetic mean,

$$c[n] = \frac{1}{n} \sum_{m=1}^{n} x[m] \tag{5.2}$$

By substitution of $x[m]$ from (5.1) into (5.2), we obtain

$$c[n] = c^* + \frac{1}{n} \sum_{m=1}^{n} \xi[n] \tag{5.3}$$

From here, it follows that the influence of the noise decreases with an increase in the number of observations, and that the estimate $c[n]$ converges to the sought value c^*.

We now modify the estimator (5.2):

$$c[n] = \frac{n-1}{n}\left(c[n-1] + \frac{1}{n-1}x[n]\right) \qquad (5.4)$$

or

$$c[n] = c[n-1] - \frac{1}{n}(c[n-1] - x[n]) \qquad (5.5)$$

Condition (5.5) shows that the influence of new information $x[n]$ decreases with increase of n, since the weighting coefficient, $1/n$, is inversely proportional to the number of observations. Therefore, $c[n]$ converges to c^*. This fact is frequently proven in practice; we must base our solutions on the past experiments, and not give excessive weight to new information which can cause great variations. The formulas of the type (5.5) have been used for a long time in the adjustments of precision devices and in firing weapons. They have the form of a rule: nth adjustment is equal to $1/n$ of the total deviation.

5.3 Another Approach

We shall now look at this problem from another point of view, and apply to it the algorithms of adaptation. It follows from (5.1) that

$$c^* = M\{x\} \qquad (5.6)$$

if the mean value of the noise is equal to zero. As in the case of the restoration (estimation) of probability density function and the moments (Section 4.16), we rewrite (5.6) as

$$M\{c - x\} = 0 \qquad (5.7)$$

We now apply to this relationship the algorithm of adaptation (3.9). By setting $\nabla_c Q(x, c) = c - x$, we find

$$c[n] = c[n-1] - \gamma[n](c[n-1] - x[n]) \qquad (5.8)$$

and for $\gamma[n] = 1/n$ we obtain the algorithm (5.5) given earlier. Therefore, using algorithms of adaptation, we obtain in particular the result which was reached in the preceding paragraph on the basis of simple physical considerations. At the same time, the values $\gamma[n] = 1/n$ are optimal with respect to the least square error criterion.

By using the modified algorithms (3.41) and (3.42) instead of (5.8), we can obtain

$$c[n] = c[n-1] - \gamma[n]\left(c[n-1] - \frac{1}{n}\sum_{m=1}^{n} x[m]\right) \qquad (5.9)$$

This algorithm gives smoother variations of $c[n]$ as n increases.

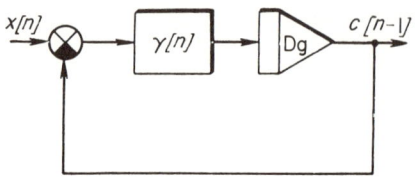

Fig. 5.1

The block diagrams of the systems which realize the algorithms of adaptation (5.8) and (5.9), i.e., which estimate the mean values, are shown in Figs. 5.1 and 5.2, respectively. They, as we already know from Section 4.16, are linear discrete systems with variable gain coefficients. The difference between these systems is that in the second system we have additional processing (averaging) of $x[n]$.

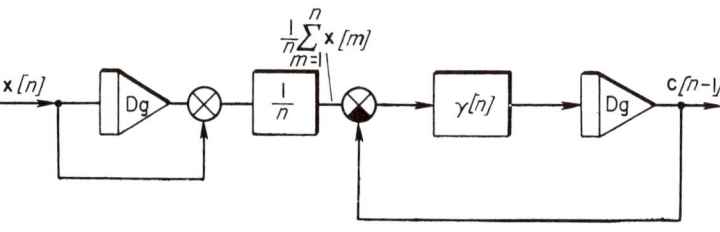

Fig. 5.2

5.4 Estimation of the Variance

The variance of a stationary random process is defined by

$$\sigma^2 = M\{(x - c^*)^2\} \tag{5.10}$$

where the mean value

$$c^* = M\{x\} \tag{5.11}$$

If the mean value is a priori known, then, since (5.10) differs from (5.6) only in notation, we can immediately use the algorithm of adaptation (5.8) with the corresponding substitution of symbols:

$$\sigma^2[n] = \sigma^2[n-1] - \gamma[n][\sigma^2[n-1] - (x[n] - c^*)^2] \tag{5.12}$$

The system which realizes this algorithm (Fig. 5.3) differs from the discrete system for estimation of the mean value (Fig. 5.1) by the presence of a

5.4 Estimation of the Variance

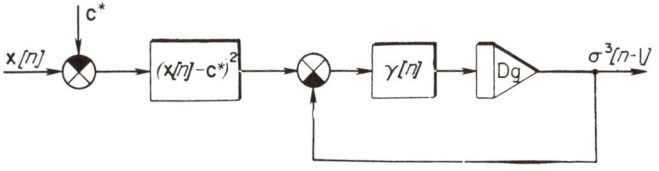

Fig. 5.3

squaring device at the input and a comparator for the algebraic addition of the signals.

But usually the mean value c^* is unknown to us. In such cases we can simultaneously form the estimates $\sigma^2[n]$ and $c[n]$ by using (5.10) and (5.11). We then obtain two algorithms which are dependent on each other:

$$\sigma^2[n] = \sigma^2[n-1] - \gamma[n][\sigma^2[n-1] - (x[n] - c[n])^2]$$
$$c[n] = c[n-1] - \gamma_1[n][c[n-1] - x[[n]]] \quad (5.13)$$

The discrete system which realizes these algorithms is shown in Fig. 5.4. This system does not require any special explanations since it represents a combination of discrete systems for the estimation of the variance (Fig. 5.3) and of the mean value (Fig. 5.2).

If the mean value is equal to zero ($c^* = 0$), we obtain a simpler algorithm from (5.12),

$$\sigma^2[n] = \sigma^2[n-1] - \gamma[n][\sigma^2[n-1] - x^2[n]] \quad (5.14)$$

which is realized by a discrete system, shown in Fig. 5.3 for $c^* = 0$. The optimal value $\gamma[n]$ is equal to $1/n$ in this case.

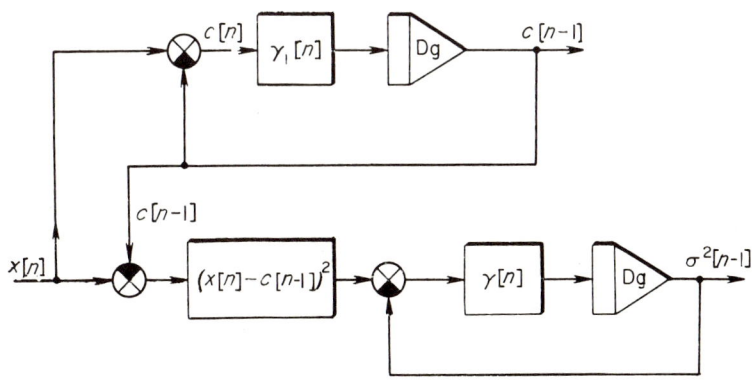

Fig. 5.4

In the continuous case, when $x(t)$ is a stationary random process, we have to use continuous algorithms of adaptation (3.28), which for $c^* = 0$ take the form

$$\frac{d\sigma^2(t)}{dt} = -\gamma(t)[\sigma^2(t) - x^2(t)] \tag{5.15}$$

A continuous system analogous to the discrete system shown in Fig. 5.3 (the digrator is replaced by a continuous integrator) corresponds to this algorithm.

5.5 Discussion

Let us now discuss briefly the physical meaning and an interpretation of the algorithms for the estimation of the variance. Of course, all this will be equally true for the estimation of the moments in general. Let us define the "current" variance

$$\sigma^2(t) = \frac{1}{t} \int_0^t x^2(t)\, dt \tag{5.16}$$

where $x(t)$ is a stationary random process. Obviously,

$$\sigma^2 = \lim_{t \to \infty} \sigma^2(t) \tag{5.17}$$

if such a limit exists. By differentiating both sides of (5.16) with respect to t, we obtain the equation

$$\frac{d\sigma^2(t)}{dt} = -\frac{1}{t}(\sigma^2(t) - x^2(t)) \tag{5.18}$$

But this equation is identical to algorithm (5.15) when $\gamma(t) = 1/t$. In a similar fashion we can obtain the discrete algorithm (5.15) if we define

$$\sigma^2[n] = \frac{1}{n} \sum_{m=1}^{n} x^2[m] \tag{5.19}$$

and take the first difference $\Delta\sigma^2[n-1] = \sigma^2[n] - \sigma^2[n-1]$. It follows that the "current" variance (5.16) or (5.19) can be considered as a solution of the corresponding differential equation (5.15) or difference equation (5.14) for a special choice of γ, which define the algorithms for estimation of the variance. This simple fact shows that the adaptive approach in the considered case has replaced an impossible operation of ensemble averaging by a possible operation of averaging with respect to time. This substitution corresponds to the processing of information as it arrives, and it is valid for ergodic stationary processes.

5.6 Estimation of Correlation Functions

The cross-correlation function, which plays an important role in modern automatic control and radiophysics, is defined for a fixed value $\tau = \tau_0$ by the following expression

$$R_{yx}(\tau_0) = M\{y(t)x(t - \tau_0)\} \quad (5.20)$$

In estimating $R_{yx}(\tau)$, we can use, for instance, the continuous algorithms of the type (5.15). Let the "current" cross-correlation be designated by $R_{yx}(\tau_0, t) = c(t)$. We then obtain

$$\frac{dc(t)}{dt} = -\gamma(t)(c(t) - y(t)x(t - \tau_0)) \quad (5.21)$$

or, in a more convenient form,

$$T_0(t)\frac{dc(t)}{dt} + c(t) = y(t)x(t - \tau_0) \quad (5.22)$$

where

$$T_0(t) = \frac{1}{\gamma(t)} \quad (5.23)$$

is the time-varying "time constant."

Algorithm (5.22) permits a simple physical realization. The signals $y(t)$ and $x(t - \tau_0)$ are the input signals to a multiplier, and at the output of the multiplier is a resistor-capacitor (RC) circuit with the resistance varying according to $1/\gamma(t)$ (Fig. 5.5).

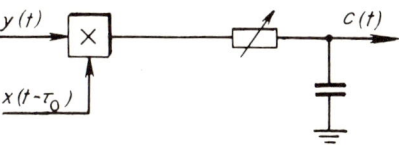

Fig. 5.5

In the case of a continuous modified algorithm for estimation of cross-correlation functions, instead of (5.21) we have

$$\frac{dc(t)}{dt} = -\gamma(t)\left[c(t) - \frac{1}{t}\int_0^t y(t)x(t - \tau_0)\, dt\right] \quad (5.24)$$

or

$$T_0(t)\frac{dc(t)}{dt} + c(t) = z(t) \quad (5.25)$$

where

$$z(t) = \frac{1}{t} \int_0^t y(t)x(t - \tau_0) \, dt \qquad (5.26)$$

By differentiating (5.26) with respect to t, we obtain

$$t \frac{dz(t)}{dt} + z(t) = y(t)x(t - \tau_0) \qquad (5.27)$$

Therefore, the modified algorithm (5.24) can be presented in the form of two simple algorithms (5.25) and (5.27). The scheme which realizes modified algorithms consists of a multiplier and a cascade of two decoupled RC circuits with variable resistances (Fig. 5.6). In this case, we have double averaging

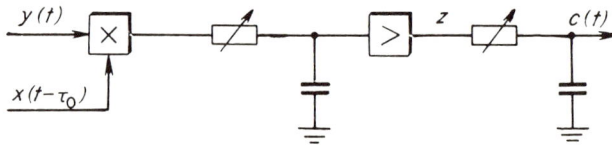

Fig. 5.6

which naturally leads to a smoother variation of $c(t)$ with increasing t. It is not difficult to show that for $\gamma(t) = 1/t$, the modified algorithm minimizes the sum of the variance of the estimate and its weighted derivative.

Although the schemes shown in Figs. 5.5 and 5.6 provide the estimates of cross-correletation functions, it is clear that similar schemes can be used for obtaining the estimates of the mean values, variances and moments. In order to achieve this, it is only necessary that the multiplier be replaced by a linear or nonlinear transformer.

5.7 Estimation of the Characteristics of Nonlinear Elements

When speaking of nonlinear elements, we have in mind nonlinear elements without inertia or the functional transformers which in general have an arbitrary number of inputs and a single output. The characteristics of a nonlinear element,

$$y = f(x) \qquad (5.28)$$

are determined by estimating the function $f(x)$ on the basis of observations of the input x and the corresponding output variable y.

If the function $f(x)$ is approximated by a sum of linearly independent functions $\mathbf{c}^T \boldsymbol{\phi}(x)$, as in earlier chapters, then the algorithms considered previously provide the solution to the posed problem. We shall not discuss such algorithms again.

5.7 Estimation of the Characteristics of Nonlinear Elements

A reader who is interested in the problem of determining the characteristics of nonlinear elements is asked to look at the algorithms listed in Tables 4.1 and 4.2 of the preceding chapter. If we assume now that y can take any arbitrary value and not only the values ± 1 as before, such algorithms can be used for finding the characteristics of the nonlinear elements. In the problem under consideration, the "perceptron" schemes which were used for the realization of the mentioned algorithms now play a role of adaptive nonlinear transformers for the estimation of nonlinear characteristics.

In a number of cases the form of the characteristics of a nonlinear element may be known except for a vector of unknown parameters. Then naturally, the approximating function should be

$$\hat{f}(x) = f_0(x, \mathbf{c}) \tag{5.29}$$

where $\mathbf{c}$ is an N-dimensional vector of parameters.

The mathematical expectation of a strictly convex function of the difference $f(x) - \hat{f}(x)$ is used as a measure of error in the approximation:

$$J(\mathbf{c}) = M_x\{F(y - f_0(x, \mathbf{c}))\} \tag{5.30}$$

The gradient is equal

$$\nabla_\mathbf{c} F(y - f_0(x, \mathbf{c})) = -F'(y - f_0(x, \mathbf{c}))\nabla_\mathbf{c} f_0(x, \mathbf{c})$$

Therefore, the algorithm of adaptation for estimating the parameters can be written

$$\mathbf{c}[n] = \mathbf{c}[n-1] + \gamma[n]F'(y[n] - f_0(x[n], \mathbf{c}[n-1]))\nabla_\mathbf{c} f_0(x[n], \mathbf{c}[n-1]) \tag{5.31}$$

or

$$\frac{d\mathbf{c}(t)}{dt} = \gamma(t)F'(y(t) - f_0(x(t), \mathbf{c}(t)))\nabla_\mathbf{c} f_0(x(t), \mathbf{c}(t)) \tag{5.32}$$

The block diagram realizing this algorithm is shown in Fig. 5.7.

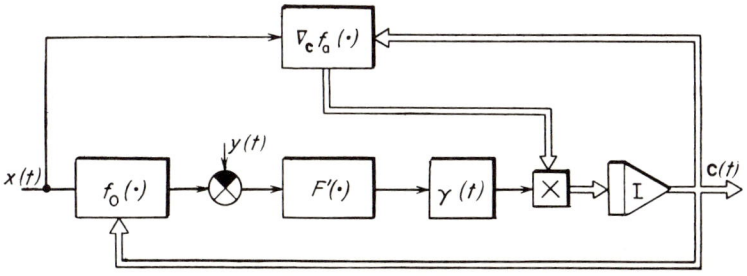

Fig. 5.7

5.8 Estimation of the Coefficient of Statistical Linearization

In the statistical analysis of nonlinear systems, a nonlinear element is frequently replaced by a linear element that is specified by an equivalent gain coefficient or a so-called coefficient of statistical linearization. In that case, the probability density of the input stationary variable is assumed to be constant. Using the algorithms of adaptation, we can remove this essential assumption and obtain a general procedure that is appropriate for any measure of approximation and not only a quadratic one.

We shall approximate the characteristic $y = f(x)$ in (5.29) by a linear function kx. The coefficient k is selected in such a way that

$$J(k) = M\{F(y - kx)\} \tag{5.33}$$

is minimized. Here, $F(\cdot)$ is a strictly convex function.

The gradient of the realization is then simply the derivative

$$\frac{d}{dk} F(y - kx) = -F'(y - kx)x \tag{5.34}$$

Using the continuous algorithms of adaptation (3.28), we obtain

$$\frac{dk(t)}{dt} = \gamma(t) F'(y(t) - k(t)x(t)) x(t) \tag{5.35}$$

When the conditions of convergence are satisfied, $k(t)$ converges almost surely to the equivalent coefficient k^*. A scheme of the continuous system realizing this algorithm is shown in Fig. 5.8.

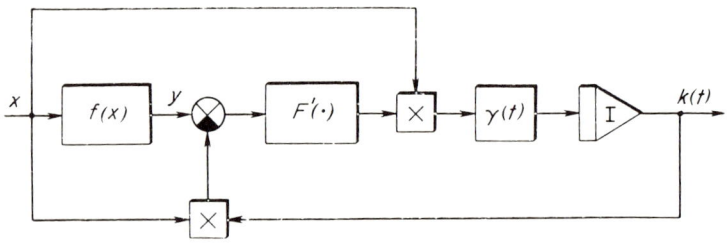

Fig. 5.8

5.9 Special Cases

For a given function $F(\cdot)$, we obtain a particular algorithm of adaptation. If $F(\cdot)$ is a quadratic function, algorithm (5.35) defines the statistical coefficient of linearization, and when $x(t) = a \sin(wt + \phi)$, where the phase angle ϕ is a random variable, we obtain the well-known harmonic coefficient of linearization.

5.9 Special Cases

We shall now select the functional

$$J(k) = M\{|y - kx|\} \tag{5.36}$$

Since

$$F'(y - kx) = \text{sign}(y - kx) \tag{5.37}$$

it follows from (5.35) that

$$\frac{dk(t)}{dt} = \gamma(t)x(t)\,\text{sign}\,(y(t) - k(t)x(t)) \tag{5.38}$$

This algorithm is realized by the block diagram shown in Fig. 5.8, where the relay characteristic corresponds to $F'(\cdot)$ defined by (5.27). The functional (5.36) characterizes a measure of the absolute error of approximation. Unlike (5.36), the functional

$$J(k) = M\left\{\left|\frac{y - kx}{x}\right|\right\} \tag{5.39}$$

defines not an absolute but a relative error of approximation; since in this case

$$\frac{d}{dk}\left|\frac{y - kx}{x}\right| = -\,\text{sign}\,x \cdot \text{sign}\,(y - kx) \tag{5.40}$$

the algorithm of adaptation has the form

$$\frac{dk(t)}{dt} = \gamma(t)\,\text{sign}\,x \cdot \text{sign}\,(y - kx) \tag{5.41}$$

The algorithm of adaptation (5.41) is much simpler than (5.38) since it does not contain the operation of multiplication by $x(t)$. This algorithm is realized using a relay, as can be seen in Fig. 5.9. Of course, the optimal values k^* obtained by these algorithms are, in general, different, i.e., they correspond to different criteria of optimality, and thus these algorithms cannot substitute for each other.

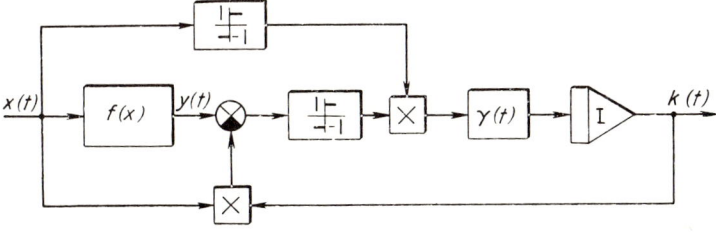

Fig. 5.9

5.10 Description of Dynamic Plants

The behavior of nonlinear dynamic plants can be described in a general case either by a nonlinear difference equation of the lth order,

$$x[n] = f(x[n-1], \ldots, x([n-l]; u[n-1], \ldots, u[n-l_1]) \quad (5.42)$$

where $x[n]$ are input and $u[n]$ are output variables (scalar functions), or by a system of nonlinear difference equations of the first order,

$$\mathbf{x}[n] = \mathbf{f}(\mathbf{x}[n-1], \mathbf{u}[n-1]) \quad (5.43)$$

where

$$\mathbf{x}[n] = (x_1[n], \ldots, x_l[n]) \qquad \mathbf{u}[n] = (u_1[n], \ldots, u_{l_1}[n])$$

are vectors of input and output variables of the plant. Although we can always obtain (5.43) from (5.42), equation (5.43) is more general, since it includes even the case when the number of control actions is greater than one.

These difference equations correspond either to continuous plants controlled by digital computers or to discrete systems. Under specific conditions, these difference equations can serve as approximate descriptions of the continuous systems.

Instead of the difference or the differential equations, it is frequently convenient to describe nonlinear dynamic objects (plants) by a Volterra series:

$$x[n] = \sum_{m=0}^{\infty} k_1[m]u[n-m]$$

$$+ \sum_{m_1=0}^{\infty} \sum_{m_2=0}^{\infty} k_2[m_1, m_2]u[n-m_1]u[n-m_2] + \cdots$$

$$+ \sum_{m_1=0}^{\infty} \cdots \sum_{m_s=0}^{\infty} k_s[m_1, \ldots, m_s]u[n-m_1] \cdots u[n-m_s] + \cdots \quad (5.44)$$

Relationship (5.44) can be considered as an approximation of the corresponding Volterra series in which the sums are replaced by the integrals and the variables are continuous. If we take only the first term of the Volterra series (5.44), we obtain the equations of a linear system.

The identification of dynamic plants consists of finding the equations of the object which relate the input and the output variables. We shall now apply the adaptive approach to the identification of various dynamic plants.

5.11 Identification of Nonlinear Plants I

The identification of nonlinear dynamic plants which are described by difference equations do not appear to be more complicated than the procedure for obtaining the characteristics of nonlinear elements without inertia.

5.12 Identification of Nonlinear Plants II

Naturally, we must now know the assumed order l of the difference equation of the plant. If l is chosen to be large, the number of computations grows faster than the accuracy of estimation. Therefore, we pose the problem in the following way: For given l we have to determine in the best possible manner the difference equations of the dynamic plant.

In solving this problem, we introduce an $(l + l_1)$-dimensional vector—the situation vector $\mathbf{z}[n]$:

$$\mathbf{z}[n] = (x[n-1], \ldots, x[n-l]; u[n-1], \ldots, u[n-l_1]) \qquad (5.45)$$

Then, the difference equation (5.42) is written in more compact form

$$x[n] = f(\mathbf{z}[n]) \qquad (5.46)$$

which was discussed in the previous chapter. Therefore, it is reasonable to use the well-known formula

$$f(\mathbf{z}) \approx \hat{f}(\mathbf{z}, \mathbf{c}) = \mathbf{c}^T \boldsymbol{\phi}(\mathbf{z}) \qquad (5.47)$$

for approximation of the right-hand side of (5.46). But this problem differs from the problems of learning in pattern recognition. As it can be seen from (5.45), the vectors $\mathbf{z}[n]$ ($n = 0, 1, 2, \ldots,$) cannot be statistically independent. We shall assume that they are stationary random sequences or processes. In this case, the algorithms of adaptation are applicable and we can determine the best estimate of $\mathbf{c}$, and thus of the function $f(\mathbf{z})$, using the algorithms of learning similar to (4.9). For instance, we can use the algorithm

$$\mathbf{c}[n] = \mathbf{c}[n-1] + \gamma[n] F'(x[n] - \mathbf{c}^T[n-1]\boldsymbol{\phi}(\mathbf{z}[n]))\boldsymbol{\phi}(\mathbf{z}[n]) \qquad (5.48)$$

and the corresponding "perceptron" scheme (Fig. 4.1).

The simple perceptrons as well as the search perceptrons can be considered to be the models of the plant; the parameters of these models are changing in such a manner that at the end, its dynamic properties differ little from the plant. Therefore, in the problem of identification of plants, the perceptrons play a role of an adjustable model.

5.12 Identification of Nonlinear Plants II

In many cases it is considerably more convenient and more natural to describe the plant by a system of nonlinear difference equations (5.43). In such a case, the output variable is not a scalar, as it was in (5.42), but a vector. This type of equation necessitates a certain modification of the identification method described above.

We shall approximate each component of the vector function $\mathbf{f}(\mathbf{x}, \mathbf{u})$ by a finite sum

$$\hat{f}_\mu(\mathbf{x}, \mathbf{u}, \mathbf{c}) = \sum_{v=1}^{N} c_v \phi_{\mu v}(\mathbf{x}, \mathbf{u}) \quad (\mu = 1, \ldots, l) \quad (5.49)$$

or in the vector form

$$\hat{\mathbf{f}}(\mathbf{x}, \mathbf{u}, \mathbf{c}) = \Phi(\mathbf{x}, \mathbf{u})\mathbf{c} \quad (5.50)$$

where

$$\Phi(\mathbf{x}, \mathbf{u}) = \|\phi_{\mu v}(\mathbf{x}, \mathbf{u})\| \quad (\mu = 1, \ldots, l; v = 1, \ldots, N)$$

is an $l \times N$ matrix of linearly independent functions $\phi_{\mu v}(\mathbf{x}, \mathbf{u})$.

The problem of identifying a plant is then reduced to one of minimizing the mathematical expectation

$$J(\mathbf{c}) = M\{F(\mathbf{x}[n] - \Phi(\mathbf{x}[n-1], \mathbf{u}[n-1])\mathbf{c})\} \quad (5.51)$$

with respect to vector $\mathbf{c}$, where $F(\cdot)$ is a strictly convex function.

We shall now apply the search algorithm of adaptation (3.15) to the functional (5.15). In the case under consideration this leads to the following algorithm:

$$\mathbf{c}[n] = \mathbf{c}[n-1] - \gamma[n]\tilde{\nabla}_{\mathbf{c}\pm} F(\mathbf{x}[n], \mathbf{u}[n-1], \mathbf{c}[n-1], a[n]) \quad (5.52)$$

In fact, since the function $F(\cdot)$ is strictly convex and usually differentiable, it is better to use the algorithm of adaptation (3.9) in the solution of the posed problem.

The gradient is

$$\nabla_{\mathbf{c}} F(\mathbf{x}[n] - \Phi(\mathbf{x}[n-1], \mathbf{u}[n-1])\mathbf{c}) = -\Phi^T(\mathbf{x}[n-1], \mathbf{u}[n-1]) \nabla F(\mathbf{x}[n] \\ - \Phi(\mathbf{x}[n-1], \mathbf{u}[n-1])\mathbf{c}) \quad (5.53)$$

By applying the algorithm of adaptation to (5.51), we find the following algorithm in the usual manner:

$$\mathbf{c}[n] = \mathbf{c}[n-1] + \gamma[n]\Phi^T(\mathbf{x}[n-1], \mathbf{u}[n-1]) \nabla F(\mathbf{x}[n] \\ - \Phi(\mathbf{x}[n-1], \mathbf{u}[n-1])\mathbf{c}[n-1]) \quad (5.54)$$

which defines the optimal value of the vector $\mathbf{c} = \mathbf{c}^*$ as $n \to \infty$. The scheme for estimating the optimal vector $\mathbf{c}^*$ and the characteristic of the plant $\mathbf{f}(\mathbf{x}, \mathbf{u})$ is presented in Fig. 5.10. This scheme is a slightly more complicated variant of the earlier perceptron scheme given in Fig. 4.1.

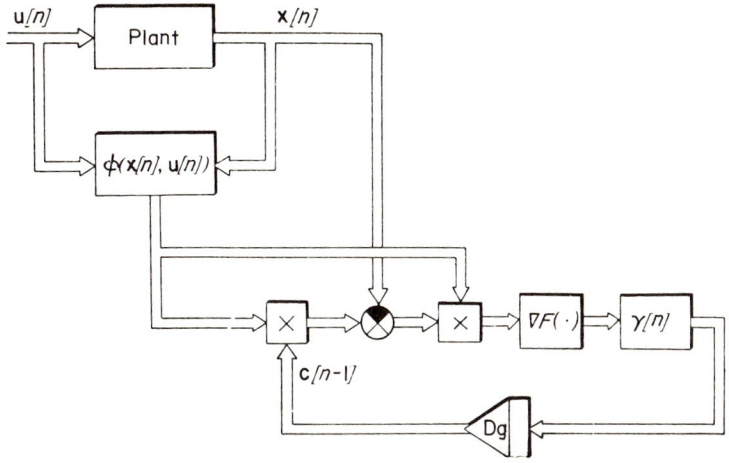

Fig. 5.10

5.13 Identification of Nonlinear Plants III

Let us finally consider the case when the dynamic plant is described by Volterra series (5.44), which can be abbreviated as

$$x[n] = \sum_{s=1}^{N_0} \sum_{m_1,\ldots,m_s=0}^{\infty} k_s[m_1, \ldots, m_s] u[n-m_1] \cdots u[n-m_s] \quad (5.55)$$

In (5.55), the internal summation sign indicates multiple summation corresponding to different indices of summation. We shall now approximate the kernel by a finite sum

$$\hat{k}_s[m_1, \ldots, m_s] = \sum_{v=1}^{N} c_{sv} \phi_v[m_1, \ldots, m_s] \quad (5.56)$$

where $\phi_v[m_1, \ldots, m_s]$ is a set of linearly independent functions. Then an estimate of $x[n]$ has the form

$$\hat{x}[n] = \sum_{s=1}^{N_0} \sum_{v=1}^{N} c_{sv} Y_{sv}(u[n]) \quad (5.57)$$

In (5.57), the quantities

$$Y_{sv}(u[n]) = \sum_{m_1,\ldots,m_s=0}^{\infty} \phi_v[m_1, \ldots, m_s] u[n-m_1] \cdots u[n-m_s] \quad (5.58)$$

can be considered as standard reactions on the input signal $u[n]$. In vector form, the relationship (5.57) looks like

$$\hat{x} = \mathbf{c}^T \mathbf{Y}(u) \quad (5.59)$$

where

$$\mathbf{c} = (c_{11}, \ldots, c_{1N}; c_{21}, \ldots, c_{2N}; \ldots; c_{N_01}, \ldots, c_{N_0N}) \quad (5.60)$$

and

$$\mathbf{Y} = (Y_{11}, \ldots, Y_{1N}; Y_{21}, \ldots, Y_{2N}; \ldots; Y_{N_01}, \ldots, Y_{N_0N})$$

For obtaining $\mathbf{c} = \mathbf{c}^*$, and thus the kernel (5.56), we construct the functional

$$J(\mathbf{c}) = M\{F(x - \hat{x})\} \quad (5.61)$$

or, due to (5.59),

$$J(\mathbf{c}) = M\{F(x - \mathbf{c}^T\mathbf{Y}(u))\} \quad (5.62)$$

If the function $F(\cdot)$ is differentiable, we can find its gradient with respect to $\mathbf{c}$:

$$\nabla_{\mathbf{c}} F(x - \mathbf{c}^T\mathbf{Y}(u)) = -F'(x - \mathbf{c}^T\mathbf{Y}(u))\mathbf{Y}(u) \quad (5.63)$$

Then, the optimal vector $\mathbf{c} = \mathbf{c}^*$ can be found using the algorithm of adaptation (3.9):

$$\mathbf{c}[n] = \mathbf{c}[n-1] + \gamma[n]F'(x[n] - \mathbf{c}^T[n-1]\mathbf{Y}(u[n]))\mathbf{Y}(u[n]) \quad (5.64)$$

5.14 A Special Case

Usually the kernels $k_s[m_1, \ldots, m_s]$ have the property

$$k_s[m_1, \ldots, m_s] = \prod_{\mu=1}^{s} k_s[m_\mu] \quad (5.65)$$

This is called the property of separability.

Let us assume that

$$k_s[m] = 0 \quad \text{for} \quad m < 0 \quad \text{and} \quad m \geq M \quad (5.66)$$

Then, in order to define the kernels $k[m_1, \ldots, m_s]$, we have to specify $(s+1)\cdots(s+M-1)/(m+1)!$ of their values at the points $m_1, \ldots, m_s = 0, 1, \ldots, M$.

The sought ordinates of the kernels will be considered as the components of an unknown vector, which is again designated by $\mathbf{c}$. We also introduce the vector

$$\overline{\mathbf{Y}}(u) = (\overline{Y}_{11}, \ldots, \overline{Y}_{1M}; \overline{Y}_{21}, \ldots, \overline{Y}_{2M}; \ldots; \overline{Y}_{N_01}, \ldots, \overline{Y}_{N_0M}) \quad (5.67)$$

where

$$\overline{Y}_{\mu_s} = u[n - \mu_1] \cdots u[n - \mu_s]$$

It is easy to see that $\overline{Y}_{\mu_s}$ is obtained from (5.58) for a special choice of ϕ_ν. Therefore, in this case we obtain the relationship analogous to (5.50), and in obtaining $\mathbf{c}^*$ we can use algorithm (5.64). A disadvantage of this case is that the dimension of the vector $\mathbf{c}$ increases. In computing the sth kernel (5.56) we needed N components, and now the number of components is increased to $(s+1) \cdots (s+N+1)/(N-1)! > N$.

The dimension of $\mathbf{c}$ can be reduced if instead of a single system of linearly independent functions $\phi_\nu(\cdot)$, we use the systems $\phi_{\nu s}(\cdot)$ as was done in Section 5.12. Then, instead of (5.56) we have

$$\hat{k}_s[m_1, \ldots, m_s] = \sum_{\nu=1}^{N} c_\nu \phi_{\nu s}[m_1, \ldots, m_s] \qquad (5.68)$$

In this case, by repeating the arguments of Section 5.12, the algorithm of adaptation has the form

$$\mathbf{c}[n] = \mathbf{c}[n-1] + \gamma[n]F'(x[n] - \mathbf{Y}_1(u[n]\mathbf{c}))\mathbf{Y}_1(u[n]) \qquad (5.69)$$

where $\mathbf{Y}_1(u[n])$ is an N-dimensional vector. The dimension of the vector $\mathbf{c}$ is now N. This is achieved at the cost of a greater variety of functions $\phi_{\nu s}(\cdot)$ defined over an infinite interval.

5.15 A Remark

After becoming familiar (in Sections 5.11–5.14) with the various possibilities of describing the objects and the various algorithms for identifying such objects, the following question immediately comes to mind: Which algorithm should be given preference?

Regardless of the fact that different descriptions of nonlinear dynamic plants are equivalent, each of them has its own advantages and, unfortunately, disadvantages. For instance, the information about the state of the object is the most complete if the system of difference equations (5.43) is used, since here $\mathbf{x}$ is a vector. In the difference equation (5.42), or in the Volterra series representation (5.44), $\mathbf{x}$ is a scalar. Also, the system of difference equations includes, in the most natural way, the case of many control actions. On the other hand, the number of degrees of freedom defines the order of the difference equation or of the system of difference equations, and we have to know it in advance. However, this is not necessary when the Volterra series is used. Therefore, the answer to the question posed above could be obtained only after collecting the experimental evidence of numerous applications of these algorithms. The algorithms can be easily extended to multivariable systems, and the reader can do this independently.

5.16 Identification of Linear Plants I

For linear dynamic plants, the algorithms of adaptation are simplified. Now we know that the plant is described by a linear difference equation

$$x[n] = \sum_{m=1}^{l} a_m x[n-m] + \sum_{m=1}^{l_1} b_m u[n-m] \qquad (5.70)$$

where some of the coefficients a_m and b_m can be equal to zero. Using the notation (5.45) for the state vector $\mathbf{z}$, and introducing the vector of coefficients

$$\mathbf{c} = (a_1, \ldots, a_l; b_1, \ldots, b_{l_1}) \qquad (5.71)$$

we write the approximating function $\hat{\mathbf{f}}(\mathbf{z}, \mathbf{c})$ as an inner product

$$\hat{\mathbf{f}}(\mathbf{z}, \mathbf{c}) = \mathbf{c}^T \mathbf{z} \qquad (5.72)$$

The dimension of the vector $\mathbf{c}$ depends upon the order of difference equations. By substituting $\phi(\mathbf{z})$ for $\mathbf{z}$ in the algorithm of adaptation, we obtain

$$\mathbf{c}[n] = \mathbf{c}[n-1] + \gamma[n] F'(x[n] - \mathbf{c}^T[n-1]\mathbf{z}[n])\mathbf{z}[n] \qquad (5.73)$$

The perceptrons are now used directly for estimation of the equations and their coefficients.

In the particular case when $F(\cdot)$ is a quadratic function and $2\gamma[n] = 1/\|\mathbf{z}[n]\|^2$, it follows from (5.73) that

$$\mathbf{c}[n] = \mathbf{c}[n-1] + \frac{1}{\|\mathbf{z}[n]\|^2}(x[n] - \mathbf{c}^T[n-1]\mathbf{z}[n])\mathbf{z}[n] \qquad (5.74)$$

The scheme which realizes this algorithm (Fig. 5.11) is naturally simpler than those considered above. This is a well-known algorithm due to Kaczmarz,

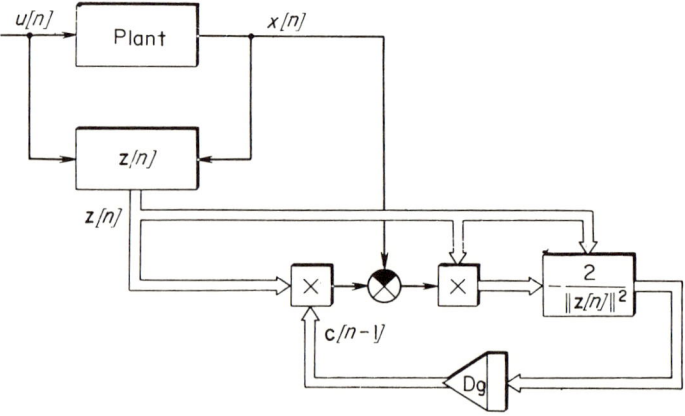

Fig. 5.11

but it was originally obtained with the assumption that the input variables are statistically independent.

All the results of this section can be easily generalized for a case when a linear plant is described by a system of linear difference equations.

5.17 Identification of Linear Plants II

If for some reason we do not know l_1 and l_2, we can use the description of a linear system in the form of convolution:

$$x[n] = \sum_{m=0}^{\infty} k[m]u[n-m] \qquad (5.75)$$

This equation can be considered as a first order approximation of the Volterra series (5.45).

We shall make the following assumptions:

(1) $k[m] \equiv 0$ when $m > M$;
(2) $u[n]$ is different from zero in the interval $[0, N]$, and outside of this interval $u[n] \equiv 0$;
(3) $x[n]$ can be observed in the interval $[0, N + M]$.

Then equation (5.75) can be written as

$$x[n] = \sum_{m=0}^{M} k[m]u[n-m] \qquad (5.76)$$

We shall determine the vectors

$$\mathbf{c} = (k[0], \ldots, k[M]) \qquad (5.77)$$

and

$$\mathbf{z} = (u[M], \ldots, u[0]) \qquad (5.78)$$

We can immediately conclude that algorithms (5.73) and (5.74) are applicable. It is also obvious that the search and continuous algorithms can be used for the estimation of the impulse characteristic.

We shall emphasize the fact that when conditions (1)–(3) are satisfied, the dimension of vector $\mathbf{c}$ does not depend upon the order of the equation of the plant. This was not true in the case of the difference equations (5.16).

In conclusion, let us note that the identification of the linear plants can be accomplished by perceptron schemes which contain a number of typical elements (delay lines, RC circuits, etc.), the outputs of which are summed with the specific weights. These weights are determined by learning, starting from the conditions of minimizing a given functional. We will discuss this in Section 6.4 while discussing the solution to a filtering problem which is similar to the problem of identification considered here.

5.18 Estimation of Parameters in Systems with Distributed Parameters

Systems with distributed parameters are described by partial differential equations. For instance, the equations describing thermal, chemical and similar processes can be written as

$$\frac{\partial \mathbf{x}(s, t)}{\partial t} = \mathscr{F}\left[s, t, \mathbf{c}, \mathbf{x}(s, t), \frac{\partial \mathbf{x}(s, t)}{\partial s}, \frac{\partial^2 \mathbf{x}(s, t)}{\partial s^2}\right] \quad (t \geq 0, 0 \leq s \leq L_0) \quad (5.79)$$

where t is time, s is a spatial coordinate, and $\mathbf{x}(s, t) = (x_1(s, t), \ldots, x_N(s, t))$ is a vector characterizing the state of the plant at an instance t and at any point of the spatial coordinate s. It is assumed that the distributive property exists only along the spatial coordinate s; $\mathbf{c}$ is the vector of unknown parameters which have to be determined.

We shall measure the state of the plant at discrete instants $n = 1, 2, \ldots$, and at a finite series of points $r = 1, 2, \ldots, R$, with a sampling interval Δr. The difference equation corresponding to the partial differential equation (5.79) is

$$\mathbf{x}[r, n] = f(\mathbf{c}, \mathbf{x}[r-1, n-1], \mathbf{x}[r, n-1], \mathbf{x}[r+1, n-1], \ldots)$$
$$(r = 1, \ldots, R; n = 0, 1, \ldots) \quad (5.80)$$

Here, $\mathbf{x}[0, n]$ and $\mathbf{x}[r, n]$ are the boundary conditions. Equation (5.80) is analogous to (5.29). Therefore, identification of the parameter $\mathbf{c}$ can be performed by the known method.

We now introduce the performance index

$$J = M\left\{\sum_{r=1}^{R} F(\mathbf{x}[r, n]; \hat{\mathbf{x}}[r, n])\right\} \quad (5.81)$$

which is the mathematical expectation of the risk function. In (5.81), $\hat{\mathbf{x}}[r, n]$ is determined by (5.80). The quantities $\mathbf{x}[r, n]$ are the results of the measurements. They can differ from the true values by the quantities determined by the measurement errors. It is further assumed that the measurement errors at each step are independent random variables with mean value equal to zero and finite variance. The algorithm of estimating the parameter $\mathbf{c}$ is then written as

$$\mathbf{c}[n] = \mathbf{c}[n-1] - \gamma[n]\nabla_\mathbf{c} \sum_{r=1}^{R} F(\mathbf{x}[r, n]; f(\mathbf{c}[n-1], \mathbf{x}[r-1, n-1],$$
$$\mathbf{x}[r, n-1], \mathbf{x}[r+1, n-1]), \ldots) \quad (5.82)$$

When c and x are scalars, and the loss function is

$$F(\mathbf{x}[r, n]; \hat{\mathbf{x}}[r, n]) = (x[r, n] - \hat{x}[r, n])^2 \quad (5.83)$$

5.19 Noise

algorithm (5.82) is simplified, and has the form

$$c[n] = c[n-1] + 2\gamma[n][x[r, n] - f(c[n-1], x[r-1, n-1],$$
$$x[r, n-1], x[r+1, n-1]) \cdots]$$
$$\times \frac{d}{dc} f(c[n-1], x[r-1, n-1], x[r, n-1], \ldots) \quad (5.84)$$

The scheme for estimating the parameter is shown in Fig. 5.12.

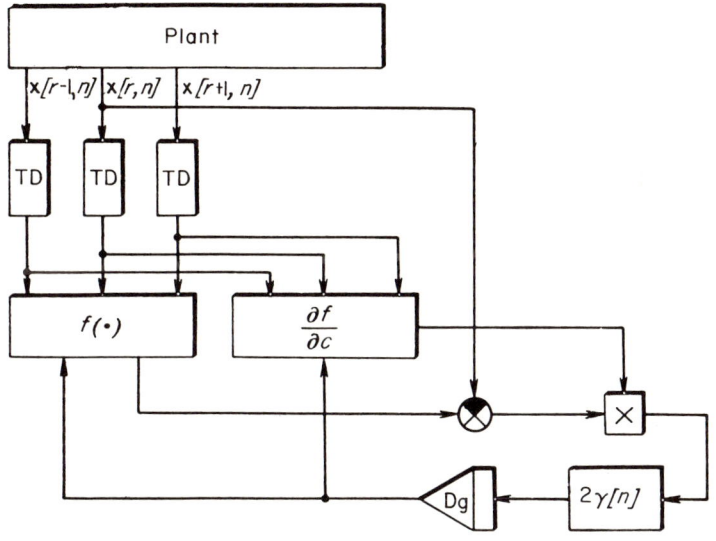

Fig. 5.12

5.19 Noise

The measurements of the input and output variables of an object (plant) are usually noisy. The existence of noise can cause biased estimates which differ from the estimates obtained when the noise is not present. Therefore, questions related to the influence of noise on the estimation procedures are very important.

In order to explain this, we shall consider an example of estimating the characteristics of nonlinear plants without inertia. We shall first explain the conditions under which noise does not influence the estimates, i.e., when the estimates of the plant characteristics are unbiased. Of course, we only consider the estimates obtained at the end of learning or adaptation.

Let us assume that the noise ξ is present at the input of the plant shown in Fig. 5.13; the mean value of the noise is zero. We write the criterion of

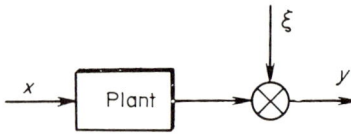

Fig. 5.13

optimality as

$$J = M\{F(y - \hat{f}(x, \mathbf{c}) + \xi)\} \qquad (5.85)$$

or

$$J(\mathbf{c}) = -M\{F(y - \mathbf{c}^T \boldsymbol{\phi}(x) + \xi)\} \qquad (5.86)$$

The condition of optimality is written as

$$\nabla J(\mathbf{c}) = M\{F'(y - \mathbf{c}^T \boldsymbol{\phi}(x) + \xi)\boldsymbol{\phi}(x)\} = 0 \qquad (5.87)$$

If $F(\cdot)$ is a quadratic function, this condition is considerably simplified, and (5.87) can be written as

$$M\{y\boldsymbol{\phi}(x)\} - M\{\mathbf{c}^T \boldsymbol{\phi}(x)\boldsymbol{\phi}(x)\} + M\{\xi\boldsymbol{\phi}(x)\} = 0 \qquad (5.88)$$

If the noise and the input signal are independent,

$$M\{\xi\boldsymbol{\phi}(x)\} = 0 \qquad (5.89)$$

Then equation (5.88) has the form

$$M\{y\boldsymbol{\phi}(x)\} = M\{\Phi(x)\mathbf{c}\} \qquad (5.90)$$

where

$$\Phi(x) = \|\phi_\nu(x)\phi_\mu(x)\| \qquad (\nu, \mu = 1, \ldots, N) \qquad (5.91)$$

is an $N \times N$ matrix. From (5.90) we obtain the unbiased estimate

$$\mathbf{c}^* = M\{\Phi(x)\}^{-1} M\{y\boldsymbol{\phi}(x)\} \qquad (5.92)$$

Therefore, the quadratic performance index, and thus the least square error method, offer unbiased estimates. This fact explains a broad popularity of the quadratic performance indices. The problems become linear, and the influence of the output noise with zero mean value is removed.

5.20 Elimination of Noise

Is it possible to eliminate the influence of noise when the performance index is not quadratic? It appears that under certain conditions, which we shall now explain, this can be done.

5.20 Elimination of Noise

Let us assume that the function $F(\cdot)$ is a polynomial of rth degree. Then when noise is not present,

$$J(\mathbf{c}) = M\left\{\sum_{v=0}^{r} a_v(y - \mathbf{c}^T\boldsymbol{\phi}(x))^v\right\} \tag{5.93}$$

Let the criterion $J(\mathbf{c})$ be minimized for $\mathbf{c} = \mathbf{c}^*$.
If the noise exists at the output of the plant,

$$J(\mathbf{c}) = M\left\{\sum_{v=0}^{r} a_v(y - \mathbf{c}^T\boldsymbol{\phi}(x) + \xi)^v\right\} \tag{5.94}$$

Obviously, by minimizing (5.94), we obtain $\mathbf{c}^{**}$ which, generally speaking, differs from $\mathbf{c}^*$, and depends upon the statistical characteristics—moments—of the noise $\overline{\xi^v}$, i.e., we obtain a biased estimate $\mathbf{c}^{**}$. If we know the moments of the noise $\overline{\xi^v}$, we can modify the criterion of optimality so that the estimate $\mathbf{c}^*$ is unbiased. Actually, let

$$J(\mathbf{c}) = M\left\{\sum_{v=0}^{r} b_v(y - \mathbf{c}^T\boldsymbol{\phi}(x) + \xi)^v\right\} \tag{5.95}$$

where b_v ($v = 1, \ldots, r$) are unknown coefficients. In order to determine these coefficients, we expand $(y - \mathbf{c}^T\boldsymbol{\phi}(x) + \xi)^v$ into a power series of ξ. For simplicity, we assume that the signal x and the noise ξ are uncorrelated, and we write the functional (5.95) in the following form:

$$J(\mathbf{c}) = M\left\{\sum_{v=0}^{r}\left(\sum_{\mu=0}^{r-v} b_{v+\mu}\binom{\mu}{v-\mu}\overline{\xi^\mu}(y - \mathbf{c}^T\boldsymbol{\phi}(x))^v\right)\right\} \tag{5.96}$$

where $\overline{\xi^\mu}$ is the μth moment of the noise.

By comparing (5.96) and (5.93), it is easy to conclude that if we choose the unknown coefficients b_k so that we satisfy the equations

$$a_v = \sum_{\mu=0}^{r-v} b_{v+\mu}\binom{\mu}{v-\mu}\overline{\xi^\mu} \tag{5.97}$$

then a minimization of the functional (5.95) with the algorithms of adaptation provides the unbiased estimate of $\mathbf{c}$. This result can be obtained even in the case when ξ and x are correlated. The coefficients b_k are calculated according to the recursive formula:

$$b_v = a_v - \sum_{\mu=1}^{r-v} b_{v+\mu}\binom{\mu}{v-\mu}\overline{\xi^\mu} \tag{5.98}$$

The moments of the noise are present in this formula. They can be computed in advance. However, the coefficients b_k can be determined by using the adaptive algorithms for estimating the moments of noise (these algorithms are analogous to those described in Sections 4.16 and 5.2–5.6). Therefore, unbiased estimates are obtained here by a proper choice of the criterion of optimality.

The problem of eliminating the influence of the noise acting at the input of a nonlinear element is much more complicated, but under certain conditions it can be solved. We shall consider this problem while discussing the adaptive filters (Chapter 6).

5.21 Certain Problems

In addition to the general problems of accelerating the convergence and of choosing the functions $\phi_\nu(\cdot)$, a very important problem in identification is one of obtaining unbiased estimates when the criterion of optimality is given. If there is a possibility of estimating the variance of the noise, we can obtain unbiased estimates of the parameters or plant characteristics. In estimating the variance, it is also efficient to use the algorithms of adaptation.

The development of the methods for obtaining unbiased estimates has begun. An important step is to remove the requirement of stationarity imposed on the external actions. Can we modify the presented methods of identification so that they can be applicable in the cases when the process is nonstationary, but converging to a stationary one?

An inverse relationship certainly exists between the identification time and the error of the obtained results. It is of great importance not only for the problems of identification but also for the problems of control, that this relationship be determined. Such a relationship would define the limiting accuracy for a given time interval of observation and for the processing of information.

A similar relationship in control theory would play a role analogous to the uncertainty principle of quantum mechanics.

5.22 Conclusion

It is probably not difficult to establish a close relationship between the problems of pattern recognition and identification. In the latter case, we have to "recognize" a plant and extract its characteristic properties, i.e., determine its characteristics. The adaptive approach enables us to solve these problems under different levels of a priori information. We have not tried to obtain explicit formulas for solving the problem under consideration. The experiments conducted thus far indicate that such an attempt would be fruitless. Instead, we obtain the algorithms, i.e., equations (difference of differential), and their solutions lead us to our goal. Since such equations are nonlinear and stochastic, they have to be solved on analog or digital computers. Therefore, the realization of the algorithms of identification leads to a construction of discrete or continuous adjustable models of the identified plants. All these models are actually different forms of perceptrons.

It is interesting that even before the pattern recognition problems were considered, the perceptrons had been used as adaptive models in the identification of the plants. But indeed, their second birth was in pattern recognition.

COMMENTS

5.1 An excellent survey of various approaches to the problems of identification is given in the paper by Eikhoff (1963), and also the survey by Eikhoff *et al.* (1966) presented at the Third International Congress on Automatic Control, IFAC.

5.2 These simple and clear considerations are also presented in the books by Chang (1961) and Wilde (1964).

5.3 The modified algorithm (5.9) was proposed by Fu and Nikolic. The authors had great hopes in that algorithm, which the algorithm deserves (see the commentary to Section 4.16).

5.5 See also the paper by the author (Tsypkin, 1965).

5.6 The nature of modified algorithms was also considered in the paper by the author. (Tsypkin, 1967).

5.7 This problem was considered by Aizerman *et al.* (1964).

5.8 For the quadratic loss function, we obtain from (5.35) the algorithm which defined the coefficients of the statistical linearization. The algorithm was introduced by Booton (1963) and Kazakov (1956, 1965).

5.9 The algorithms of the type (5.41) were proposed by Fabian (1960) and H. Cruz (Wilde, 1964) as a simpler modification of the algorithms of the type (5.35). However, as it was shown by Avedian (1967), these algorithms cannot substitute for each other.

5.11 The idea of using the solution of pattern recognition problems for identification of linear plants described by differential equations of a high order was presented in the paper by Braverman (1966). Useful considerations in connection with the identification of nonlinear plants, and also a description of the experiment was given in the work by Kirvaitis and Fu (1966).

5.13 and 5.14 On discrete and continuous functional series of Volterra, see the works of Alper (1964, 1965) and the monograph by Van Trees (1962).

5.16 Algorithm (5.74) is equivalent to the algorithm proposed by Kaczmarz (1937). It was obtained (for the special case of independent input signals) and successfully applied in the solution of identification problems by Chadeev (1964, 1965). Such constraints are removed here. As we have already mentioned, a similar algorithm for the restoration of a differential, and not difference equation, was obtained by Braverman (1966) for correlated external actions. Naturally, the difficulties of measuring the state vector which is defined by the derivatives of a higher order are present here. Similar algorithms of Kaczmarz's type were again "discovered" by Nagumo and Noda (1966), and were applied by them in the solution of an identification problem.

5.17 A similar problem on the basis of another approach was considered by Levin (1960).

5.18 The results obtained by Zhivogliadov and Kaipov (1966) are presented here.

5.19 A similar method of eliminating the influence of noise was described by Sakrison (1961).

BIBLIOGRAPHY

Aizerman, M.A., Braverman, E.M., and Rozonoer, L.I. (1964). The probabilistic problem of training automata to recognize patterns and the potential function method, *Automat. Remote Contr.* (*USSR*) **25** (9).

Alper, P. (1964). Higher-dimensional z-transforms and non-linear discrete systems, *Rev. Quart. A* **6** (4).

Alper, P. (1965). A consideration of the discrete Volterra series, *IEEE Trans. Automat. Contr.* **AC-10** (3).

Avedian, E.D. (1967). On a modification of Robbins-Monro algorithm, *Automat. Remote Contr.* (*USSR*) **28** (4).

Barker, H.A., and Hawley, D.A. (1966). Orthogonalising techniques for system identification and optimization. *In* "Automatic and Remote Control. Proceedings of the Third Congress of the International Federation of Automatic Control." Butterworths, London.

Bellman, R. (1966). Dynamic programming, system identification and suboptimization, *J. SIAM Contr.* **4** (1).

Bellman, R., Kalaba, R., and Sridhar, R. (1966). Adaptive control via quasilinearization and differential approximation, *Computing* **1** (1).

Booton, R.C. (1963). The analysis of nonlinear control systems with random inputs. *In* "Proceedings of the Symposium on Nonlinear Circuit Analysis." Polytechnic Institute of Brooklyn, New York.

Braverman, E.M. (1966). Estimation of the plant differential equation in the process of normal exploitation, *Automat. Remote Contr.* (*USSR*) **27** (3).

Butler, R.E., and Bohn, E.V. (1966). An automatic identification technique for a class of nonlinear systems, *IEEE Trans. Automat. Contr.* **AC-11** (2).

Chadeev, V.M. (1964). Estimation of the dynamic characteristics of the plants in the process of their normal exploitation for the purpose of self-adaptation, *Automat. Remote Contr.* (*USSR*) **25** (9).

Chadeev, V.M. (1965). Certain questions regarding the estimation of the characteristics using an adjustable model. *In* "Tekhnicheskaia Kybernetika," Nauka, Moscow. (In Russian.)

Chang, S.S.L. (1961). "Synthesis of Optimum Control Systems." McGraw-Hill, New York.

Eikhoff, P. (1963). Some fundamental aspects of process parameter estimation, *IEEE Trans. Automat. Contr.* **AC-8** (4).

Eikhoff, P., van der Grinten, P., Kvakernaak, H., and Weltman, B. (1966). System modelling and identification. *In* "Automatic and Remote Control. Proceedings of the Third Congress of the International Federation of Automatic Control." Butterworths, London.

Elistratov, M.G. (1963). Estimation of dynamic characteristics of non-stationary plants in the process of normal operation, *Automat. Remote Contr.* (*USSR*) **24** (7).

Elistratov, M.G. (1965). On a question of accuracy in the estimation of non-stationary plant characteristics, *Automat. Remote Contr.* (*USSR*) **26** (7).

Fabian, V. (1960). Stochastic approximation methods, *Czech. Math. J.* **10** (1).
Flake, R.H. (1963a). Volterra series representation of nonlinear systems, *IEEE Trans. Appl. Ind.* **82** (64).
Flake, R.H. (1963b). Theory of Volterra series and its applications to nonlinear systems with variable coefficients, *Theory Self-Adapt. Contr. Syst. Proc. IFAC Symp. 2nd* **1963**.
Ho, Y.C., and Lee, R.C.K. (1964). Identification of linear dynamic systems, *Proc. Nat. Electron. Conf.* **20**.
Hsieh, H.C. (1964a). Least square estimation of linear and nonlinear weighting function matrices, *Inform. Contr.* **7** (1).
Hsieh, H.C. (1964b). An on-line identification scheme for multivariable nonlinear systems, *Proc. Conf. Comp. Methods Optimizations* **1964**.
Kaczmarz, S. (1937). Angenäherte Auflösung von Systemen linearer Gleichungen, *Bull. Inter. Acad. Pol. Sci. Lett. Cl. Sci. Math. Natur.* **1937**.
Kazakov, I.E. (1956). Approximate probabilistic analysis of the accuracy of operation of essentially nonlinear automatic systems, *Automat. Remote Contr. (USSR)* **17** (5).
Kazakov, I.E. (1965). A generalization of the method of statistical linearization to multichannel systems, *Automat. Remote Contr. (USSR)* **26** (7).
Kirvaitis, K., and Fu, K.S. (1966). Identification of nonlinear systems by stochastic approximation. *In* "Joint Automatic Control Conference, Seattle, Washington, 1966."
Kitamori, G. (1961). Application of the orthogonal functions for obtaining dynamic characteristics of the objects and the design of self-optimizing systems of automatic control. *In* "Proceedings of the First IFAC." Butterworths, London.
Kitchatov, Yu.F. (1965). Estimation of nonlinear plant characteristics with gaussian input signals, *Automat. Remote Contr. (USSR)* **26** (3).
Kumar, K.S.P., and Sridhar, R. (1964). On the identification of control systems by the quasilinearization method, *IEEE Trans. Automat. Contr.* **AC-9** (2).
Kushner, H. (1963). A simple iterative procedure for identification of the unknown parameters of a linear time-varying discrete system, *J. Basic Eng.* **85** (2).
Labbok, J.K., and Barker, H.A. (1964). A solution of the identification problem, *IEEE Trans. Appl. Ind.* **83** (72).
Lelashvili, S.G. (1965). Application of an iterative method to analysis of multivariable automatic systems. *In* "Skhemi Automaticheskogo Upravlenia." Tbilisi. (In Russian.)
Levin, M.J. (1960). Optimum estimation of impulse response in the presence of noise, *IRE Trans. Circuit Theory* **CT-7** (1).
Levin, M.J. (1964). Estimation of a system pulse transfer function in the present noise, *IEEE Trans. Automat. Contr.* **AC-9** (3).
Magil, D.T. (1966). A sufficient statistical approach to adaptive estimation. *In* "Automatic and Remote Control. Proceedings of the Third Congress of the International Federation of Automatic Control." Butterworths, London.
Merchav, S.T. (1962). Equivalent gain of single valued nonlinearities with multiple inputs, *IRE Trans. Automat. Contr.* **AC-7** (5).
Meyerhoff, H.T. (1966). Automatic process identification. *In* "Automatic and Remote Control. Proceedings of the Third Congress of the International Federation of Automatic Control." Butterworths, London.
Miller, R.W., and Roy, R. (1964). Nonlinear process identification using statistical pattern matrices, *Joint Automat. Contr. Conf.* **1964**.
Nagumo, J., and Noda, A. (1966). A learning method for system identification, *Proc. Nat. Electron. Conf.* **22**.
Pearson, A.E. (1966). A regression analysis in functional space for the identification of multivariable linear systems. *In* "Automatic and Remote Control. Proceedings of the

Third Congress of the International Federation of Automatic Control." Butterworths, London.

Perdreaville, F.J., and Goodson, R.F. (1966). Identification of systems described by partial differential equations, *J. Basic Eng.* **88** (2).

Raibman, N.S., and Chadeev, V.M. (1966). "Adaptive Models in Control Systems." Sovetskoe Radio, Moscow. (In Russian.)

Roberts, P.D. (1966). The application of orthogonal functions to self-optimizing control systems containing a digital computer. *In* "Automatic and Remote Control. Proceedings of the Third Congress of the International Federation of Automatic Control." Butterworths, London.

Roy, R.J. (1964). A hybrid computer for adaptive nonlinear process identification, *AFIPS Conf. Proc.* **26**.

Rucker, R.A. (1963). Real time system identification in the presence of noise, *WESCON Conv. Rec.* **7** (4).

Sakrison, D.T. (1961). Application of stochastic approximation methods to optimal filter design, *IRE Inter. Conv. Rec.* **9**.

Skakala, J., and Sutekl, L. (1966). Adaptive model for solving identification problems. *In* "Automatic and Remote Control. Proceedings of the Third Congress of the International Federation of Automatic Control." Butterworths, Londons.

Tsypkin, Ya.Z. (1965). On estimation of the characteristic of a functional transformer using randomly observed points, *Automat. Remote Contr.* (*USSR*) **26** (11).

Tsypkin, Ya.Z. (1967) On algorithms for restoration of probability density functions and moments using observations, *Automat. Remote Contr.* (*USSR*) **28** (7).

Van Trees, H. (1962). "Synthesis of Nonlinear Control Systems." MIT Press, Cambridge, Mass.

Wild, D.J. (1964). "Optimum Seeking Methods." Prentice-Hall, Englewood Cliffs, N.J.

Zhivoglladov, V.P., and Kaipov, V.H. (1966). On application of stochastic approximation method in the problem of identification, *Automat. Remote Contr.* (*USSR*) **27** (10).

6

FILTERING

6.1 Introduction

Filtering is usually considered to be a process of separating a signal from noise. In this chapter, the concept of filtering is broader, since it also includes the detection, extraction, recovery and transformation of the input signals.

If the signal is known and invariant, the design of a filter which performs filtering in a certain optimal manner is not difficult. However, when the operating conditions are changed, the optimality is destroyed. Furthermore, an optimal filter cannot be designed in advance if a priori information about input signals is insufficient. Such cases have indeed created a necessity for the design of adaptive filters, which can adjust themselves to unknown or varying operating conditions. This adjustment can be caused by modifications of either external signals or the internal structure of the filter alone. Adaptation is accomplished by a variation of filter parameters (or if it is necessary, even by modifying the structure of the filter) so that a certain criterion of optimality which characterizes the operation of the filter is minimized.

This chapter is devoted to a discussion of possible principles for designing adaptive filters, where the filters are designated to solve various problems of filtering in the broadest sense of that word.

6.2 Criterion of Optimality

Let the input signal to the filter be

$$u(t) = s(t) + \xi(t) \qquad (6.1)$$

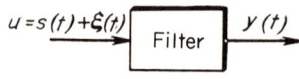

Fig. 6.1

where $s(t)$ is the signal and $\xi(t)$ is noise; both the signal and the noise are stationary random processes with known probability distributions. It is assumed that the mean value of noise is zero (Fig. 6.1).

The output of the filter $y(t)$ is the response of the filter on the input signal $u(t)$. The output signal also depends upon the filter structure and, particularly, on a parameter vector $\mathbf{c} = (c_1, \ldots, c_N)$. It is usually required that the output signal differs slightly from a desired function $y_0(t)$. The desired function $y_0(t)$ represents a certain ideal transformation of the desired signal $s(t)$; i.e., it is a result of a certain operation acting upon $s(t)$. This transformation, performed by the operator, can be integration, differentiation, prediction, smoothing, etc.

As a measure of deviation of the output signal $y(t)$ from the desired signal $y_0(t)$, we can again select the familiar mathematical expectation of a certain convex function F of the difference $y_0(t) - y(t)$. Then

$$J = M\{F(y_0(t) - y(t))\} \tag{6.2}$$

The problem now is to determine the structure of the adaptive filter and the algorithm of adaptation which ensures the minimization of the measure of deviation—the functional (6.2).

Let us be more specific in describing this problem. We consider the block diagram shown in Fig. 6.2. Here, the outputs of the adaptive and the ideal filter are compared when their inputs are the signals $u(t)$ and $s(t)$, respectively. The deviation $\varepsilon = y_0(t) - y(t)$ is the input signal to a nonlinear transformer with the characteristic $F(\varepsilon)$. By applying the sample values $F(y_0(t) - y(t))$ to a special device which is modifying the parameter vector $\mathbf{c}$ in accordance with

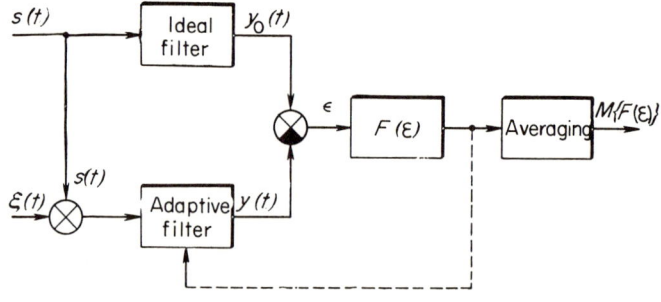

Fig. 6.2

the algorithm of adaptation, we have to find the minimum of the measure of deviation. This problem will be solved if the corresponding algorithms of adaptation are found.

6.3 Adaptive Filter

Let us represent the input variable of an adaptive filter in the form of a linear combination of linearly independent functions of the input signal $u(t) = s(t)$, i.e.,

$$y(t) = \mathbf{c}^T \boldsymbol{\phi}(s(t)) \tag{6.3}$$

So far we have assumed that there is no noise at the input. Using (6.3), expression (6.2) becomes

$$J(\mathbf{c}) = M\{F(y_0(t) - \mathbf{c}^T \boldsymbol{\phi}(s(t)))\} \tag{6.4}$$

In the problems of pattern recognition and identification, the argument of the vector function $\boldsymbol{\phi}$ is a vector function, but here it is a scalar function of time. This fact only simplifies the problem. Since the gradient is

$$\nabla_{\mathbf{c}} F(y_0(t) - \mathbf{c}^T \boldsymbol{\phi}(s(t))) = -F'(y_0(t) - \mathbf{c}^T \boldsymbol{\phi}(s(t))) \boldsymbol{\phi}(s(t)) \tag{6.5}$$

on the basis of the results obtained in Section 2.8 we can immediately write a continuous algorithm of adaptation:

$$\frac{d\mathbf{c}(t)}{dt} = \gamma(t) F'(y_0(t) - \mathbf{c}^T(t) \boldsymbol{\phi}(s(t))) \boldsymbol{\phi}(s(t)) \tag{6.6}$$

This algorithm indeed specifies the structure of an adaptive filter, which is presented in Fig. 6.3. This diagram differs only slightly from the diagram of the perceptron shown in Fig. 4.1.

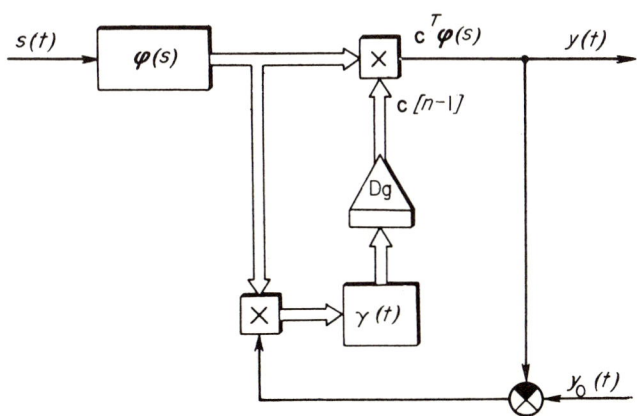

Fig. 6.3

6.4 Special Cases

A choice of the system of linearly independent functions frequently does not permit any simplification in the structure of the adaptive filter, but it can add certain interesting and useful properties. Let us show the algorithm of adaptation (6.6) in the component form

$$\frac{dc_\mu(t)}{dt} = \gamma(t)F'\left(y_0(t) - \sum_{v=1}^{N} c_v(t)\phi_v(s(t))\right)\phi_\mu(s(t)) \qquad (\mu = 1, 2, \ldots, N) \quad (6.7)$$

Let

$$\phi_v(s(t)) = s(t - vT) \qquad (v = 1, \ldots, N) \quad (6.8)$$

which can be simply realized with the delay lines (TD). Then,

$$\frac{dc_\mu(t)}{dt} = \gamma(t)F'\left(y_0(t) - \sum_{v=1}^{N} c_v(t)s(t - vT)\right)s(t - \mu T)$$

$$(\mu = 1, 2, \ldots, N) \quad (6.9)$$

A diagram of the adaptive filter which realizes this algorithm is shown in Fig. 6.4. The adaptive filter can be further simplified if $F(\cdot)$ is chosen to be a quadratic function. Then the derivative $F'(\cdot)$ is a linear function, and a need for nonlinear transformers ceases to exist. As we shall see later, similar filters based on delay lines are successfully used in various communication systems, television and control. Of course, instead of using delay lines, such filters can also use integrators, resonant circuits, etc.

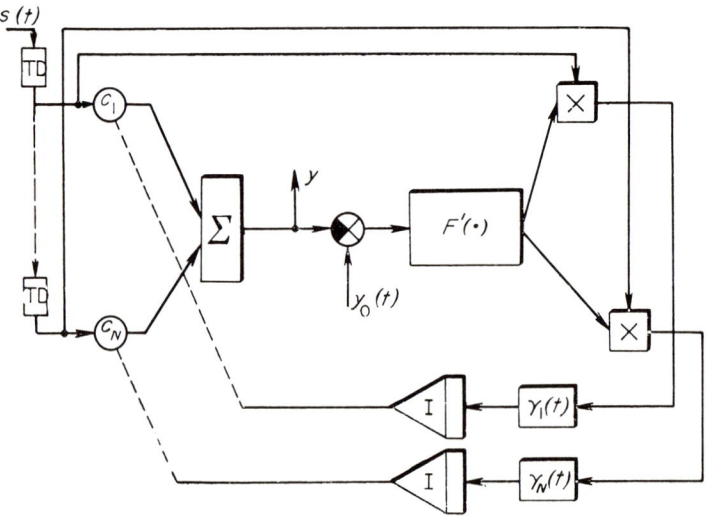

Fig. 6.4

6.4 Special Cases

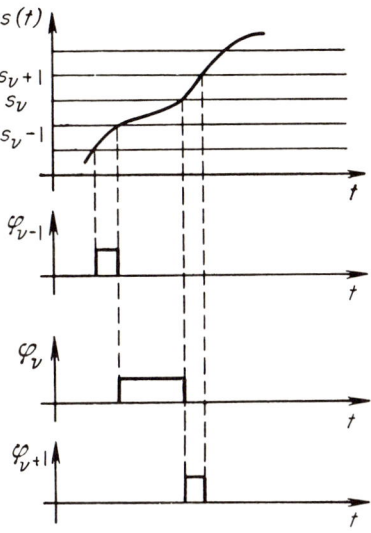

Fig. 6.5

In all the algorithms of adaptation discussed thus far, the coefficients $c_\nu(t)$ are interrelated, i.e., they have to be determined simultaneously. By selecting a special system of functions $\phi_\nu(s(t))$, we can "decouple" these coefficients, and determine them sequentially. For instance, let

$$\phi_\nu(s(t)) = \begin{cases} 1 & \text{when } s_{\nu-1} \leq s(t) \leq s_\nu \\ 0 & \text{when } s(t) < s_{\nu-1}; \ s_\nu < s(t) \end{cases} \quad (\nu = 1, \ldots, N) \quad (6.10)$$

These are threshold functions, and their form is shown in Fig. 6.5. They are orthogonal, i.e.,

$$\phi_\nu(s(t))\phi_\mu(s(t)) = \begin{cases} 1 & \text{when } \nu = \mu \\ 0 & \text{when } \nu \neq \mu \end{cases} \quad (6.11)$$

Therefore, the algorithm of adaptation (6.7) can be simplified. Considering (6.10) and (6.11), we obtain the following equation:

$$\frac{dc_\mu(t)}{dt} = \gamma(t)F'(y_0(t) - c_\mu(t)\phi_\mu(s(t)))\phi_\mu(s(t)) \quad (\mu = 1, \ldots, N) \quad (6.12)$$

Knowing $c_\mu(t)$, we obtain

$$y(t) = \sum_{\nu=1}^{N} c_\nu(t)\phi_\nu(s(t)) \quad (6.13)$$

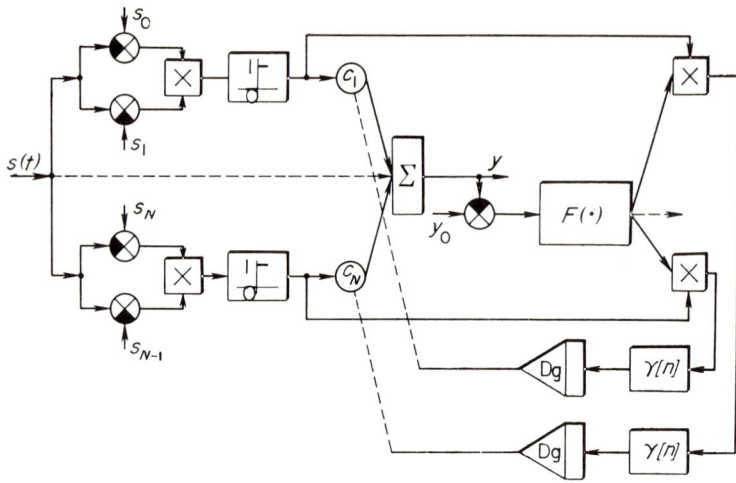

Fig. 6.6

The structural scheme of such an adaptive filter is shown in Fig. 6.6. From equation (6.12) it follows that each coefficient c_μ is determined independently from the others. Such a property permits us to consider only a simple one-dimensional adaptive filter. By varying the thresholds s_{v-1}, s_v we can sequentially determine the coefficients $c_v = c_v^*$ ($v = 1, \ldots, N$) and then $y(t) = \sum_{v=1}^{N} c_v^* \phi_v(s(t))$, which is a staircase approximation of the signal $s(t)$ (Fig. 6.7). This possibility is especially convenient for an initial choice of the optimal parameters of the filters based on experimental data.

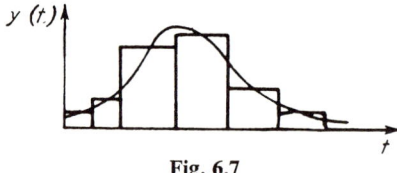

Fig. 6.7

6.5 Adaptive Equalizer

The rate of signal transmission, particularly in the case of television signals, is limited by distortions which are caused by the transients in the transmission channels. In order to remove these distortions at the receiving end of the channel, we have an equalizer. An equalizer is a filter based on delay lines which has weights set in time through a process of training or learning while some special test signals are transmitted through the channel. After a period of learning, the characteristics of the equalizer are such that the distortions in the channel are minimized.

6.5 Adaptive Equalizer

The impulse characteristic of the system "communication channel–equalizer," $k[n]$, is defined as a convolution of the impulse characteristic of the communication channel, $w[n]$, and the equalizer, c_n ($n = 1, \ldots, N$), so that

$$k[n] = \sum_{m=0}^{n} c_m w[n-m] \quad \text{when } n \leq N \tag{6.14}$$

$$k[n] = \sum_{m=0}^{N} c_m w[n-m] \quad \text{when } n > N$$

The distortion would not exist if

$$k[n] = \begin{cases} k(0) & \text{when } n = 0 \\ 0 & \text{when } n > 0 \end{cases} \tag{6.15}$$

i.e., if the communication channel does not have a memory. Therefore, we can quantitatively evaluate the distortion by the functional

$$J(\mathbf{c}) = M\left\{ \sum_{n=1}^{\infty} |k[n]| \right\} \tag{6.16}$$

Let us determine the optimal impulse characteristic of the equalizer by minimizing the functional (6.16). We shall compute the gradient of $Q = \sum_{n=1}^{\infty} |k[n]|$ with respect to c_m. Its components, according to (6.14), are defined by the expression

$$\frac{\partial Q}{\partial c_\mu} = \sum_{n=\mu}^{\infty} w[n-\mu] \, \text{sign} \sum_{m=0}^{n} c_m w[n-m] \quad (\mu = 1, \ldots, N) \tag{6.17}$$

If we assume that

$$w[n] \ll w[0] = 1 \tag{6.18}$$

then the expressions for the components of the gradient are simplified:

$$\frac{\partial Q}{\partial c_\mu} \approx \text{sign} \sum_{m=0}^{\mu} c_m w[\mu - m] \quad (\mu = 1, \ldots, N) \tag{6.19}$$

The algorithm of adaptation can now be written as

$$c_\mu[n] = c_\mu[n-1] - \gamma[n] \, \text{sign} \sum_{m=0}^{\mu} c_m[n-1] w[\mu - m]$$
$$(\mu = 1, \ldots, N) \tag{6.20}$$

This algorithm differs from those encountered earlier since it accumulates the preceding estimates using the impulse characteristic of the equalizer. The scheme of such an adaptive equalizer is shown in Fig. 6.8. It should be noticed

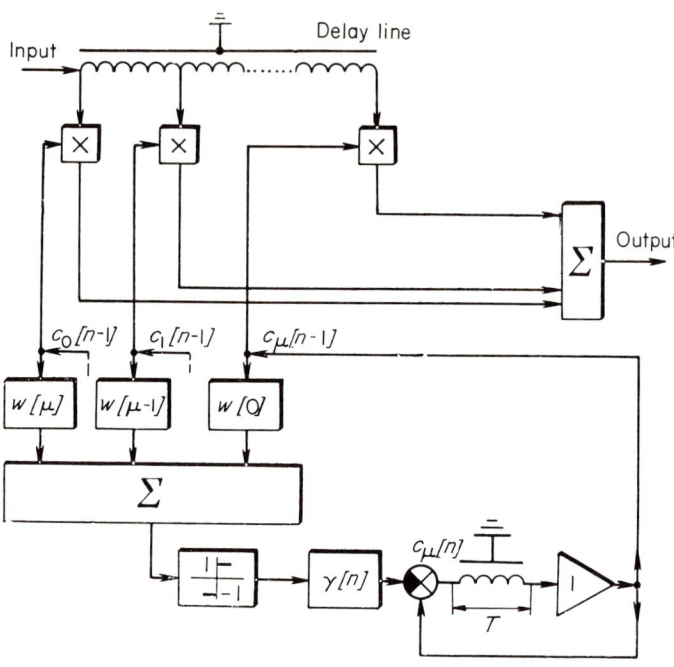

Fig. 6.8

that the algorithm becomes considerably more complicated when (6.18) is not satisfied.

6.6 Adaptive Filters Based upon Search

Let us designate the quantity

$$Y(t) = F(y_0(t) - \mathbf{c}^T \boldsymbol{\phi}(s(t))) \tag{6.21}$$

The observations of such a quantity can be used in adaptation. When the coefficient is incremented by $\pm a$ and all the other coefficients are kept constant, the values of $Y(t)$ are designated respectively by

$$Y_+^\mu \quad \text{for } c_\mu + a$$

and (6.22)

$$Y_-^\mu \quad \text{for } c_\mu - a$$

Then, the quantity which approximately defines the $Y_c(t)$ component of the gradient is equal to

$$Y_{c_\mu \pm} = \frac{Y_+^\mu - Y_-^\mu}{2a} \tag{6.23}$$

6.6 Adaptive Filters Based upon Search

Using the search algorithm of adaptation (3.15), we obtain

$$\mathbf{c}[n] = \mathbf{c}[n-1] - \gamma[n]\tilde{\nabla}_{\mathbf{c}\pm} Y \quad (6.24)$$

where the observed gradient

$$\tilde{\nabla}_{\mathbf{c}\pm} Y = \{Y_{c_1\pm}, \ldots, Y_{c_N\pm}\} \quad (6.25)$$

has the components defined by relationship (6.23). In the case of sequential search, at each time interval of duration $2NT$ there are $2N$ observations of the quantities Y_+^μ and Y_-^μ.

In order to reduce the sensitivity to noise, the averaging over the time intervals of duration T can be performed so that

$$\overline{Y}_+^\mu[n] = \frac{1}{T}\int_{\tau_n+(k-1)T}^{\tau_n+kT} Y_+^\mu(t)\,dt \quad (6.26)$$

and

$$\overline{Y}_-^\mu[n] = \frac{1}{T}\int_{\tau_n+kT}^{\tau_n+(k+1)T} Y_-^\mu(t)\,dt \quad (6.27)$$

Such quantities should then be used in formula (6.23) and in algorithm (6.24). The averaging can be accomplished either in a continuous or in a discrete manner. In the latter case, the integrals in (6.26) and (6.27) are replaced by the sums. The adaptation time can be reduced in the discrete case if all the parameters are sought simultaneously and not sequentially.

The simultaneous search can be accomplished even in the case of continuous adaptation. In this case, the discrete algorithm is replaced by the continuous one:

$$\frac{d\mathbf{c}(t)}{dt} = -\gamma(t)\tilde{\nabla}_{\mathbf{c}\pm} Y(t) \quad (6.28)$$

or

$$\mathbf{c}(t) = \mathbf{c}(1) - \int_1^t \gamma(\tau)\tilde{\nabla}_{\mathbf{c}\pm} Y(\tau)\,d\tau \quad (6.29)$$

The components $\tilde{\nabla}_{\mathbf{c}\pm} Y(t)$ are now

$$Y_{c_\mu\pm} = \frac{Y_+^\mu(t) - Y_-^\mu(t)}{2a(t)} \quad (6.30)$$

It should be noticed that expression (6.30) does not represent an approximation of the gradient, but a random process with a mean value which approximates the gradient. A block diagram of the filter which realizes the continuous adaptation is presented in Fig. 6.9.

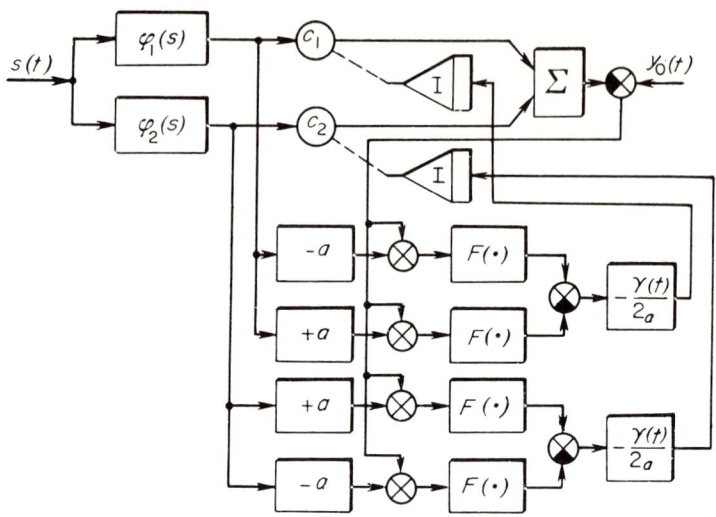

Fig. 6.9

6.7 Adaptive Filter Predictor

Gabor's well-known adaptive filter predictor has a number of interesting properties which will be discussed briefly here. The scheme of the filter is shown in Fig. 6.10. Its operation is based on a particular search algorithm of adaptation. The criterion of optimality here is the mean square deviation of the input signal from the desired reference signal:

$$J = M[(y_0 - y)^2] \tag{6.31}$$

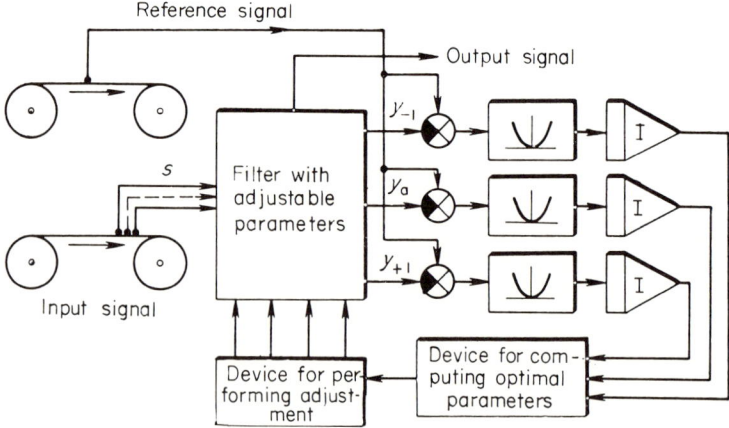

Fig. 6.10

The filter is trained by using a recording of a certain input signal and the corresponding recording of the reference signal.

The optimal value of a certain parameter is computed after measuring the square deviation during a finite interval of time. Three values of the quantity J are simultaneously obtained, and they correspond to three values of the sought parameter: the smallest (y_{-1}), the intermediate (y_0), and the largest (y_{+1}). Because the performance index is quadratic, these three values can be used to compute J_{min} and the corresponding value of the parameter c_{opt}. After adjusting this parameter, the process is repeated again and again for similar adjustment of the remaining parameters, i.e., there exists a search. Empirically established, the number of learning cycles necessary for complete adaptation of N adjustable parameters is on the average equal to $N^2/2$.

The described algorithm can be related to the relaxation algorithms, since every step in the modification of each parameter is based upon the conditions of minimization of the performance index–mean square deviation.

6.8 Kolmogorov-Wiener Filters

If we assume that the performance index is a quadratic function of the difference between the desired signal and the output signal of the filter, and that the filter is linear and described by a convolution, we then face the problem of designing an optimal Kolmogorov-Wiener filter. The solution of this problem is then reduced to minimization of the functional

$$J = M\left\{\left(y_0(t) - \int_0^t k(\tau)u(t-\tau)\,d\tau\right)^2\right\} \tag{6.32}$$

The analytical approach, in which the calculus of variation and the correlation theory are used, leads to Wiener-Hopf equations or to an equivalent boundary value problem which is solved by the method of factorization. A relatively compact form of the final result, which defines either the impulse response $k(\tau)$ or its corresponding transfer function $K(p)$ of the optimal filter, gives an illusion of simplicity in computing these optimal characteristics. However, this is not the case. A large number of computations is necessary to find the corresponding correlation functions and the spectral densities from the sample realizations. The forms of the autocorrelations or the spectral densities are somehow always assumed to be known and, as a rule, to be relatively simple. It is possible to bypass that step and directly determine the optimal characteristic of the filter using a sample realization and the adaptive approach.

We shall assume that the impulse characteristic of the filter has the form $k(\tau) = \mathbf{c}^T \boldsymbol{\phi}(\tau)$. By substituting it into (6.32), we obtain

$$J(\mathbf{c}) = M(y_0(t) - \mathbf{c}^T \boldsymbol{\psi}(t))^2 \tag{6.33}$$

where, obviously,

$$\psi(t) = \int_0^t \phi(\tau) u(t - \tau) \, d\tau \qquad (6.34)$$

We can now minimize $J(\mathbf{c})$ with respect to $\mathbf{c}$ by the usual procedure. A system of linear equations with respect to $\mathbf{c}$ is obtained. The coefficients and the elements on the right-hand side of the equations are the correlation functions. The continuous solution of this system of equations defines the optimal value with the passage of time, and thus

$$k^*(\tau) = \mathbf{c}^{*T} \phi(\tau) \qquad (6.35)$$

The application of an algorithm of adaptation, for instance,

$$\frac{d\mathbf{c}(t)}{dt} = \gamma(t)(y_0(t) - \mathbf{c}^T \psi(t)) \psi(t) \qquad (6.36)$$

greatly simplifies the device of adaptation. In an analogous manner we can consider a very interesting approach for the solution of similar problems which was successfully developed by Kalman.

6.9 Statistical Theory of Signal Reception

The basic problems of the statistical theory of signal reception are those of detection or extraction of the signal from the background noise. In the detection problem, on the basis of the observation

$$\mathbf{y} = Y(\mathbf{s}, \xi) \qquad (6.37)$$

one has to decide whether the signal $\mathbf{s}$ is present. In the problem of signal extraction, it is a priori known that the signal $\mathbf{s}$ is present in the observation $\mathbf{y}$, and it is then necessary to determine the parameters or, in a general case, the characteristics of the signal $\mathbf{s}$. The problems of detection and extraction of the signal depends on one parameter c only, i.e., $\mathbf{s}(t) = \mathbf{s}(c, t)$ and that the parameter c can take only two values: $c = c_1$, with probability P_1, and $c = c_0$ with probability P_0, where $P_1 + P_0 = 1$. In particular,

$$\mathbf{s}(c, t) = c\mathbf{f}(t) \qquad (6.38)$$

In the detection problem, one of the values c, for instance c_0, is chosen to be equal to zero.

The statistical theory of signal reception or, as it is sometimes called, statistical decision theory, provides an effective solution for the detection or extraction of a signal from the noisy background if we have sufficient a priori information. According to this theory, the solutions regarding the parameters of the signal are optimal, i.e., they minimize a certain criterion of optimality in signal reception.

6.10 Criterion of Optimality in Signal Reception

In the observation space, to each observed point, **y**, there corresponds a particular decision. Therefore, all the points in the observation space have to be partitioned into two regions: Λ_1, which corresponds to decision c_1, and Λ_0, which corresponds to decision c_0. Two types of errors accompany each decision:

(1) The error of the first kind, when actually $c = c_0$, and the decision is $c = c_1$. In the detection problem, this is a false alarm; there is no signal, but the decision is that there is one.

(2) The error of the second kind, when actually $c = c_1$, and the decision is that $c = c_0$. In the detection problem, such an error corresponds to a miss; the signal exists, but the decision is that there is no signal.

Let us designate by α and β the corresponding error probabilities of the first and the second kind, and by $p(\mathbf{y}|c)$, the conditional probability of observing a particular signal **y** when the parameter is equal to c. Then, the error probabilities of the first and second kind are

$$\alpha = \int_{\Lambda_1} p(\mathbf{y}|c_0) \, d\mathbf{y} \tag{6.39}$$

and

$$\beta = \int_{\Lambda_0} p(\mathbf{y}|c_1) \, d\mathbf{y} \tag{6.40}$$

The total unconditional error probability is equal to

$$J = P_0 \alpha + P_1 \beta \tag{6.41}$$

Relationships (6.39) and (6.40) are basic for the definitions of various optimal decisions. A decision is optimal if it minimizes a particular criterion of optimality.

The criterion of an ideal observer (or Kotelnikov's criterion) is to minimize the unconditional error probability (6.41). The Neyman-Pearson criterion is to minimize the probability of a miss

$$J = P_1 \beta \tag{6.42}$$

for a given constant probability of false alarm

$$P_0 \alpha = \text{const} \tag{6.43}$$

160   6 Filtering

Instead of the unconditional error probability when the errors of the first and second kind are not equally costly, we can consider the criterion of optimality

$$J_\lambda = \lambda P_0 \alpha + P_1 \beta \tag{6.44}$$

where λ is a certain constant multiplier that characterizes the weight of the errors of the first kind. For $\lambda = 1$ in (6.44), we obtain Kotelnikov's criterion (6.41). If λ is considered to be a variable which is determined by (6.43) after minimizing (6.44), we have Neyman-Pearson's criterion of optimality.

6.11 A Decision Rule

Classical methods of obtaining a decision rule which minimizes a criterion of optimal signal reception (for instance, Neyman-Pearson's or Kotelnikov's criteria) are based on the computations of a posteriori or conditional probabilities using the results of signal reception. However, in order to compute a posteriori probabilities, we need a priori probabilities which are usually unknown. We either do not possess necessary statistics to estimate a priori probabilities, or such statistics have not been studied for the simple reason that similar situations did not exist in the past.

The difficulty caused by insufficient a priori information always follows the classical approach. However, such "a priori difficulty" is not insurmountable. In those cases where the estimated functions or quantities do not depend or depend little upon the form of the probability density function, we can choose any probability density function suitable for computations, and then use well-known methods based upon the Bayesian approach. Unfortunately, such cases are not frequent. They only include the problems of estimating the moments. In all the other cases, the problem is resolved by the adaptive approach.

We consider here the general problem reduced to the minimization of J_λ. By considering (6.39) and (6.40),

$$J_\lambda = \lambda P_0 \int_{\Lambda_1} p(\mathbf{y}|c_0)\, d\mathbf{c} + P_1 \int_{\Lambda_0} p(\mathbf{y}|c_1)\, d\mathbf{y} \tag{6.45}$$

or

$$J_\lambda = \lambda \int_\Lambda \{P_0 d_1(\mathbf{y}) p(\mathbf{y}|c_0) + P_1 d_0(\mathbf{y}) p(\mathbf{y}|c_1)\}\, d\mathbf{y} \tag{6.46}$$

where $d_\mu(\mathbf{y})$ ($\mu = 0, 1$) is the decision rule,

$$d_\mu(\mathbf{y}) = \begin{cases} 1 & \text{if } \mathbf{y} \in \Lambda_\mu \\ 0 & \text{if } \mathbf{y} \notin \Lambda_\mu \end{cases} \tag{6.47}$$

An analogous representation was already encountered in Sections 1.2 and 4.8.

Let us assume that the boundary between regions Λ_1 and Λ_0 is defined by

$$\hat{f}(\mathbf{y}, \mathbf{c}) = \mathbf{c}^T \boldsymbol{\phi}(\mathbf{y}) = 0 \tag{6.48}$$

where $\hat{f}(\mathbf{y}, \mathbf{c}) > 0$ in the region Λ_1, and $\hat{f}(\mathbf{y}, \mathbf{c}) < 0$ in the region Λ_2. The boundary is then determined by finding the vector $\mathbf{c}$. In order to accomplish this, we introduce the function

$$\theta(\mathbf{y}, \mathbf{c}) = \begin{cases} 1 & \text{if an error of the first kind is made} \\ \lambda & \text{if an error of the second kind is made} \\ 0 & \text{if there is no error} \end{cases} \tag{6.49}$$

Then

$$J_\lambda = \lambda P_0 \alpha + P_1 \beta = M_\mathbf{y}\{\theta(\mathbf{y}, \mathbf{c})\} \tag{6.50}$$

and to find the optimal vector or the parameters, we can use the search algorithm

$$\mathbf{c}[n] = \mathbf{c}[n-1] - \gamma[n]\tilde{\nabla}_{\mathbf{c}\pm}\,\theta(\mathbf{y}, \mathbf{c}, a) \tag{6.51}$$

which indeed solves the posed problem.

6.12 Signal Detection in Background Noise I

Since examples are frequently more educational than rules, we shall consider a detection problem on the basis of the adaptive approach. We shall consider that a signal is present when the observation $\mathbf{y}$ belongs to the region Λ_1, and that it is not present when $\mathbf{y}$ does not belong to Λ_1. We shall characterize the region Λ_1 by a vector of threshold values $\mathbf{c} = (c_1, \ldots, c_N)$. The problem is then to find the best estimate of the vector $\mathbf{c}$.

We shall first select the probability of error (either $\mathbf{y} \in \Lambda_1$, when there is no signal, or $\mathbf{y} \notin \Lambda_1$, when a signal is present) as a criterion of optimality. Such a criterion can be written as

$$J = P\{\mathbf{y} \in \Lambda_1, \mathbf{s} = 0 \text{ when } \mathbf{y} \notin \Lambda_1, \mathbf{s} \neq 0\} \tag{6.52}$$

The best estimate of $\mathbf{c}$ corresponds to the minimum of this functional. We introduce the following notation of the characteristic functions:

$$\theta(\mathbf{y}, \mathbf{c}) = \begin{cases} 1 & \text{if } \mathbf{y} \in \Lambda_1 \\ 0 & \text{if } \mathbf{y} \notin \Lambda_1 \end{cases} \tag{6.53}$$

$$y_0 = \begin{cases} 1 & \text{if } \mathbf{s} \neq 0 \\ 0 & \text{if } \mathbf{s} = 0 \end{cases} \tag{6.54}$$

and finally,

$$\Omega(z) = \begin{cases} 1 & \text{if } z = 1 \\ 0 & \text{if } z \neq 0 \end{cases} \tag{6.55}$$

The criterion of optimality (6.52) can then be given in a more convenient form:

$$J(\mathbf{c}) = P\{\theta(\mathbf{y}, \mathbf{c}) + y_0 = 1\} \tag{6.56}$$

or, considering (6.55),

$$J(\mathbf{c}) = M\{\Omega[\theta(\mathbf{y}, \mathbf{c}) + y_0]\} \tag{6.57}$$

In finding the optimal vector $\mathbf{c} = \mathbf{c}^*$ which minimizes this functional, we apply the search algorithm of learning (2.21)

$$\mathbf{c}[n] = \mathbf{c}[n-1] - \gamma[n]\tilde{\nabla}_{\mathbf{c}\pm}\Omega[\theta(\mathbf{y}[n], \mathbf{c}[n-1], a[n]) + y_0] \tag{6.58}$$

A block diagram of the receiver which operates according to this algorithm is shown in Fig. 6.11.

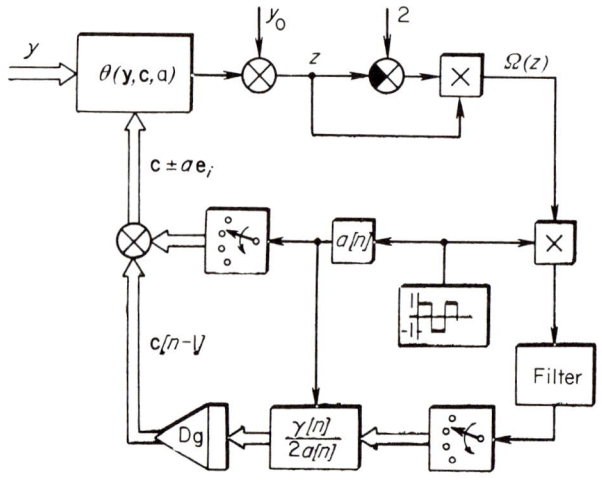

Fig. 6.11

6.13 Signal Detection in Background Noise II

Let us now use another criterion of optimality, the equality of errors of the first and the second kind:

$$P\{\mathbf{y} \in \Lambda_1, \mathbf{s} = 0\} = P\{\mathbf{y} \notin \Lambda_1, \mathbf{s} \neq 0\} \tag{6.59}$$

But, according to the formula of total probability,

$$P\{\mathbf{y} \in \Lambda_1\} = P\{\mathbf{y} \in \Lambda_1, \mathbf{s} = 0\} + P\{\mathbf{y} \in \Lambda_1, \mathbf{s} \neq 0\} \tag{6.60}$$

and

$$P\{\mathbf{s} \neq 0\} = P\{\mathbf{y} \notin \Lambda_1, \mathbf{s} \neq 0\} + P\{\mathbf{y} \in \Lambda_1, \mathbf{s} \neq 0\}$$

Therefore, equality (6.59) is equivalent to

$$P\{\mathbf{y} \in \Lambda_1\} = P\{\mathbf{s} \neq 0\} \quad (6.61)$$

which means that the probability of the decision that indicates the presence of the signal is equal to the probability that the signal is actually present.

It is not difficult now to write the criterion of optimality as

$$M_\mathbf{y}\{\theta(\mathbf{y}, \mathbf{c}) - y_0\} = 0 \quad (6.62)$$

where the characteristic functions are defined in (6.53) and (6.54). In order to determine the optimal estimate $\mathbf{c}$, we can use the algorithm

$$\mathbf{c}[n] = \mathbf{c}[n-1] - \gamma[n][\theta(\mathbf{y}[n], \mathbf{c}[n]) - y_0[n]] \quad (6.63)$$

If the signal and the threshold vectors are one-dimensional,

$$\theta(y, c) = \operatorname{sgn}(y - c) \equiv \begin{cases} 1 & \text{if } y > c \\ 0 & \text{if } y \leq c \end{cases} \quad (6.64)$$

and we obtain a simple scalar algorithm of learning

$$c[n] = c[n-1] - \gamma[n](\operatorname{sgn}(y[n] - c[n-1]) - y[_0 n]) \quad (6.65)$$

A block diagram of the receiver which operates according to such algorithms is shown in Fig. 6.12. These receivers estimate optimal threshold values by

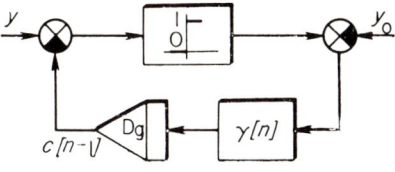

Fig. 6.12

learning and adaptation. We shall consider similar algorithms again in Chapter 10.

The detection problem is very closely related to the problem of learning in pattern recognition, however, we have decided to present it here. We hope that the reader will not consider this an inappropriate arrangement since the problems of filtering differ slightly from the pattern recognition problems.

6.14 Signal Extraction from Background Noise

In the problems of signal extraction from background noise, it is a priori known that the signal is present in the observation y, and the parameters of the signal have to be determined by processing the observed value y. We

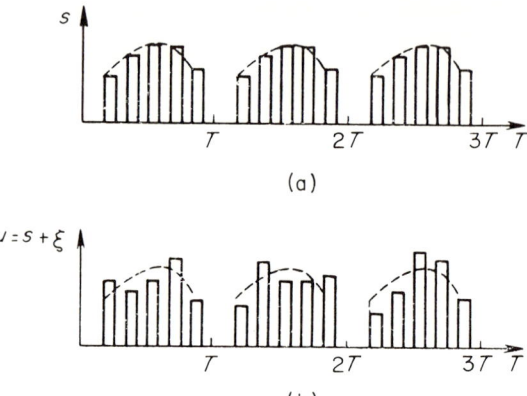

Fig. 6.13

shall consider a "constant" signal $\mathbf{s} = (s_1, \ldots, s_M)$ which represents a short sequence of impulses which are repeated with a period T (Fig. 6.13). The components of the vector $\mathbf{s}$ are equal to the amplitudes of the pulses in the sequence. At the input of the receiver, we have an action $\mathbf{u}$ which is the sum of the signal $\mathbf{s}$ and the stationary noise $\xi = (\xi_1[n], \ldots, \xi_M[n])$ dependent on the index of the sequence. The mean value of the noise is assumed to be equal to zero, and the correlation matrix B is assumed to be known. The output signal of the receiver is then a linear combination of the input signal,

$$y = \mathbf{c}^T \mathbf{u} = \mathbf{c}^T \mathbf{s} + \mathbf{c}^T \xi \tag{6.66}$$

It is a sum of the signal component $\mathbf{c}^T\mathbf{s}$ and the noise component $\mathbf{c}^T\xi$.

6.15 Criterion of Optimal Signal Extraction

As a criterion of optimality we can select the ratio of the signal power to noise power, which is briefly (but inaccurately) called the *signal-to-noise* ratio. When the signal-to-noise ratio is large, its selection is traditional. But in general, it cannot be claimed that the maximization of this ratio guarantees the greatest amount of information about the signal. It should always be remembered that the selected criterion is only good for a small noise level, i.e., for the large signal-to-noise ratio. The criterion of optimality (the signal-to-noise ratio) can be written as

$$J(\mathbf{c}) = \frac{M\{(\mathbf{c}^T \mathbf{s})^2\}}{M\{(\mathbf{c}^T \xi)^2\}} \tag{6.67}$$

6.16 Algorithm of Signal Extraction

The next problem is to determine the maximum of this expression under the conditions of constant noise power,

$$M\{(\mathbf{c}^T\boldsymbol{\xi})^2\} = \mathbf{c}^T B\mathbf{c} = A \tag{6.68}$$

6.16 Algorithm of Signal Extraction

Let us form the Lagrange function

$$J_\lambda(\mathbf{c}) = M\{(\mathbf{c}^T\mathbf{s})^2\} + \lambda(M\{(\mathbf{c}^T\boldsymbol{\xi}^2)\} - A) \tag{6.69}$$

If the signal and the noise are uncorrelated, it follows from (6.66) that

$$M\{(\mathbf{c}^T\mathbf{u})^2\} = M\{(\mathbf{c}^T\mathbf{s})^2\} + M\{(\mathbf{c}^T\boldsymbol{\xi})^2\} \tag{6.70}$$

and the Lagrange function (6.69) can be expressed through the measured and known variables $y = \mathbf{c}^T\mathbf{u}$, B, A, i.e.,

$$J_\lambda(\mathbf{c}) = M\{(\mathbf{c}^T\mathbf{u})^2 + (\lambda - 1)(\mathbf{c}^T B\mathbf{c}) - \lambda A\} \tag{6.71}$$

The gradient of the Lagrange function is

$$\nabla_\mathbf{c} J_\lambda(\mathbf{c}) = 2M\{(\mathbf{c}^T\mathbf{u})\mathbf{u} + (\lambda - 1)B\mathbf{c}\} \tag{6.72}$$

By applying the algorithms which determine the conditional extremum under the constraints of the type (3.19), we obtain

$$\begin{aligned}\mathbf{c}[n] &= \mathbf{c}[n-1] - \gamma_1[n][(\mathbf{c}^T[n-1]\mathbf{u}[n])\mathbf{u}[n] + (\lambda[n-1]-1)B\mathbf{c}[n-1]]\\ \lambda[n] &= \lambda[n-1] - \gamma_2[n][\mathbf{c}^T[n-1]B\mathbf{c}[n-1] - A]\end{aligned} \tag{6.73}$$

The diagram of the receiver which performs signal extraction on the basis of this algorithm is shown in Fig. 6.14. This diagram differs from the previous

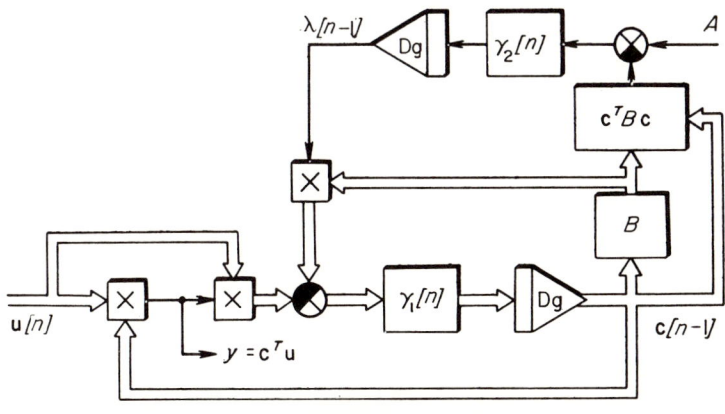

Fig. 6.14

perceptron schemes by the presence of an additional loop which guarantees that condition (6.73) is satisfied.

6.17 More about Signal Extraction from Background Noise

Let us now assume that we do not have any a priori knowledge about the signal $s(t)$. Regarding the noise, we assume that it is uncorrelated with the signal, and that it is white noise with zero mean value and with the variance equal to σ_ξ^2. The input signal to the receiver $u(t)$ is an additive mixture of the unknown signal and noise. Let us consider the receiver (based upon delay lines) which is designed to extract the signal $s(t)$. Let the input signal be defined in the following fashion:

$$y(t) = \sum_{v=0}^{2N} c_v u(t - vT) \tag{6.74}$$

where c_v are still the unknown parameters of our receiver. The output signal consists of two parts:

$$y_s(t) = \sum_{v=0}^{2N} c_v s(t - vT) \tag{6.75}$$

related to the signal, and

$$y_\xi(t) = \sum_{v=0}^{2N} c_v \xi(t - vT) \tag{6.76}$$

called noise. Since the signal $s(t)$ is not known a priori, we cannot define the desired signal $y_0(t)$ and, therefore, we cannot now use the solutions presented in Sections 6.3–6.8.

6.18 Another Criterion of Optimality

For the problem under consideration, we have to select another criterion of optimality which must take into consideration the known a priori information. Such a criterion can be

$$J = M\{y^2(t)\} - \lambda M\{y_\xi^2(t)\} \tag{6.77}$$

where λ is a constant weighting multiplier. The functional (6.77) represents the excess of power in the output signal over the noise power. This functional is maximal for an optimal receiver. Before we get involved with the maximization, we want to direct the attention of the reader to the close relationship of this criterion and the signal-to-noise ratio.

Due to the properties of noise $\xi(t)$,

$$M(y_\xi^2) = \sigma_\xi^2 \sum_{v=0}^{2N} c_v^2 \tag{6.78}$$

Therefore, (6.77) can be written as

$$J(\mathbf{c}) = M\left\{y^2(t) - \lambda\sigma_\xi^2 \sum_{\nu=1}^{2N} c_\nu^2\right\} \qquad (6.79)$$

6.19 Optimal Receiver

The condition of the extremum is obtained by differentiating (6.79) with respect to c_μ, and by setting the partial derivatives equal to zero:

$$\frac{\partial J(\mathbf{c})}{\partial c_\mu} = M\left[2y(t)\frac{\partial y(t)}{\partial c_\mu} - 2\lambda\sigma_\xi^2 c_\mu\right] = 0 \qquad (\mu = 0, 1, \ldots, 2N) \quad (6.80)$$

As it can be seen from (6.74),

$$\frac{\partial y(t)}{\partial c_\mu} = u(t - \mu T) \qquad (\mu = 0, 1, \ldots, 2N) \quad (6.81)$$

and we find from (6.80) that

$$c_\mu^* = \frac{1}{\lambda\sigma_\xi^2} M[y(t)u(t - \mu T)] \qquad (\mu = 0, 1, \ldots, 2N) \quad (6.82)$$

Equation (6.82) is very clear. It indicates that the coefficients c_μ^* are proportional to the values of the cross-correlation function of the input and output signals of the receiver,

$$c_\mu^* = \frac{1}{\lambda\sigma_\xi^2} R_{yu}(\mu T) \qquad (6.83)$$

One can obtain a system of linear equations which defines the optimal value c_μ^* ($\mu = 0, 1, \ldots, 2N$) using $y(t)$ in (6.74) and (6.80). But is this necessary?

We can use (6.82) directly, and thus avoid the tedious solutions of the systems of linear equations. The problem of finding c_μ^* is then reduced to the computations of the cross-correlation functions using sample functions. Therefore, we can use the results of Section 5.6 to estimate the cross-correlation function, and obtain the continuous algorithm

$$\frac{dc_\mu(t)}{dt} = \gamma(t)\left[\frac{1}{\lambda\sigma_\xi^2} y(t)u(t - \mu T) - c_\mu(t)\right] \qquad (6.84)$$

or, in a more convenient form,

$$T_0(t)\frac{dc_\mu(t)}{dt} + c_\mu(t) = \frac{1}{\lambda\sigma_\xi^2} y(t)u(t - \mu T) \qquad (\mu = 0, 1, \ldots, 2N) \quad (6.85)$$

where

$$T_0(t) = \frac{1}{\gamma(t)} \qquad (6.86)$$

These equations define the variations of the coefficients in the block diagram of the receiver. The diagram of the optimal adaptive receiver which defines the output signal according to equation (6.84) is shown in Fig. 6.15.

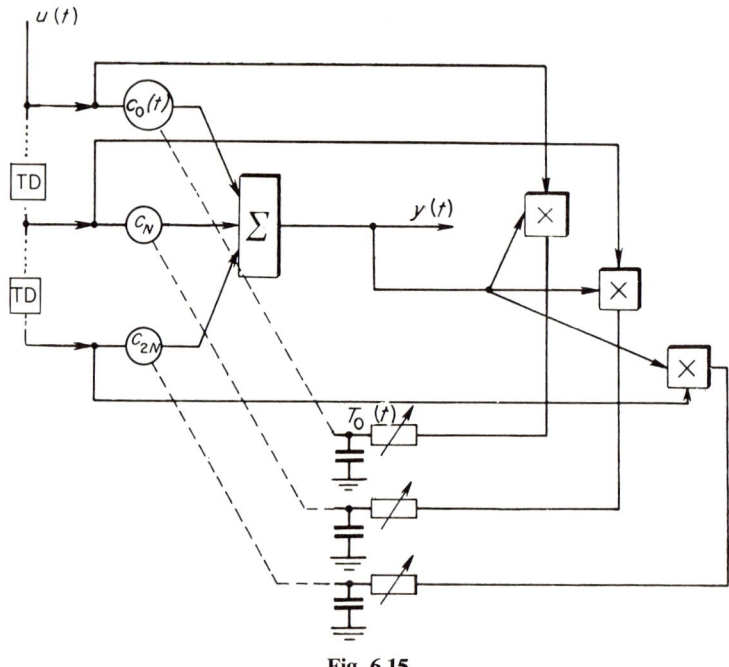

Fig. 6.15

6.20 Possible Simplifications

The adaptive receiver shown in Fig. 6.15 is converted into an optimal correlation receiver after a period of learning. Its characteristic is that the weighting coefficients are proportional to the cross-correlation function of the input and output signals. The diagram of the receiver can be simplified if instead of the usual correlation function,

$$R_{yu}(\mu T) = M\{y(t)u(t - \mu T)\} \tag{6.87}$$

we use a so-called "relay" cross-correlation function,

$$R_{\text{sign } y, u}(\mu T) = M\{\text{sign } y(t)u(t - \mu T)\} \tag{6.88}$$

which takes into consideration only the sign of the input signal. Moreover, let the time constant T_0 be invariant and define λ such that $\lambda \sigma_\xi^2 = 1$. Then instead of (6.85), we obtain

$$T_0 \frac{dc_\mu(t)}{dt} + c_\mu(t) = \text{sign } y(t)u(t - \mu T) \quad (\mu = 0, 1, \ldots, 2N) \tag{6.89}$$

6.21 Recovery of Input Signals

The diagram of the receiver which operates according to this algorithm (Fig. 6.16) differs from the previous one by the presence of an additional relay in the path of the signal, and by the time-invariant quantity T_0.

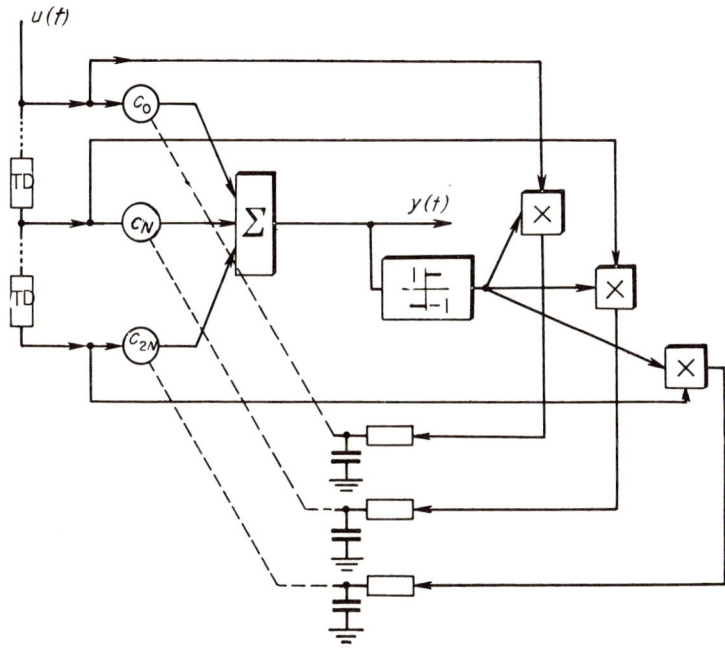

Fig. 6.16

6.21 Recovery of Input Signals

Adaptive filters are very useful in the solution of problems which recover the input signals $s(t)$ and which act at the input of a certain known system. Such a problem appears in the decoding of recorded signals which were deformed by the measuring devices. This problem is very closely related to the identification problem. Let the system be described by

$$y(t) = As(t) \tag{6.90}$$

where A is a certain nonlinear operator, and $s(t)$ is the input signal. For simplicity, we assume that y and s are the scalar quantities. The output signal cannot be measured exactly. Instead of $y(t)$, we measure

$$v(t) = y(t) + \xi \tag{6.91}$$

where ξ is a noise with the mean value equal to zero and with a finite variance.

The problem is to recover the signal $s(t)$. We assume that $s(t)$ can be accurately described by

$$\hat{s}(t) = \mathbf{c}^T \boldsymbol{\phi}(\mathbf{L}\zeta) \tag{6.92}$$

where $\mathbf{L} = (L_1, \ldots, L_N)$ is a vector of linearly independent operators, and ζ is a set of signals created by a special generator. We shall determine the optimal vector $\mathbf{c} = \mathbf{c}^*$ so that a certain functional, for instance,

$$J(\mathbf{c}) = M\{[As(t) - A\mathbf{c}^T \boldsymbol{\phi}(\mathbf{L}\zeta)]^2\} \tag{6.93}$$

is minimized. Considering (6.90) and (6.91), this functional can be written in another form:

$$J(\mathbf{c}) = M\{[v(t) - \xi - A\mathbf{c}^T \boldsymbol{\phi}(\mathbf{L}\zeta)]^2\} \tag{6.94}$$

6.22 Algorithm of Signal Recovery

The condition specifying the optimal vector of the parameters has the form

$$\nabla J(\mathbf{c}) = -2M\{[v(t) - \xi - A\mathbf{c}^T \boldsymbol{\phi}(\mathbf{L}\zeta)]A'(\mathbf{c}^T \boldsymbol{\phi}(\mathbf{L}\zeta))\boldsymbol{\phi}(\mathbf{L}\zeta)\} = 0 \tag{6.95}$$

where A' is the derivative of the operator A (in Frechet sense, for instance). Since ξ and ζ are correlated, and the mean value of ξ is equal to zero, we obtain from (6.95),

$$M\{[v(t) - A\mathbf{c}^T \boldsymbol{\phi}(\mathbf{L}\zeta)]A'(\mathbf{c}^T \boldsymbol{\phi}(\mathbf{L}\zeta))\boldsymbol{\phi}(\mathbf{L}\zeta)\} = 0 \tag{6.96}$$

and then find the algorithm for defining the optimal vector $\mathbf{c} = \mathbf{c}^*$, and thus the recovery of the input signal:

$$\frac{d\mathbf{c}(t)}{dt} = \gamma(t)[v(t) - A\mathbf{c}^T(t)\boldsymbol{\phi}(\mathbf{L}\zeta)]A'(\mathbf{c}^T\boldsymbol{\phi}(\mathbf{L}\zeta))\boldsymbol{\phi}(\mathbf{L}\zeta) \tag{6.97}$$

In the special case when the operator A is linear, algorithm (6.97) is considerably simpler and has the form

$$\frac{d\mathbf{c}(t)}{dt} = \gamma(t)[v(t) - \mathbf{c}^T(t)\boldsymbol{\psi}(t)]\boldsymbol{\psi}(t) \tag{6.98}$$

where

$$\boldsymbol{\psi}(t) = A\boldsymbol{\phi}(\mathbf{L}\zeta) \tag{6.99}$$

A diagram of the adaptive device for the recovery of the input signal according to algorithm (6.98) is shown in Fig. 6.17. By substituting discrete

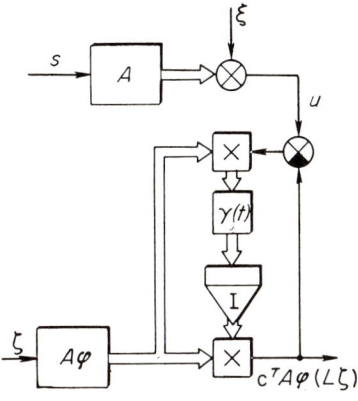

Fig. 6.17

integrators for the continuous integrators in this diagram, we can recover discrete input signals.

6.23 The Influence of Noise

The influence of the measurement errors in measuring the output signals (noise with a mean value equal to zero and with a finite variance) is averaged in time when the loss function is quadratic. This can be seen from the equivalence of the optimality conditions (6.95) and (6.96). The effect of noise removal in the process of adaptation is preserved even when the noise is present not only at the output, but also at the input of the linear adaptive filter. As it was shown in Section 5.20, noise can be removed or the unbiased estimates of the signals can be obtained even in the case of a more general loss function, but only when the input noise is not present. This case is characteristic for the measuring devices used to recover the measured signals. But what happens when the loss function is not quadratic, and the adaptive filter is nonlinear? An attempt to solve this problem is met with many difficulties. Obviously, one possible way is to use an additional linear adaptive filter to extract the signal and to reduce the noise, i.e., to obtain an estimate of the signal. The signal obtained by such smoothing is then applied at the input of a nonlinear adaptive filter for necessary transformations.

6.24 Certain Problems

The structure and properties of adaptive filters and optimal receivers for detection and extraction of signals depends considerably on the choice of the system of functions $\phi_\nu(\cdot)$. Therefore, the investigations related to the choice of these functions would be very useful.

Since the adaptive filters have to operate on variable external signals, it is important to determine a relationship between the rate of change in such external signals and the duration of the adaptation process. In order to decrease the duration of adaptation, i.e., the learning time, it would be valuable to use a priori information about the distribution of random variables. It would also be important to develop the procedures for effective usage of this a priori information for an improvement of the algorithms of adaptation.

The schemes of adaptive filters considered in this chapter frequently contain a desired signal. It is important that such a desired signal be replaced by its probabilistic characteristics. It would also be important to establish a relationship between the Bayesian approach, the statistical decision theory, and the conditional Markov processes.

6.25 Conclusion

By broadening the concept of "filtering," in this chapter we have considered different problems: filtering in the usual sense, detection, extraction and recovery of the signals which are deformed due to noise or to the properties of the devices used for signal processing. We have tried to show that the applications of the adaptive approach not only serve in the solution for a number of new problems, but that they considerably simplify the solutions of well-known classical problems (for instance, the problems of Kolmogorov-Wiener filter design). This approach permits us to look at the statistical theory of reception from an unusual viewpoint, and to develop a method for solving the problems of this theory which does not require complete a priori information about signals and noise.

COMMENTS

6.4 The algorithms of the type (6.9) are realized with the help of digital filters with adjustable weighting coefficients. The examples of such filters are described in the papers by Gersho (1966) and Prouza (1957).

The threshold functions (6.10) are broadly used by Schetzen (1965) for design of optimal Weiner filters. See also Lee and Schetzen (1965).

6.5 This problem was investigated by Lucky (1966), although the result presented here differs slightly from the one obtained by him.

6.6 The content of this section is based on the results obtained by Sakrison (1961, 1963, 1965a). It would be interesting to use the parallel methods of search developed by Fitsner (1964) in the solution of similar problems.

6.7 The reader can find the details about the filter predictor in the paper by Gabor *et al.* (1961). The theory of adaptive filter predictors which are optimal with respect to a quadratic criterion were developed by Gardner (1963a, b).

6.8 The continuous adaptive filter based on the continuous solution of the system of equations was considered by Šefl (1960).

The approach presented in this section has simplified the device of adaptation. Instead of $N(N+s)/2$ operations of multiplication, and the same number of time-averaging operations necessary to obtain the required correlation functions, only N operations of multiplication are necessary to determine the gradient. There is also no need for a device which solves a system of linear equations.

The discrete variant of the analogous filter was described earlier by Prouza (1957). Some interesting results regarding the relationship between stochastic approximation and the method proposed by Kalman are presented in the paper by Ho (1962).

6.9 and 6.10 The books by Gutkin (1961), Falkovich (1961) and Middleton (1960) are devoted to the statistical theory of reception. They also contain broad bibliographies related to these questions. In these books, it is assumed that a priori information is sufficiently complete.

6.11 A similar approach was also described by Leonov (1966).

6.13 Algorithm (6.65) was first obtained for $\gamma = $ const by Kac (1962) in a completely different manner. This algorithm corresponds to the criterion of optimality in which the probabilities of the first and the second kind are equal. Such a criterion is applicable only when these errors are small. Therefore, the algorithm without search proposed by Kac is obtained at a rather high cost. Another possibility of detecting signals with the help of adaptive receivers was described by Jacowatz *et al.* (1961) and Groginsky *et al.* (1966). See also the paper by Tong and Liu (1966). A survey about threshold detectors has been written by Levin and Bonch-Bruyevich (1965).

6.14–6.16 This problem was posed by Kushner (1963). We have chosen the same criterion of optimality as that used by him. Although Kushner considered the problem of optimal detection using the stochastic approximation method, the solution presented here is different from Kushner's solution. It should be noted that our attempts to establish a relationship between these solutions were not successful. We were also neither able to understand the method of solution nor the result obtained by Kushner.

6.20 It is interesting to note that the scheme of the adaptive receiver obtained by us is identical to the scheme proposed by Morishita (1965) for extraction of narrow band signals from background noise, or from other narrow band signals of smaller power. However, the last possibility was not theoretically established by the author. We were also not able to establish such a property for the considered adaptive receiver.

6.21 and 6.22 The road toward the design of adaptive devices for the restoration of input signals is very closely related to the analytical theory of nonlinear systems originated by Wiener (1949), and also with the methods of signal estimation developed by Kalman (1960).

BIBLIOGRAPHY

Chelpanov, I.B. (1963). Design of optimal filters under incomplete statistical properties of the signals. *In* "Tekhnicheskaia Kybernetika." Academy of Sciences USSR, Moscow. (In Russian.)

Falkovich, S.E. (1961). "Reception of Radar Signals in Noise." Sovetskoe Radio, Moscow. (In Russian.)

Fitsner, L.N. (1964). Automatic optimization of complex distributive systems, *Automat. Remote Contr. (USSR)* **25** (5).

Gabor, D., Wilby, W.P.Z., and Woodcock, R. (1961). An universal nonlinear filter, predictor and simulator which optimizes itself by a learning process, *Proc. IEE* **108** (40).

Gardner, L.A. (1963a) Adaptive predictors. *In* "Transactions of the Third Prague Conference on Information Theory, Statistical Decision Functions, Random Process, Prague, 1963."

Gardner, L.A. (1963b). Stochastic approximation and its application to problems of prediction and control systems. *In* "International Symposium on Nonlinear Differential Equations and Nonlinear Mechanics." Academic Press, New York.

Gersho, A. (1966). Automatic time-domain equalization with transversal filters, *Proc. Nat. Electron. Conf.* **22**.

Glaser, E.M. (1961). Signal detection by adaptive filters, *IRE Trans. Inform. Theory* **IT-7** (2).

Groginsky, H.L., Wilson, L.R., and Middleton, D. (1966). Adaptive detection of statistical signal in noise, *IEEE Trans. Inform. Theory* **IT-12** (3).

Gutkin, L.S. (1961). "Theory of Optimal Radio Receivers." Gosenergoizdat, Moscow. (In Russian.)

Hinich, M.T. (1962). A model for a self-adapting filter, *Inform. Contr.* **5** (3).

Ho, Y.C. (1962). On stochastic approximation and optimal filtering methods, *J. Math. Anal. Appl.* **6** (1).

Jakowatz, C.K., Shuey, R.L., and White, G.M. (1961). Adaptive waveform recognition. *In* "Information Theory, Fourth London Symposium" (C. Cherry, ed.). London.

Kac, M. (1962). A note on learning signal detection, *IRE Trans. Inform. Theory* **IT-8** (2).

Kalman, R. (1960). New approach to the linear filtering and prediction problems, *J. Basic Eng.* **82** (1).

Kalman, R., and Bucy, R.S. (1961). New result in linear filtering and prediction theory, *J. Basic Eng.* **83** (1).

Kolmogorov, A.N. (1941). Interpolation and Extrapolation of Stationary Random Series, *Akad. Nauk SSSR Ser. Mat.* **5** (1). (In Russian.)

Kushner, H. (1963). Adaptive techniques for the optimization of binary detections signals, *IEEE Inter. Conv. Rec.* **4**.

Lee, Y.W., and Schetzen, M. (1965). Some aspects of the Wiener theory of nonlinear systems, *Proc. Nat. Electron. Conf.* **21**

Leonov, Yu.P. (1966). Classification and statistical hypothesis testing, *Automat. Remote Contr. (USSR)* **27** (12).

Levin, G.A., and Bonch-Bruyevich, A.M. (1965). The ways to design detectors with self-optimizing threshold level in the receivers, *Elektrosvyaz* **19** (5). (In Russian.)

Lucky, R.W. (1966). Techniques for adaptive equalization of digital communication systems, *Bell Syst. Tech. J.* **45** (2).

Middleton, D. (1960). "An Introduction to Statistical Communication Theory." McGraw-Hill, New York.

Middleton, D. (1965). "Topics in Communication Theory." McGraw-Hill, New York.

Morishita, I. (1965). An adaptive filter for extracting unknown signals from noise, *Trans. Soc. Instrument Contr. Eng.* **1** (3).

Prouza, L. (1957). Bemerkung zur linear Prediktion mittels eines lernenden Filters. *In* "Transactions of the First Prague Conference on Information Theory." Czechoslovak Academy of Sciences, Prague.

Sakrison, D.T. (1961). Application of stochastic approximation methods to optimal filter design, *IRE Inter. Conv. Rec.* **9**.

Sakrison, D.T. (1963). Iterative design of optimum filters for non-meansquare error performance criteria, *IEEE Trans. Inform. Theory* **IT-9** (3).

Sakrison, D.T. (1965a). Efficient recursive estimation of the parameters of a radar or radio astronomy target, *IEEE Trans. Inform. Theory* **IT-12** (1).

Sakrison, D.T. (1965b). Estimation of signals containing unknown parameters; comparison of linear and arbitrary unbiased estimates, *J. SIAM Contr.* **13** (3).

Schalkwijk, J.P.M., and Kailath, T. (1966). A coding scheme for additive noise channels with feedback—Part 1; No bandwidth constrain. Part II: Band-limited signals, *IEEE Trans. Inform. Theory* **IT-12** (2).

Schetzen, M. (1965). Determination of optimum nonlinear systems for generalized error criterion based on the use of gate functions, *IEEE Trans. Inform. Theory* **IT-11** (1).

Scudder, H.T. (1965). Adaptive communication receivers, *IEEE Trans. Inform. Theory* **IT-11** (2).

Šefl, O. (1960). Filters and predictors which adapt their values to unknown parameters of the input process. *In* "Transactions of the Second Prague Conference on Information Theory." Czechoslovak Academy of Sciences, Prague.

Tong, P.S., and Liu, B. (1966). Application of stochastic approximation techniques to two signal detection problem, *Proc. Nat. Electron. Conf.* **22**.

Wiener, N. (1949). "Extrapolation, Interpolation and Smoothing of Stationary Time Series with Engineering Applications." MIT Press, Cambridge, Mass.

Wiener, N. (1958). "Nonlinear Problems in Random Theory." MIT Press, Cambridge, Mass.

7

CONTROL

7.1 Introduction

Until now we have studied only the class of open-loop systems. This explains the relative simplicity of the solutions of pattern recognition, identification and filtering problems. We shall now consider a more complicated class of systems—those of automatic control. The automatic control systems described in this chapter are closed-loop nonlinear systems which cannot be reduced to open-loop systems. First we shall select a group of problems which cannot be solved by the ordinary methods of strong negative feedback. These problems will then be solved by using the adaptive approach, which becomes traditional for this book. It will further be shown that a number of ordinary and realistic constraints, before unmanageable by the classical methods, can now be taken into consideration. The reader will find that the theory of sensitivity plays an important role in the realization of adaptive control systems without search. This chapter is the most complicated and therefore the most interesting one in this book.

7.2 When Is Adaptation Necessary?

The basic principle underlying the design of automatic control systems is the principle of control by error signals or the principle of negative feedback. If an error signal reflects different uncontrollable external actions (disturbances), then it can serve as a measure of deviation of the system from a prescribed regime for which such an error signal would not exist. It was noticed long ago, first in radio engineering and then in automatic control, that an increase of the corresponding gain coefficients in the closed loops can reduce the influence of the external uncontrollable disturbances and variations in the characteristics of the controlled object (plant characteristics).

Many results of control theory are related to this fact. The effect of strong negative feedback can be obtained either by a direct increase of the gain coefficients or, indirectly by creating the so-called sliding modes in the relay systems of automatic control and in the systems with variable structure. This effect can also be found in the ideas of absolute invariance but in a slightly camouflaged form. Why then cannot such an effect of strong negative feedback sometimes remove the influence of external disturbances and plant variations? The answer to this question has been known for a long time. The remarkable properties of strong negative feedback can only be used when the stability of the closed-loop systems is guaranteed simultaneously with the increase of the gain coefficients. However, this cannot always be achieved.

The presence of time delays, inertias, and occasionally of nonlinearities, does not permit an unlimited increase of gain coefficients since that can destroy the stability of the system. Without an increase in a priori information, it is not possible to increase the gain coefficients and not affect the stability of the system at the same time. For some reason, this is not often obvious. However, with the larger gain coefficients, certain parameters and nonlinearities which were not considered earlier in the analysis of the system could become more apparent. But if they are unknown, then the boundaries of stability could easily be crossed. This tool, which is effective under deterministic conditions, cannot be used when a priori information is insufficient due to the existence of the "barriers of stability."

Some other possibilities based on the effects of compensation and the broadly advertised theory of invariance are also inapplicable, but for a different reason. They are based on the complete knowledge of the plant characteristics and these are frequently unknown. If it is also assumed that the unknown plant characteristics can vary with time, then one can easily see the difficulties which confront these ordinary methods when the problem is to control in an optimal fashion.

In all these cases for which the insufficiency of a priori information is characteristic, the ordinary classical methods are not applicable and we are forced to use adaptation.

It should also be mentioned that in a number of cases it is worthwhile to employ adaptation for obtaining a priori information through the experiments, and later to apply the usual classical methods.

7.3 Statement of the Problem

As we have seen in Chapter 5, the controlled objects (plants) can be described by various equations. Here we accept the description in the form of a system of nonlinear difference equations in vector form

$$\mathbf{x}[n] = \mathbf{f}(\mathbf{x}[n-1], \mathbf{u}[n-1]) \qquad (7.1)$$

where $\mathbf{f}(x, u) = (f_1(x, u), \ldots, f_l(x, u))$ is an l-dimensional vector of unknown functions, $\mathbf{x} = (x_1, \ldots, x_l)$ is the state vector of the plant, and $\mathbf{u} = (u_1, \ldots, u_l)$ is the vector of control actions.

The controller is characterized by the control law in the general form

$$\mathbf{u}[n] = \mathbf{k}(\mathbf{x}[n]) \qquad (7.2)$$

where $\mathbf{k}(x) = (k_1(x), \ldots, k_{l_1}(x))$ is an l_1-dimensional vector of unknown functions. The state variables and the control actions can (or must) satisfy certain additional constraints. For instance,

$$h_v^I(\mathbf{x}[n]) = 0 \qquad (v = 1, \ldots, M_I) \quad (7.3)$$

and

$$h_v^{II}(\mathbf{u}[n]) \leq 0 \qquad (v = 1, \ldots, M_{II}) \quad (7.4)$$

The basic problem consists in finding the control law (7.2) such that the state variables and the control actions satisfy the constraints, and that the given criterion of optimality is minimized.

Let the desired state vector of the system be a stationary random vector $\mathbf{x}^0[n]$. The functional

$$J_1 = M\{F(\mathbf{x}^0[n] - \mathbf{x}[n])\} \qquad (7.5)$$

where $F(\cdot)$ is a certain complex function, can then serve as a performance index.

This problem differs from the problems in optimal system theory (both deterministic and stochastic). Here the plant equations are not known and a priori information is insufficient to compute the optimal control law in advance. Within the scope of the theory of optimal systems this problem is not only unsolvable, but its formulation in such a form is also awkward. The adaptive approach provides the solution to the problem. The solution is related to the simultaneous identification and control of the plant.

7.4 Dual Control

Insufficient a priori information necessitates a certain combination of plant identification and control. We cannot control a plant in an optimal fashion without knowing its characteristics, but we can study a plant by controlling it. This can then give us an opportunity to improve the control and eventually to reach an optimal control law. In this situation the control actions have a dual character. They serve as the tools for investigation, plant identification, and also as the tools for steering the plant to a desired (i.e., optimal) state. Such a control in which the control actions have a dual character is called dual control.

In dual control systems there is always a conflict between the identification and the control aspects of a control action. Successful control is possible if the properties of a plant are well-known and the controller can quickly react to the variations in the state of the plant. However, each plant identification requires a certain definite period of time. One could then expect that a short period of control without sufficient information about the properties of the plant followed by a more careful control based on gathered information (even when such a control is no longer necessary) can lead to a successful result.

Duality of knowledge and control, as it was emphasized by Shannon, is closely related to the duality of the past and the future. It is possible to possess knowledge of the past, but we cannot control the past. It is also possible to control the future without knowing the future. In this lies the charm and the meaning of control.

Dual control was discovered and considerably developed by Feldbaum on the basis of statistical theory. It is obvious that such an approach is the best one in those situations when a priori probability density functions of the external disturbances and the plant parameters are given, and when the performance index is the average risk. Unfortunately, this way of obtaining the solution is frequently so complex that it could be used only in certain simple cases. Since the lack of a priori information is also related to the probability density functions, it is then logical to search for some other ways to solve the dual control problems which do not require knowledge of a priori probability density functions. The reader may guess that one of such possibilities is related to the application of adaptation.

7.5 Algorithms of Dual Control

Let us start with the familiar problem of learning what the plant is or, as it is called in Chapter 5, the problem of identification. This problem is reduced to one of minimizing an index of the quality of approximation to the plant characteristics. For a plant described by a system of equations, this index is given in the form of (5.51)

$$J_2(\mathbf{c}) = M\{F_2(\mathbf{x}[n] - \Phi(\mathbf{x}[n-1], \mathbf{u}[n-1])\mathbf{c})\} \qquad (7.6)$$

As we already know, the minimization of $J_2(\mathbf{c})$ is achieved either with the help of the search algorithm of identification (5.52),

$$\mathbf{c}[n] = \mathbf{c}[n-1] - \gamma_2[n]\tilde{\nabla}_{\mathbf{c}\pm} F_2(\mathbf{x}[n], \mathbf{u}[n-1], \mathbf{c}[n-1], a[n]) \qquad (7.7)$$

or by using the algorithm without search (5.53),

$$c[n] = c[n-1] - \gamma_2[n]\nabla_c F_2(x[n] - \Phi(x[n-1], u[n-1])c[n-1]) \quad (7.8)$$

In algorithms (7.7) and (7.8), which could be called the algorithms of "learning," the control action is not arbitrary. It is defined by the control law which is acting in the system at that particular instant. Let us write this control law in the general form

$$u[n] = \hat{k}(x[n], b) \quad (7.9)$$

The right-hand side in (7.9) is a known scalar function of the vector of unknown parameters, b. In particular, the function $u[n]$ can have the form

$$u[n] = \Psi(x[n])b \quad (7.10)$$

Here, $\Psi = \|\psi_{\nu\mu}\|$ ($\nu = 1, \ldots, N_1$; $\mu = 1, \ldots, l_1$) is an $N_1 \times l_1$ matrix of linearly independent functions $\psi_{\nu\mu}$ and b is an M_1-dimensional vector of unknown coefficients in the controller. The choice of $\hat{k}(x[n], b)$ or $\psi_{\nu\mu}$ depends on the realizability of the system. Very often a decisive factor is the simplicity of control or the existence of standard elements used as the building blocks for the controllers. The criterion of optimality (7.5) can now be written in the explicit form

$$J_1(b, c) = M\{F_1(x^0[n] - \Phi(x[n-1], \Psi(x[n-1])b)c)\} \quad (7.11)$$

The optimal vector $b = b^*$ corresponds to the minimum $J_1(b, c)$. Naturally, this optimal vector b^* depends essentially on the vector c which defines the plant characteristics. This is emphasized by the arguments of the functional $J(b, c)$. In finding the optimal vector b^*, we can proceed in the same manner as in determining the optimal vector of the plant characteristic c^*. The minimization of the function (7.11) is obtained either with the help of the search algorithm,

$$b[n] = b[n-1] - \gamma_1[n]\tilde{\nabla}_{b\pm} F_1(x^0[n], x[n], b[n-1], a[n], c[n-1]) \quad (7.12)$$

or using the algorithm without search,

$$b[n] = b[n-1] - \gamma_1[n]\nabla_b F_1(x^0[n] - \Phi(x[n-1],$$
$$\Psi(x[n-1])b[n-1])c[n-1]) \quad (7.13)$$

The values $c[n-1]$ are determined by algorithm (7.7) or (7.8). It is appropriate to name (7.12) and (7.13) the algorithms of "control." In the algorithm of control (7.13) for determining optimal value b it is necessary to know the gradient of the performance index J_1 with respect to c.

The algorithms of "learning" and "control" are closely related to each other, i.e., they depend on each other. This is proof that the processes of identification and control are inseparable, and that is the essence of dual control.

7.6 Adaptive Control Systems I

Let us use the search algorithms of dual control (algorithm of learning (7.7) and algorithm of control (7.12)) for determining the block diagram of an adaptive control system. We have already noticed (see Section 3.6) that an estimate of the gradient based on divided differences can be obtained by synchronous detection. The block diagram of the adaptive system is given in Fig. 7.1. As it is not difficult to see, the block diagram represents a unification

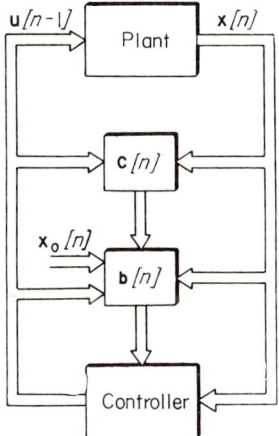

Fig. 7.1

of a plant identification scheme and an adaptive filter which were based on the search algorithms of dual control. In this adaptive system, each new value of the state of the plant $x[n]$ causes a variation in the parameters $c[n]$ and $b[n]$. Some other strategies of learning and control are also possible. For instance, the variations of $c[n]$ and $b[n]$ can be introduced after each series of l steps (where l is the assumed order of the system, i.e., the number of essential state variables), or after a certain number of steps when it can be assumed with certainty that we have found actual optimal values of the vectors b^* and c^* under given conditions.

7.7 Adaptive Control Systems II

If it is impossible or for some reason undesirable to apply the method of search, an adaptive control system without search can then be designed on the basis of the algorithms of dual control (7.8) and (7.13).

Let us consider the algorithm of control (7.13) in more detail. It could be noticed that the gradient $\nabla_b F(\cdot)$ contains not only the known functions

$$\frac{\partial F(\cdot)}{\partial x_i} \quad \frac{\partial \phi_{\mu\nu}}{\partial x_i} \quad \frac{\partial \psi_{\nu\mu}}{\partial x_i} \quad \frac{\partial \psi_{\nu\mu}}{\partial u_i}$$

(since the functions $F(\cdot)$, $\phi_{\mu\nu}$, and $\psi_{\nu\mu}$ are known), but that it also includes the unknown functions

$$\frac{\partial x_\mu}{\partial b_\nu} = v_\nu{}^\mu \quad (\nu = 1, \ldots, N_1; \mu = 1, \ldots, l_1) \quad (7.14)$$

which are the partial derivatives of the plant output with respect to the parameters **b**. These functions are called sensitivity functions. They characterize the influence of the variations in system parameters upon the processes in the system, and can be obtained by the special sensitivity models. We shall consider them briefly later.

In order to explain the basic idea and not complicate its essence by the details (which in time could become even essential), we shall limit ourselves to the linear plants described by the difference equations of the form

$$x[n] = \sum_{m=1}^{l} a_m x[n-m] + hu[n-1] \quad (7.15)$$

Introducing the vector of coefficients,

$$\mathbf{c} = (a_1, \ldots, a_l; h) \quad (7.16)$$

and the vector of situations,

$$\mathbf{z}[n] = (x[n-1], \ldots, x[n-1]; u[n-1]) \quad (7.17)$$

we write an estimate of $x[n]$ as

$$x[n] = \mathbf{c}^T \mathbf{z}[n] \quad (7.18)$$

We assume that the control law is linear,

$$u[n-1] = \sum_{\mu=1}^{l_1} b_\mu x[n-\mu] \quad (7.19)$$

or, briefly,

$$u[n-1] = \mathbf{b}^T \mathbf{Y}[n] \quad (7.20)$$

where

$$\mathbf{b} = (b_1, \ldots, b_{l_1}) \quad (7.21)$$

7.7 Adaptive Control Systems II

is the vector of the controller parameters, and

$$\mathbf{Y}[n] = (x[n-1], \ldots, x[n-l_1]) \qquad (7.22)$$

is the vector of input variables for the controller.

Plant identification is performed with the help of the algorithm of learning,

$$\mathbf{c}[n] = \mathbf{c}[n-1] + \gamma_2[n]F_2'(x[n] - \mathbf{c}^T[n-1]\mathbf{z}[n])\mathbf{z}[n] \qquad (7.23)$$

which minimizes the functional of the quality of identification

$$J_2(c) = M\{F_2(x[n] - \mathbf{c}^T\mathbf{z}[n])\} \qquad (7.24)$$

If we assume that

$$d\mathbf{c}[n]/d\mathbf{b}[n] \approx 0$$

then the algorithm of control is of the following form:

$$\mathbf{b}[n] = \mathbf{b}[n-1] + \gamma_1[n]F_1'(x^0[n] - \mathbf{c}^T[n-1]\mathbf{z}[n])V_b[n]\mathbf{c}[n] \qquad (7.25)$$

where

$$V_b[n] = \begin{pmatrix} \dfrac{\partial x[n-1]}{\partial b_1[n-1]}, & \cdots, & \dfrac{\partial x[n-l]}{\partial b_1[n-1]}, & \dfrac{\partial u[n-1]}{\partial b_1[n-1]} \\ \vdots & & \vdots & \vdots \\ \dfrac{\partial x[n-1]}{\partial b_{l_1}[n-1]}, & \cdots, & \dfrac{\partial x[n-1]}{\partial b_{l_1}[n-1]}, & \dfrac{\partial u[n-1]}{\partial b_{l_1}[n-1]} \end{pmatrix} \qquad (7.26)$$

is an $(l_1 \times l_1 + 1)$-dimensional sensitivity matrix.

If we introduce the matrices

$$V_{bl}[n] = \left\| \dfrac{\partial x[n-v]}{\partial b_\mu[n-1]} \right\| \qquad (v = 1, \ldots, l; \ \mu = 1, \ldots, l_1)$$

$$V_{bl_1}[n] = \left\| \dfrac{\partial x[n-v]}{\partial b_\mu[n-1]} \right\| \qquad (v = 1, \ldots, l_1; \ \mu = 1, \ldots, l_1) \qquad (7.27)$$

and the vector $\mathbf{a} = (a_1, \ldots, a_l)$, then algorithm (7.27) can be written as

$$\mathbf{b}[n] = \mathbf{b}[n-1] + \gamma_1[n]F_1'(x^0[n] - \mathbf{c}^T[n-1]\mathbf{z}[n]) \\ \times (V_{bl}[n]\mathbf{a}[n] + h[n]V_{bl_1}[n]\mathbf{b}[n-1] + h[n]\mathbf{Y}[n]) \qquad (7.28)$$

The block diagram of an adaptive control system based on algorithms (7.25) and (7.28) for the case of $l - l_1 = 2$ is given with all the details in Fig. 7.2. This scheme includes the sensitivity model which will be discussed in the next section.

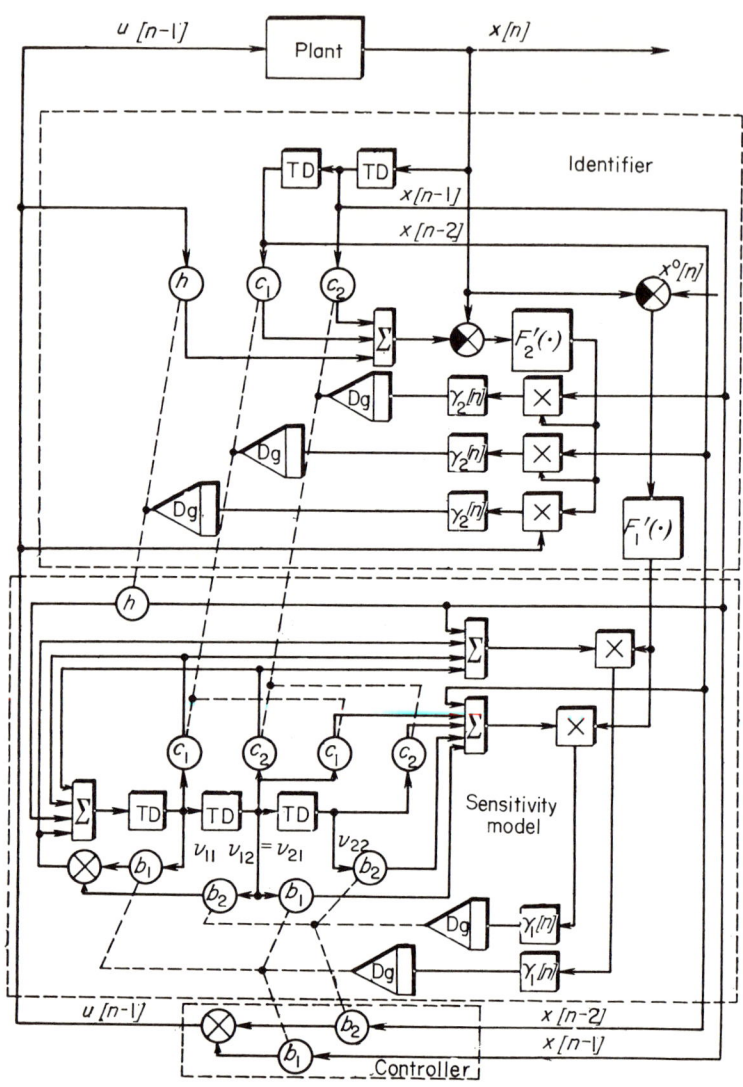

Fig. 7.2

7.8 Sensitivity Model

The sensitivity model offers a direct way to determine the sensitivity functions $v_v = \partial x/\partial a_v$. The input to the model is the plant output, and the signals measured at different points in the model correspond to the sought sensitivity functions.

7.8 Sensitivity Model

In order to determine the block diagram of the sensitivity model, we differentiate both sides of the plant equation (7.15) with respect to a_v. We then obtain an equation related to the sensitivity functions:

$$u_v[n] = \sum_{m=1}^{l} a_m u_v[n-m] + x[n-v] \tag{7.29}$$

This equation is similar to the plant equation (7.15); the input $x[n]$ in (7.15) corresponds to the sensitivity function in (7.29), and the control action $hu[n-1]$ corresponds to the plant output delayed by v time intervals. The block diagram of the sensitivity model described by equation (7.29) is presented in Fig. 7.3a. It is a discrete filter with the input signal $x[n-v]$, and the

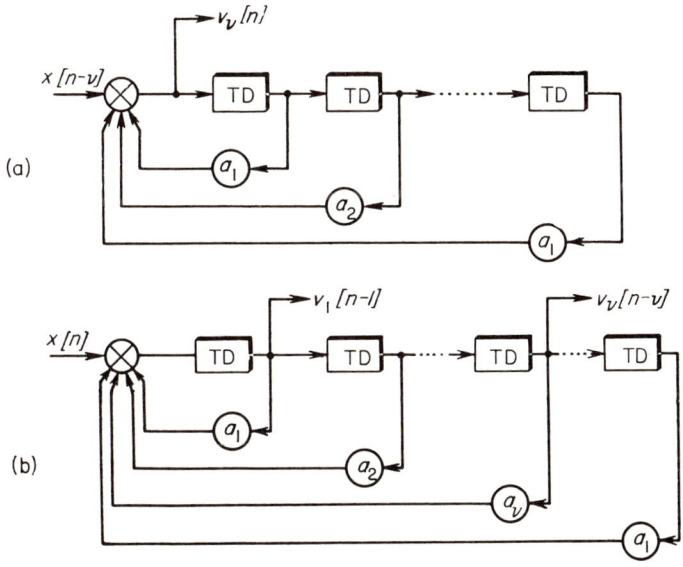

Fig. 7.3

error represents a corresponding sensitivity function. This block diagram of the sensitivity model is similar to the block diagram of the model of a plant described by linear difference equations. In general, l sensitivity models are needed for obtaining sensitivity functions $v_v[n]$, ($v = 1, \ldots, l$). However, if the plant is assumed to be linear with constant coefficients, the equations of sensitivity are also linear with constant coefficients, and only one model can be used for obtaining all sensitivity functions. This possibility is obvious from Fig. 7.3b. The input signal to the discrete filter is the plant output $x[n]$, and the signals which correspond to the sensitivity functions $v_v[n-m]$ are measured through a delay line.

Similarly, we can construct the sensitivity models for nonlinear plants. But in that case we are not able to use a single model as in the case of a linear plant. If it were possible to provide the conditions under which the input and output signals of the plant are decoupled or independent, then we could use the plant above as a sensitivity model.

7.9 Adaptive Control Systems III

Let us assume that the vector **c** of "uncontrollable" parameters is randomly varying, and that we can act upon the vector of "control" parameters **b** so as to minimize the influence of the uncontrollable parameters which cause a deviation of the system from the normal regime of operation. If these vectors were not varying, we could define a certain performance index using well-known results of automatic control theory,

$$J^0 = S(\mathbf{c}, \mathbf{b}) \qquad (7.30)$$

In the case of stationary random variations of the vector **c**, the performance index J^0 is also a random variable; therefore, we shall try to find a control, i.e., vector **b**, for which the mathematical expectation of the performance index

$$J(\mathbf{b}) = M\{S(\mathbf{c}, \mathbf{b})\} \qquad (7.31)$$

is minimized.

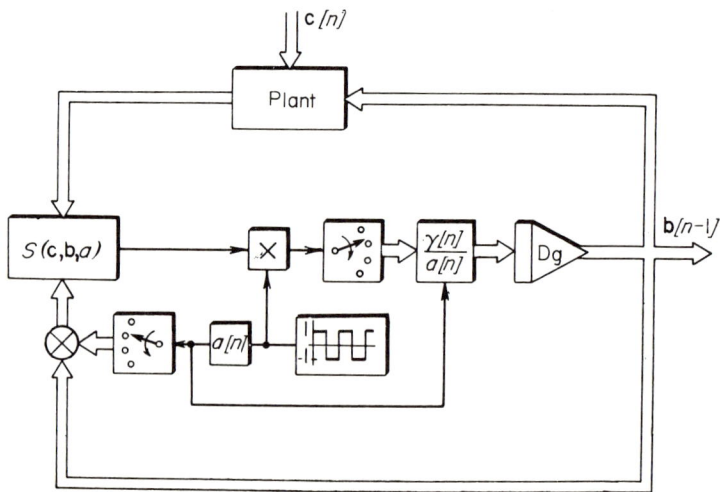

Fig. 7.4

The algorithm of control could be a search algorithm, for instance,

$$\mathbf{b}[n] = \mathbf{b}[n-1] - \gamma[n]\tilde{\nabla}_{\mathbf{b}+}S(\mathbf{c}[n], \mathbf{b}[n-1], a[n]) \qquad (7.32)$$

or an algorithm without search,

$$\mathbf{b}[n] = \mathbf{b}[n-1] - \gamma[n]\nabla_b S(\mathbf{c}[n], \mathbf{b}[n-1]) \qquad (7.33)$$

Here, $\mathbf{c}[n]$ is a realization of the random vector $\mathbf{c}$.

The block diagram which corresponds to algorithm (7.32) is shown in Fig. 7.4. In order to reduce the adaptation time, we can construct another internal loop of search for selecting the optimal values of the quantity $a[n]$ which in this case has the form $a[n] = A/n^{\alpha}$.

7.10 Simplified Adaptive Systems

Adaptive systems without search, which use the sensitivity models, as we have seen in Section 7.6, are very complicated even in the simplest cases. But the adaptive systems can be simplified if we do not attempt to identify the plant, i.e., if we remove the plant identification algorithm. This can be accomplished in the following way. We shall request that the system "imitates" another system which is considered to be a model reference. Naturally, the choice of this model reference system is based on specific a priori information if we ask only that the properties of the adaptive system approach a model reference system. But let us assume that the model reference is in some way chosen. This means that we have created a model of a desired system. Then this model can serve as a teacher that trains our system to have properties similar to its own. The block diagram of such a simplified adaptive system (or a model reference adaptive system) is shown in Fig. 7.5. From this block diagram, it is apparent that our system plays the role of a perceptron which identifies the model—the desired system. The quality of such an adaptive system is lower than that of the systems considered in Section 7.7. This verifies that the simplicity is not obtained without a price.

7.11 Systems of Control by Perturbation

In these systems the control is accomplished on the basis of disturbance measurements which cause the deviation of the state of the system from a desired one. The control by perturbation, if this perturbation can be measured, is a very effective way of improving the dynamic characteristics of the control systems. An extension of the adaptive approach to this type of system does not present any difficulties, and the reader can develop the algorithms of control for the cases of his interest when the need occurs.

We shall now touch upon a very interesting problem of control by perturbation of the so-called overdetermined plants. The number of controlled

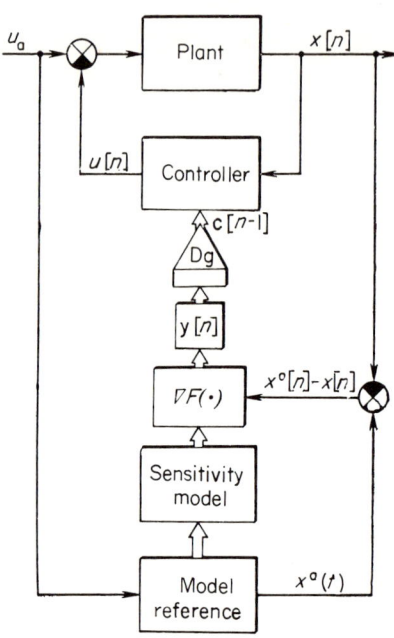

Fig. 7.5

variables (states) in these plants exceeds the number of control actions. The equation of overdetermined plants is written as

$$y = f(x, u) \tag{7.34}$$

where

$$x = (x_1, \ldots, x_{l_1}) \tag{7.35}$$

is an l_1-dimensional vector of perturbations,

$$u = (u_1, \ldots, u_{l_2}) \tag{7.36}$$

is an l_2-dimensional vector of control actions and

$$y = (y_1, \ldots, y_l) \tag{7.37}$$

is an l-dimensional vector of output variables.

For overdetermined systems, $l_1 > l_2$. Therefore, it is not possible to guarantee the maintenance of all the output variables on the desired levels. However, it is possible to keep all the output variables within certain limits,

$$a_k' \le y_k \le a_k'' \qquad (k = 1, \ldots, l) \tag{7.38}$$

where a_k' and a_k'' are preselected on the basis of technical constraints imposed on the plant under consideration.

In order to accomplish this, we use the following control law based on perturbation:

$$\mathbf{u} = \Phi(\mathbf{x})\mathbf{c} \tag{7.39}$$

where $\Phi(\mathbf{x}) = \|\phi_{\mu\nu}(\mathbf{x})\|$ ($\mu = 1, \ldots, M$; $\nu = 1, \ldots, N$) is the matrix of known functions $\phi_{\mu\nu}(\mathbf{x})$. Equation (7.39) defines the structure of the controller.

7.12 Algorithms of Optimal Control

The probability that the kth component of the vector lies within the permissible limits defined by the interval $[a_k', a_k'']$ depends on the choice of the parameter vector $\mathbf{c}$, i.e.,

$$P_k(\mathbf{c}) = P\{y_k \notin [a_k', a_k'']\} \tag{7.40}$$

where

$$y_k = f_k(\mathbf{x}, \Phi(\mathbf{x})\mathbf{c})$$

The quality of control is completely defined by the vector function $P(\mathbf{c}) = (P_1(\mathbf{c}), \ldots, P_l(\mathbf{c}))$, and the problem of optimal control of an overdetermined plant can be formulated as a problem of finding such a vector $\mathbf{c} = \mathbf{c}^*$ for which

$$J(\mathbf{c}) = M_k\{P_k(\mathbf{c})\} \tag{7.41}$$

attains the minimum. Here k is a random index which takes the value of the index of one of the components of the output vector $\mathbf{y}$ which falls outside the permissible limits. The rules by which only one of the components which simultaneously fall outside the permissible limits is chosen may be different. For instance, if the components with the indices $\mu_1, \ldots, \mu_s$ are simultaneously outside the permissible limits, then the random index k can take one of these values with probability $1/s$. If all the components lie within the permissible limits, at that instant of time the random index k is considered to have the value zero. The probability distribution of k is completely determined by the probability distributions of the components of the output vector $\mathbf{y}$, and by the rule for selecting one of a number of components which at the given instant of time fall outside the permissible limits.

We introduce our well-known characteristic function,

$$\theta_k(\mathbf{c}) = \begin{cases} 0 & \text{if } y_k \notin [a_k', a_k''] \\ 1 & \text{if } y_k \in [a_k', a_k''] \end{cases} \tag{7.42}$$

and the criterion of optimality (7.41) can be written as

$$J(\mathbf{c}) = M_{\theta, k}\{\theta_k(\mathbf{c})\} \tag{7.43}$$

where mathematical expectation is taken with respect to θ and k. For finding the optimal value $\mathbf{c} = \mathbf{c}^*$ which minimizes $J(\mathbf{c})$, we use the search algorithm of adaptation. Then

$$\mathbf{c}[n] = \mathbf{c}[n-1] - \gamma[n]\tilde{\nabla}_{\mathbf{c}^+}\theta_m(\mathbf{c}[n-1], a[n]) \qquad (7.44)$$

where m is the value of the random index at the nth step.

If only one of the components in the output vector can fall outside the permissible boundaries at each instant of time, then the criterion of optimality (7.43) can be written as

$$J(\mathbf{c}) = M_k\{P_k(\mathbf{c})\} = \sum_{v=1}^{l} P_v^2(\mathbf{c})$$

or

$$J(\mathbf{c}) = \|P(\mathbf{c})\|_2 \qquad (7.45)$$

where $\|P(\mathbf{c})\|$ is the euclidean norm of the vector $P(\mathbf{c})$. Therefore, the search algorithm of adaptation provides the minimum of the criterion of optimality (7.43), and in certain cases even of the criterion (7.45).

It might frequently be necessary to minimize the probability that even one component of the output vector does not leave the region Ω_l, i.e., to minimize

$$J(\mathbf{c}) = P[\mathbf{y} \in \Omega_l] \qquad (7.46)$$

where

$$\Omega_l = \{z; a_v' \leq z_v \leq a_v''\} \qquad (7.47)$$

By introducing the characteristic function

$$\theta(\mathbf{c}) = \begin{cases} 0 & \text{if } \mathbf{y} \in \Omega_l \\ 1 & \text{if } \mathbf{y} \notin \Omega_l \end{cases} \qquad (7.48)$$

we can write the functional (7.46) in another form:

$$J(\mathbf{c}) = M\{\theta(\mathbf{c})\} \qquad (7.49)$$

In this case, the search algorithm of adaptation which minimizes the criterion of optimality (7.46) can be written as

$$\mathbf{c}[n] = \mathbf{c}[n-1] - \gamma[n]\tilde{\nabla}_{\mathbf{c}^+}\theta(\mathbf{c}[n-1], a[n]) \qquad (7.50)$$

7.13 Another Possibility

If for a certain component y_m of the output vector $\mathbf{y}$ we could select a value $\mathbf{c}_m^0$ of the vector $\mathbf{c}$ for which $P_m(\mathbf{c}_m^0) = 0$, then instead of the search algorithm, we could use the relaxation algorithms, such as

$$\mathbf{c}[n] = \mathbf{c}[n-1] - \gamma[n](\mathbf{c}[n-1] - \mathbf{c}_m^0) \qquad (7.51)$$

where m is the value of the random index k at the nth step. A similar approach to the solution of the problem can be based on the following.

If at the nth step the mth component of the output vector falls outside the permissible limits, then it is logical to assume that, for $\mathbf{c} = \mathbf{c}[n-1]$,

$$\max_{\nu} P_{\nu}(\mathbf{c}[n-1]) = P_m(\mathbf{c}[n-1]) \tag{7.52}$$

and, therefore, there is a reason to change $\mathbf{c}[n-1]$ in order to satisfy the condition

$$P_m(\mathbf{c}[n]) \leq P_m(\mathbf{c}[n-1]) \tag{7.53}$$

If $P_k(\mathbf{c})$ ($k = 1, \ldots, l$), is "strongly unimodal" (for any permissible values of $\mathbf{c}$, m and λ such that $0 < \lambda < 1$, $P_m[\mathbf{c} - \lambda(\mathbf{c}_m^0 - \mathbf{c})] \leq P_m(\mathbf{c})$), then algorithm (7.51) acts to satisfy condition (7.53) at each step of its operation. Such an algorithm can be applied, for instance, in the solution of the problem of controlling an overdetermined system which is optimal according to the performance criterion

$$J(\mathbf{c}) = \max_{\nu} P_{\nu}(\mathbf{c}) \tag{7.54}$$

7.14 Extremal Control Systems

Test signals of one kind or another are usually employed in the construction of extremal control systems. These test signals can define the direction of motion along the extremal characteristic. If for some reason it is impossible or undesirable to perform the test directly on the plant, extremal control can be accomplished with the help of an adaptive model of the plant. This model is used for estimating the extremal characteristic.

In the general case, the extremal plant is described by the equation

$$x = A(u, \zeta) + \xi \tag{7.55}$$

where A is a nonlinear and complex operator, u is the control action, ζ is the uncontrollable disturbance, and ξ is the perturbation at the input of the plant.

The basic problem of extremal control is to define a control action for which the criterion of optimality

$$J(u) = M\{x\} \tag{7.56}$$

is maximal. The simplest way to do this is to perform the search directly on the plant. For this, one can use search algorithms of adaptation,

$$u[n] = u[n-1] - \gamma[n]\tilde{\nabla}_{u\pm} x[n] \tag{7.57}$$

where

$$\tilde{\nabla}_{u\pm} x[n] = \nabla_{u\pm}[A(u[n-1], a[n], \zeta[n]) + \xi[n]] \tag{7.58}$$

Of course, instead of $\tilde{\nabla}_{u\pm} x[n]$ we can use here $\tilde{\nabla}_{u+} x[n]$. This is a traditional way of constructing the extremal systems. However, we are frequently not allowed to perturb the plant by the test signals. In such cases we can use algorithms of control and identification without search.

7.15 Algorithms of Extremal Control

The algorithms without search, such as

$$u[n] = u[n-1] - \gamma_1[n]\nabla_u x[n] \tag{7.59}$$

can be used in order to avoid the applications of the test signals. The gradient

$$\nabla_u x[n] = \nabla_u [A(u, \zeta) - \xi] \tag{7.60}$$

is computed with the help of a self-adjusting model. A typical diagram of this model is a cascade of linear and nonlinear parts (Fig. 7.6) which are assumed

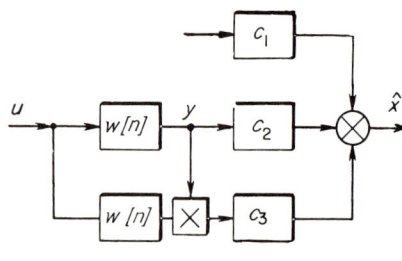

Fig. 7.6

to be described by the following equations:

$$y[n] = \sum_{m=1}^{n} w[m]u[n-m] \tag{7.61}$$

and

$$\hat{x}[n] = c_1 + c_2 y + c_3 y^2 \tag{7.62}$$

where $w[n]$ is the impulse response of the linear part with $w[0] = 0$, and c_1, c_2, c_3 are the coefficients of the approximating extremal characteristic. For simplicity we assume that the impulse response $w[n]$ is known and that the coefficients c_1, c_2, c_3 are unknown. We shall mention, however, that the general case in which $w[n]$ is also unknown should not present any difficulties to the reader who has both sufficient time and patience.

By setting (7.61) into (7.62), we obtain the equation of the model for the extremal plant

$$x[n] = c_1 + c_2 \sum_{m=1}^{n} w[m]u[n-m]$$

$$+ c_3 \sum_{m_1, m_2 = 1}^{n} w[m_1]w[m_2]u[n-m_1]u[n-m_2] \tag{7.63}$$

The derivative of $\hat{x}[n]$ with respect to $u[n-1]$ for fixed $u[n-m]$ ($m = 2, 3, \ldots, n-1$) is

$$\frac{\partial \hat{x}[n]}{\partial u[n-1]} = \left(c_2 + 2c_3 \sum_{m=1}^{n} w[m]u[n-m]\right) w[1] \tag{7.64}$$

7.16 Algorithms of Learning

In the realization of the algorithm we do not have to pay any attention to the multiplier $w[1]$ which influences only the scale of the derivative (7.64) if we select the sign of $\gamma[n]$ so that

$$\gamma[n]w[1] > 0 \tag{7.65}$$

Then the algorithm of control can be written in the form

$$u[n] = u[n-1] - \gamma_1[n](c_2[n-1] + 2c_3[n-1]\sum_{m=1}^{n} w[n]u[n-m]) \tag{7.66}$$

7.16 Algorithms of Learning

The algorithm of control (7.66) contains the unknown coefficients c_2, c_3, which can be determined by solving the problem of identification. Let the performance index of identification be

$$J(\mathbf{c}) = M\{F(x - \hat{x})\} \tag{7.67}$$

The algorithm of identification or learning can then be written as

$$c_\nu[n] = c_\nu[n-1] - \gamma_2[n]\frac{\partial}{\partial c_\nu} F(x[n] - \hat{x}[n]) \qquad (\nu = 1, 2, 3) \tag{7.68}$$

where $\hat{x}[n]$ is defined by (7.63).

It is not difficult to compute the components of the gradients which appear in algorithm (7.68):

$$\frac{\partial F(\cdot)}{\partial c_1} = -\frac{\partial F(\cdot)}{\partial \hat{x}}\left(1 + c_2[n-1]w[1]\frac{\partial u[n-1]}{\partial c_1}\right.$$

$$\left. + 2c_3[n-1]w[1]\sum_{m=1}^{n} w[m]u[n-m]\frac{\partial u[n-1]}{\partial c_1}\right)$$

$$\frac{\partial F(\cdot)}{\partial c_2} = -\frac{\partial F(\cdot)}{\partial \hat{x}}\left(\sum_{m=1}^{n} w[m]u[n-m] + c_2[n-1]w[1]\frac{\partial u[n-1]}{\partial c_2}\right.$$

$$\left. + 2c_3[n-1]w[1]\sum_{m=1}^{n} w[m]u[n-m]\frac{\partial u[n-1]}{\partial c_2}\right)$$

$$\frac{\partial F(\cdot)}{\partial c_3} = -\frac{\partial F(\cdot)}{\partial \hat{x}}\left(\sum_{m_1,m_2=1}^{n} w[m_1]w[m_2]u[n-m_1]u[n-m_2]\right. \tag{7.69}$$

$$+ c_2w[1]\frac{\partial u[n-1]}{\partial c_3}$$

$$\left. + 2c_3w[1]\sum_{m=1}^{n} w[m]u[n-m]\frac{\partial u[n-1]}{\partial c_3}\right)$$

7.17 Continuous Algorithms

In the case of a continuous control of an extremal plant, algorithms (7.66) and (7.68) are replaced by the following:

$$\frac{du(t)}{dt} = -\gamma_1(t)\left[c_2(t) - 2c_3(t)\int_0^t w(\tau)u(t-\tau)\,d\tau\right]$$

$$\frac{dc_\nu}{dt} = -\gamma_2(t)\frac{\partial F(x(t) - \hat{x}(t))}{\partial c_\nu} \qquad (\nu = 1, 2, 3) \tag{7.70}$$

where now

$$\hat{x}(t) = c_1 + c_2\int_0^t w(t)u(t-\tau) + c_3\int_0^t\int_0^t w(\tau_1)w(\tau_2)u(t-\tau_1)u(t-\tau_2)\,d\tau_1\,d\tau_2 \tag{7.71}$$

We shall not write explicitly the components of the gradient appearing in the second of the algorithms (7.70). They can be determined very simply by analogy to (7.69). We shall mention that these components, as the algorithms alone, are obtained from (7.69), (7.66) and (7.68) by a limiting process.

7.18 Structural Scheme

The components of the gradient which appear in the algorithm of learning depend on the sensitivity functions of the algorithm of control. These sensitivity functions can be determined using the sensitivity models. The final structural scheme of the extremal control is shown in Fig. 7.7.

7.19 Possible Simplifications

Let us assume that the control action is constant during an interval of duration M_0. Then, from (7.69) we obtain simpler relations:

$$\frac{\partial F(\cdot)}{\partial c_1} = -\frac{\partial F(\cdot)}{\partial \hat{x}}$$

$$\frac{\partial F(\cdot)}{\partial c_2} = -\frac{\partial F(\cdot)}{\partial \hat{x}}\sum_{m=1}^n w[m]u[n-m] \tag{7.72}$$

$$\frac{\partial F(\cdot)}{\partial c_3} = -\frac{\partial F(\cdot)}{\partial \hat{x}}\sum_{m_1, m_2=1}^n w[m_1]w[m_2]u[n-m_1]u[n-m_2]$$

and the algorithm of learning becomes

$$c_\nu[r] = c_\nu[r-1] - \gamma_2[r]F'(x[r] - \hat{x}[r]) \qquad (\nu = 1, 2, 3) \tag{7.73}$$

7.19 Possible Simplifications

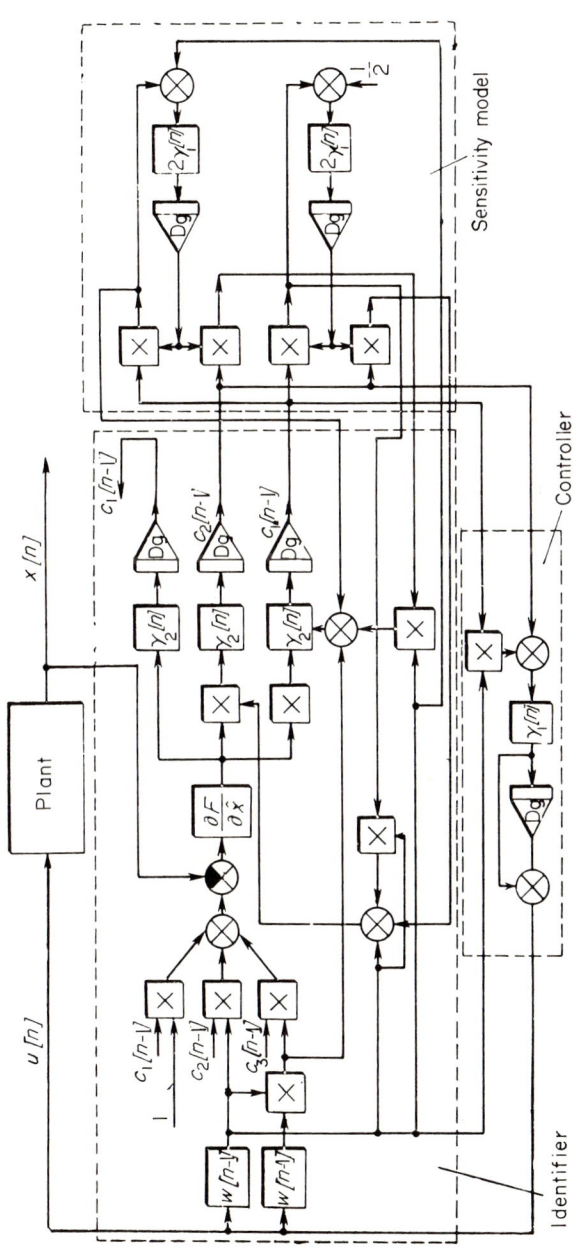

Fig. 7.7

where $n = M_0 r$, i.e., during the interval of time when a constant control action is applied and there are M_0 steps of plant identification. We can go even further by discarding the algorithm of control. Actually, if we can define c_1, c_2, c_3, then it is not difficult to determine the optimal control in an analytic way, at least in those cases where we are convinced that the extremal characteristic can be represented with sufficient accuracy by a second-order parabola. In the case that we are considering,

$$u_{\text{opt}} = c_2^*/2c_3^* \sum_{m=1}^{\infty} w[m] \qquad (7.74)$$

We have met a similar situation in Section 6.7. Such a simplified block diagram of the extremal system is shown in Fig. 7.8. In addition to the familiar elements, a divider indicated by the symbol : is also used.

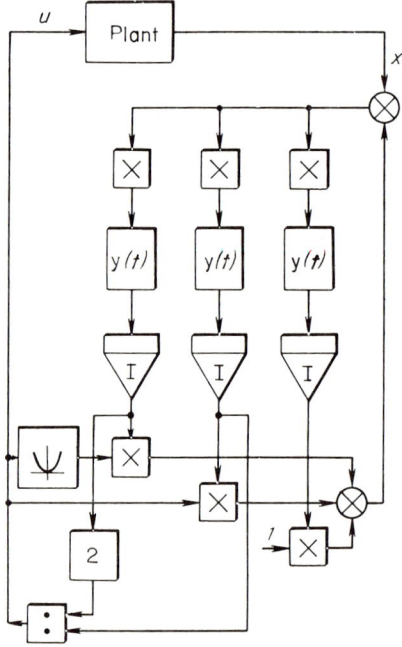

Fig. 7.8

When the extremal characteristic differs from a parabolic one, and when other nonlinearities exist, the construction of a sensitivity model is very complicated and sometimes even impossible. The answer could be found by using the plant alone as a sensitivity model. It is also possible to test the model and then transfer the control actions to the plant.

7.20 The Synthesis of Optimal Systems

The problem of synthesis of optimal control systems has for many years attracted the attention of scientists and engineers. We shall try to show one possibility of solving a similar problem on the basis of the adaptive approach. So that the reader will not get the impression that the adaptive approach is applicable only to the discrete systems described by the difference or sum equations, we shall consider here a plant described by the vector differential equation

$$\dot{\mathbf{x}}(t) = \mathbf{f}(\mathbf{x}(t), \mathbf{u}(t)) \tag{7.75}$$

where

$$\mathbf{x}(t) = (x_1(t), \ldots, x_l(t)) \tag{7.76}$$

is the state vector and

$$\mathbf{u}(t) = (u_1(t), \ldots, u_{l_1}(t)) \tag{7.77}$$

is the vector of control actions.

The performance of control is evaluated by the functional

$$J^0 = \int_0^T f_0(\mathbf{x}(t))\, dt \tag{7.78}$$

In the problem of synthesis it is necessary to find such a control law for which the functional J^0 is minimal under the constraint (7.75). When u is scalar, this law has the form

$$u(t) = k(\mathbf{x}(t), t) \tag{7.79}$$

The equation describing the controller will be sought in the form

$$u(t) = \hat{k}(\mathbf{x}(t), t) = \mathbf{c}^T \boldsymbol{\phi}(\mathbf{x}(t), t) \tag{7.80}$$

which is assumed to be known to us. In this case the problem is similar to the problem of estimating an unknown function of k (Section 7.5). It should be mentioned that the form of the vector function $\boldsymbol{\phi}(\mathbf{x}, t)$ is a priori defined on the basis of practical considerations of the design, and not only by a desire to approximate a function $\hat{k}(\mathbf{x}, t)$ in a certain "best" sense.

Equation (7.75) can be replaced by the equation,

$$\dot{\mathbf{x}}(t) = \mathbf{f}(x(t), \mathbf{c}^T \boldsymbol{\phi}(\mathbf{x}(t), t)) = \mathbf{f}_1(x(t), \mathbf{c}) \tag{7.81}$$

The solution of this equation,

$$\mathbf{x}(t) = \mathbf{x}(\mathbf{x}(0), \mathbf{c}, t) \tag{7.82}$$

depends on the initial state vector $\mathbf{x}(0)$ and the parameter vector $\mathbf{c}$. Then, the functional (7.79),

$$J^0(\mathbf{c}) = \int_0^T f_0(\mathbf{x}(\mathbf{x}(0), \mathbf{c}, t))\, dt = \Phi(\mathbf{x}(0), \mathbf{c}) \qquad (7.83)$$

also depends on the initial condition. This means that it is impossible to uniquely determine the optimal vector $\mathbf{c}$ since it depends on the initial state $\mathbf{x}(0)$. In order to avoid this nonuniqueness, it is natural to consider $\mathbf{c} = \mathbf{c}^*$ as optimal if it minimizes (7.78) in an average sense with respect to an unknown but nevertheless existing, probability density function of the initial conditions, $p(\mathbf{x}(0))$.

Therefore, we have to minimize the functional

$$J(\mathbf{c}) = M_x\{J^0(\mathbf{c})\} = M_x\left\{\int_0^T f_0(\mathbf{x}(\mathbf{x}(0), \mathbf{c}, t))\, dt\right\} \qquad (7.84)$$

7.21 Application of Adaptation Algorithms

In order to apply adaptation algorithms, it is first of all necessary to find the gradient of the performance index $J(\mathbf{c})$ with respect to $\mathbf{c}$. The simplest way to accomplish this is to use the conjugate system. The equation describing the conjugate system is

$$\begin{aligned}\psi(t) &= -\nabla f_0(\mathbf{x}) - F(\mathbf{x}, \mathbf{c})\psi(t) \\ \psi(T) &= 0\end{aligned} \qquad (7.85)$$

where

$$F(\mathbf{x}, \mathbf{c}) = \left\|\frac{\partial f_v(\mathbf{x}, \mathbf{c})}{\partial x_\mu}\right\| \qquad (v, \mu = 1, \ldots, l) \quad (7.86)$$

The Hamiltonian of the system is

$$H(\psi, \mathbf{x}, \mathbf{c}) = -f_0(\mathbf{x}) - \psi^T \mathbf{f}(\mathbf{x}, \mathbf{c}) \qquad (7.87)$$

The differential of $J^0(\mathbf{c})$ defined in (7.83) is

$$\delta J^0(\mathbf{c}) = \int_0^T \nabla_c^T H\, \delta \mathbf{c}\, dt = \delta^T \mathbf{c} \int_0^T \nabla_c H\, dt \qquad (7.88)$$

where

$$\nabla_c H = \left(\frac{\partial H}{\partial c_1}, \ldots, \frac{\partial H}{\partial c_N}\right) \qquad (7.89)$$

Therefore, we can see that the vector $Q(\mathbf{x}(0), \mathbf{c}) = \int_0^T \nabla_c H\, dt$ plays the role of the gradient of the performance index in the parameter space

$$\nabla J^0(\mathbf{c}) = Q(\mathbf{x}(0), \mathbf{c}) = \int_0^T \nabla_c H(\psi, \mathbf{x}, \mathbf{c})\, dt \qquad (7.90)$$

Now, using the adaptive approach, we can propose the following algorithm for finding the optimal vector $c = c^*$:

$$c[n] = c[n-1] - \gamma[n]Q(x[n], c[n-1]) \tag{7.91}$$

This algorithm "works" in the following fashion. First we select an arbitrary value $c[0]$ and measure the initial state $x_0[0]$. Knowing $c[0]$ and $x_0[0]$, and using relations (7.85), (7.87), (7.89) and (7.90) we find $Q(x_0[0], c[0])$, and according to (7.91) we determine $c[1]$; then the procedure repeats. For every iteration we have to measure the output of the system $x(t)$ during the time interval equal to T while $c = c[n]$ is kept constant.

It should be mentioned that in the general case of a nonlinear system (7.75), the performance index surface is multimodal, and it is therefore difficult to state the conditions under which $J(c)$ is convex, i.e., with a single extremum.

7.22 The Synthesis of Optimal Systems in the Presence of Noise

In the previous section we have assumed that the state vector x can be measured exactly. Unfortunately, very often another vector, say y, is measured instead of the state vector x. The observation vector y depends on the state vector x, noise ξ, which characterizes measurement accuracy and, perhaps, time t.

How, then, can we synthesize an optimal system under these conditions? The methods for solving this problem in the case of nonlinear plants are still unknown. A single, and thus perhaps well-known, solution consists of the following. The plant is assumed to be linear, the noise is assumed to be gaussian, and the performance index is assumed to be quadratic. These assumptions ensure the complete success of the solution of a problem not actually posed but obtained as a result of linearization and "gaussianization." But even if we close our eyes to such a substitution, or, in the best case, if such a substitution is proper, the traditional methods would not give us satisfactory results. Plants described by equations of a higher order would require enormous computational difficulties caused by a necessity to solve a large number of nonlinear Ricatti equations.

In order to solve the synthesis problem considered here, we shall try to find the equation of the controller in the form

$$u(t) = k(y(t), c) \tag{7.92}$$

where k is a certain given function,

$$y(t) = (y(t), y^{(i)}(t), \ldots, y^{(l)}(t))$$

or

$$y(t) = (y(t), y(t-\tau), \ldots, y(t-l\tau)) \tag{7.93}$$

is a vector of measurable variables, and

$$\mathbf{c} = (c_1, \ldots, c_N) \tag{7.94}$$

is the vector of still unknown parameters, where, as it was already mentioned earlier,

$$\mathbf{y}(t) = \mathbf{r}(\mathbf{x}(t), \xi, t) \tag{7.95}$$

We shall assume that either the optimal control law (a function of state variables) or the optimal control function (a function of time) is known. By constructing a model of the plant, we make $\mathbf{x}(t) = (x(t), x^{(1)}(t), \ldots, x^{(l)}(t))$ or $\mathbf{x}(t) = (x(t), x(t-\tau), \ldots, x(t-l\tau))$ available for measurement. This model can also be used to test different optimal control actions $u(t)$, i.e., for different initial states. With this we also find the optimal values of the vector $\mathbf{y}(t)$, and then a table of behavior for an "ideal" optimal system can be formed. The problem now is to select a parameter vector $\mathbf{c} = \mathbf{c}^*$ in (7.92) for which the ideal system is approximated in a certain "best" sense.

This approach is especially efficient for the minimum-time optimal control system. In this case, as it is known, $u(t) = \pm 1$. It is then necessary to determine $\mathbf{c}$ such that for small $\varepsilon > 0$ the inequalities

$$\begin{aligned} u(t) + \varepsilon - k(\mathbf{y}(t), \mathbf{c}) > 0 \\ u(t) - \varepsilon - k(\mathbf{y}(t), \mathbf{c}) < 0 \end{aligned} \tag{7.96}$$

are satisfied. If k is linearly dependent on $\mathbf{c}$, we obtain a system of linear inequalities already found in Section 4.11. The method presented there solves the problem in this case too. In the general case, this problem becomes one of solving a system of nonlinear inequalities (7.96), and under certain conditions we can apply analogous, but necessarily more general, algorithms.

7.23 Control and Pattern Recognition

In Sections 7.21 and 7.22 it was explained that a minimum-time optimal control problem is reduced to a solution of a system of inequalities. The problem is therefore similar to a pattern recognition problem. This relationship will be discussed using a particular example. For the simplest conservative system described by an equation of the second order, the relay type of control law ensures minimal-time operation if the switching boundary is a semicircle (Fig. 7.9).

Let us assume that we measure

$$y = x_1^2 + x_2^2 + \xi \tag{7.97}$$

This variable is measured, for example, by observing a satellite through a telescope if x_1 and x_2 are the coordinates of the satellite, and ξ is noise. We

7.24 Generalization of the Methods of Synthesis

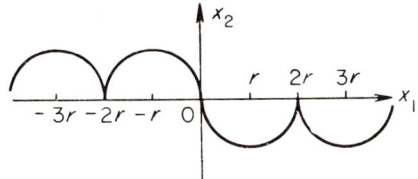

Fig. 7.9

shall also assume that y is measured at discrete instants of time, $t, t - \tau, t - 2\tau, \ldots$ Let us construct a vector of l measured values,

$$\mathbf{y} = (y(t), y(t - \tau), \ldots, y(t - l\tau)) \tag{7.98}$$

For switching $u(t)$, we choose the following rule: If $\mathbf{y}^T\mathbf{a} - a_0 > 0$, the sign of $u(t)$ is changed; if $\mathbf{y}^T\mathbf{a} - a_0 < 0$, the sign of $u(t)$ is not changed. Here, $\mathbf{a} = (a_1, \ldots, a_l)$. Optimal values $(\mathbf{a}, a_0)$ can be determined in the same way as in the problem of pattern recognition.

An optimal system will be modeled on a digital computer. It is then possible to find the corresponding optimal trajectories for different initial conditions. Those vectors $\mathbf{y}$ for which $u(t) = -1$ will be classified into the pattern class A, and the others into pattern class B. Therefore,

$$\begin{aligned} \mathbf{y}_A{}^T\mathbf{a} - a_0 > 0 \\ \mathbf{y}_B{}^T\mathbf{a} - a_0 < 0 \end{aligned} \tag{7.99}$$

and the problem is to solve the system of linear inequalities. This could be accomplished by any of the algorithms discussed in Sections 4.11 and 4.12.

7.24 Generalization of the Methods of Synthesis

The method of synthesizing optimal systems presented in Section 7.22 can be slightly generalized if it is reduced to a direct minimization of a certain functional, bypassing the system of inequalities. We shall assume that an "ideal" optimal system which not only operates in an optimal fashion, but in which all the necessary variables can be measured, is at our disposal. This ideal system can be considered to be a "teacher." The role of a "student" is performed by typical controllers which have several adjustable parameters. These controllers have to "learn" to perform the functions usually performed by expensive "real" optimal systems in the best way possible.

Let us proceed in the following way: We shall first determine the correspondence between the initial conditions $\mathbf{x}_\mu(0)$ ($\mu = 1, \ldots, M_0$) and the optimal control $u_\mu{}^*(t)$; this in turn determines the optimal process $\mathbf{x}_\mu{}^*(t)$. Therefore, the number of samples must be equal to M_0.

Let us choose the control law in the form of

$$u(t) = \mathbf{c}^T \boldsymbol{\phi}(\mathbf{x}(t)) \tag{7.100}$$

We shall then determine the optimal parameter vector so that it minimizes the performance index

$$J(\mathbf{c}) = \frac{1}{M_0} \sum_{\mu=1}^{M_0} F(u_\mu{}^* - \mathbf{c}^T \boldsymbol{\phi}(\mathbf{x}_\mu)) \tag{7.101}$$

which represents a measure of deviation of the control law (7.100) from the ideal optimal control law $u_\mu{}^*(t)$. This can be accomplished simply by an adaptive algorithm such as

$$c[n] = \mathbf{c}[n-1] - \gamma[n]\nabla_c F(u^*[n] - \mathbf{c}^T[n-1]\boldsymbol{\phi}(\mathbf{x}[n-1])) \tag{7.102}$$

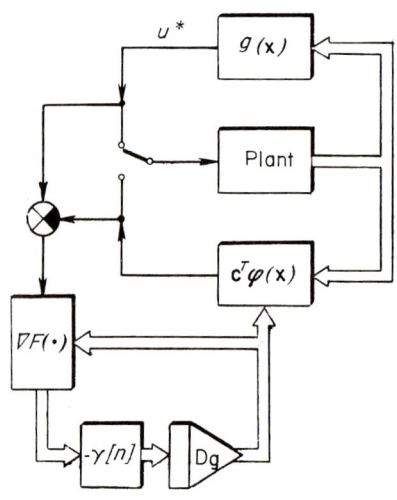

Fig. 7.10

The scheme of "learning" is shown in Fig. 7.10. Here, $k(\mathbf{x})$ is the characteristic of the ideal controller ("teacher"), and $\mathbf{c}^T \boldsymbol{\phi}(\mathbf{x})$ is the sought characteristic of a typical controller ("student").

7.25 Certain Problems

We shall now formulate a number of problems which could be solved using the adaptive approach. In addition to such problems, we shall also formulate some other problems for which the solutions are not yet known.

The behavior of closed-loop control systems is more complicated than the behavior of the open-loop systems which were considered in the previous chapters. The questions of interaction between the algorithms of learning and those of control occurred while studying the problems of learning and adaptation in the closed-loop system and the question related to the influence of this interaction on the convergence properties of such algorithms. But let us assume that the conditions of convergence are established; then, as we have already mentioned in Section 7.6, different strategies of learning and control are possible. All these strategies are equivalent when the adaptation time is infinite. The situation is drastically changed when the adaptation time is finite. In such a case, the question of selecting an appropriate relationship between the algorithms of learning and algorithms of control is of primary importance. When the plant characteristic is unknown to a certain specified degree, the problem of optimal control arises.

Frequently we face the difficulty of selecting the performance index. It must usually contain several components such as the cost of the raw materials, energy resources, volume of the achieved production, etc. The performance index components must come with definite coefficients (weights) which are not usually known a priori, and which must be determined during the process of exploitation. In other words, part of the problem is the adaptation of the performance index. The solution of this problem is also possible on the basis of the described approach.

A whole series of problems in optimization of automatic systems is related to the minimization of functionals such as

$$J_1 = \int_0^T F(\mathbf{x}^0(t) - \mathbf{x}(t)) \, dt \qquad (7.103)$$

These problems can also be reduced to the problems considered above. The functional (7.103) can indeed be replaced by

$$J_2 = \frac{1}{T} \int_0^T F(\mathbf{x}^0(t) - \mathbf{x}(t)) \, dt \qquad (7.104)$$

These functionals reach the minimum for the same function $x(t)$. Therefore, if the condition of ergodicity is fulfilled, the minimization of this functional is reduced to the minimization of

$$J = M\{F(\mathbf{x}^0(t) - \mathbf{x}(t))\} \qquad (7.105)$$

There are some problems in which we must minimize the functionals of the form

$$J_3 = M\left\{ \int_0^\infty F(\mathbf{x}^0(t) - \mathbf{x}(t)) \, dt \right\} \qquad (7.106)$$

Minimization of the functional (7.106) by a direct application of the algorithms considered above is impossible since at t (current instant of time), the value of the function $x(t)$ is unknown over the whole interval $0 \leq t < \infty$. Therefore, there is a question of whether we can minimize the performance indices (7.105) and (7.106) by observing, at each instant of time,

$$y(t) = \int_0^t F(\mathbf{x}^0(t) - \mathbf{x}(t))\, dt \qquad (7.107)$$

and using the stochastic and regular algorithms similar to those considered above.

There is some difficulty in determining $y(t)$ according to (7.107), since it requires infinite memory. How close can we come to the optimum if instead of (7.107) we use the variable

$$y_T(t) = \int_{t-T}^t F(\mathbf{x}^0(t) - \mathbf{x}(t))\, dt \qquad (7.108)$$

where T is a fixed quantity?

7.26 Conclusion

In the closed-loop systems, the deviation (error with respect to the desired or prescribed state) is used to form a control action which would remove this error. In contrast to the problems of pattern recognition, identification and filtering, we are confronted with two types of algorithms in the problems of control solved by closed-loop systems. The first type includes those known algorithms of learning whose role is to "learn" or identify the object of control (plant). The second type includes all algorithms which have to be implemented by the controller; we have not yet studied these algorithms.

The interaction between the algorithms of learning and the algorithms of control is an indication of the fact that the plant and the controller form an inseparable unit in the closed-loop automatic systems. This makes the processes in the closed-loop systems much more complex than the processes in the open-loop systems. Although certain concrete results related to adaptive systems are already obtained here, many questions are still far from being answered and we are now just at the beginning of a period of expanding the theory and the principles of design of such systems.

COMMENTS

7.2 The idea of strong negative feedback was expressed by Black (1934) in connection with the improvement of the characteristics of an amplifier by using negative feedback. The effect of negative feedback is also broadly used in the design of operational amplifiers in analog computers (Kogan, 1963).

The stability conditions of the system which theoretically permit an unlimited increase in the gain coefficient were established by Meerov (1947, 1959). These conditions can be considered as the conditions of realizability of strong negative feedback. The effect of strong negative feedback, as it was shown by Pospelov (1955) and the author (Tsypkin, 1966), is achieved in the case of the sliding modes in the relay automatic systems. It was recently discovered (Bermant and Emelyanov, 1963) that a similar effect may be found in the systems with variable structure. It is interesting to note that regardless of the differences in the systems, and the methods for eliminating the variations in their characteristics, they are all actually based on the same effect of strong negative feedback. About invariance, see the work by Kukhtenko (1963). The conditions which require application of adaptation are presented in the book by Krasovskii (1963).

7.3 The first papers describing the applications of the stochastic approximation method to the simplest control systems were written by Bertram (1959) and Ulrikh (1964). See also the papers by Schultz (1965) and the book by Kulikovski (1966).

7.4 On dual control, see the basic papers by Feldbaum (1960–1961, 1961, 1964) and the book by Feldbaum (1965). The theory of dual control proposed by Feldbaum defines an optimal system on the basis of statistical decision theory. No better system can be obtained using the same information. However, the great number of computations prohibits a physical realization of the optimal system at this time and, most probably, in the near future. We have considered the term dual control in a more general sense, and not only in relationship with the statistical decision theory.

7.5 The strategy of learning and control in l stages was used by Ho and Whalen (1963) in the control of linear plants. As it can now be understood, this represents a special case of the problem that we considered. The reader can find certain evaluations of the possible strategies of learning and control in the paper by Kishi (1964).

7.7 The interesting results in the area of adaptive systems without search (and when noise is not present), were obtained by Bikhovskii (1964) and Kokotović (1964). See also the papers by Rissanen (1963, 1966a, b), Evlanov (1963), Kazakov (1962) and Kazakov and Evlanov (1963).

7.8 On the sensitivity model, see the papers by Bikhovskii (1964) and Kokotović (1964), and the book by Meisinger (1962). The idea of using a single model for obtaining (simultaneously) all sensitivity functions is due to Kokotović (1964).

An attempt to use the plant as the sensitivity model was described by Narendra and Streer (1964). Obviously, this idea cannot be realized completely.

7.9 The results of this section are related to the work by Hill and Fu (1965). The idea of adaptively reducing the adaptation time is due to them.

7.10 Adaptive systems with models are extensively used since they are very simple. One of the first works describing such adaptive systems is by Margolis and Leondes (1961).

The questions regarding the design of adaptive systems with models were considered by Rissanen (1963, 1966a), Krutova and Rutkovskii (1964a, b) and others.

The surveys by Donalson and Kishi (1965) and Krutova and Rutkovskii (1964a) contain expositions of the design principles and applications of adaptive systems with models, and also bibliographies.

7.11 The statement of the problem of controlling an overdetermined plan is due to Thomas and Wilde (1963).

7.13 The possibility of using the relaxation algorithm was described by Thomas and Wilde (1963). They have also proved the convergence of the relaxation algorithm (7.51).

7.14 The reader can learn about extremal systems from the book by Krasovskii (1963).

7.15 A similar description of an extremal plant was used by Popkov (1966, 1967).

7.19 See also the paper by Kostyuk (1965). Such a design of extremal systems was proposed for a particular case. Interesting results have also been given in the works by Kulikovski (1963, 1966).

7.20 We have used the form of the differential (7.88) obtained in an interesting article by Rozonoer (1959).

7.21 These results are based on the work by Propoi and Tsypkin (1967).

7.22 Such a statement of the problem of synthesis of optimal systems is due to Ho and Whalen (1963).

7.23 The example was borrowed from the paper by Knoll (1965). In the experiment carried out by Knoll, $l = 10$ and $\tau = 0.03$ sec were chosen. The initial values for x were uniformly distributed within the interval $1.0 \leq x \leq 18.25$. Therefore, each vector from A or B consisted of ten points; there was a total of 36 vectors. After y_A and y_B were determined, the optimal values a_1^* and a_0^* were found in three iterations. These optimal values were then verified on fifteen other initial conditions $x(0)$, and the separation was complete.

BIBLIOGRAPHY

Bermant, M.A., and Emelyanov, S.V. (1963). On a method of control for a certain class of plants with time-varying parameters, *Automat. Remote Contr.* (*USSR*) **24** (5).
Bertram, J.E. (1959). Control by stochastic adjustment, *Trans. AIEE* **78** (11).
Bikhovskii, M.L. (1964). Sensitivity and dynamic accuracy of the control systems, *Izv. Akad. Nauk SSSR Tekh. Kibern.* **1964** (6). (Engl. transl.: *Eng. Cybern.* (*USSR*).)
Black, H.S. (1934). Stabilized feedback amplifiers, *Bell Syst. Tech. J.* **14** (1).
Comer, J.P. (1965). Application of stochastic approximation to process control, *J. Roy. Stat. Soc.* **B-27** (3).
Donalson, D.E., and Kishi, F.H. (1965). Review of adaptive control system theories and techniques. *In* "Modern Control Systems Theory" (C.T. Leondes, ed.). McGraw-Hill, New York.
Emelyanov, S.V., and Bermant, M.A. (1962). On a question of design of high-quality control systems with time-varying parameters, *Dokl. Akad. Nauk SSSR* **45** (1).
Evlanov, L.G. (1963). A self-organizing system with the search gradient. *In* "Tekhnicheskaia Kybernetika," No. 1. Academy of Sciences USSR, Moscow.
Feldbaum, A.A. (1960–1961). Theory of dual control I–IV, *Automat. Remote Contr.* (*USSR*) **21** (9) and (11); **22** (1) and (2).
Feldbaum, A.A. (1961). On accumulation of information in closed loop control systems. *In* "Energetika i Avtomatika," No. 4. Academy of Sciences USSR, Moscow.
Feldbaum, A.A. (1964). On a class of self-learning systems with dual control, *Automat. Remote Contr.* (*USSR*) **25** (4).
Feldbaum, A.A. (1965). "Optimal Control Systems." Academic Press, New York.
Hill, J.D., and Fu, K.S. (1965). A learning control system using stochastic approximation for hill-climbing. *In* "Joint Automatic Control Conference, New York, 1965."
Hill, J.D., McMurtry, G.J., and Fu, K.S. (1964). A computer-simulated on-line experiment in learning control, *AFIPS Proc.* **25**.

Ho, Y.C., and Whalen, B. (1963). An approach to the identification and control of linear dynamic systems with unknown coefficients, *AIEE Trans. Automat. Contr.* **AC-8** (3).
Hsu, J.C., and Meserve, W.E. (1962). Design-making in adaptive control systems, *IRE Trans. Automat. Contr.* **AC-7** (1).
Janáč, K., Skrivánek, T., and Šefl, O. (1965). Adaptivni, řizeni systému s promennyin zesilenim v připade náhodnych poruch a jeho modelováni. *In* " Problemy Kibernetiky." Praha.
Kazakov, I.E. (1962). On a statistical theory of continuous self-adaptive systems. *In* "Energetika i Avtomatika," No. 6. Academy of Sciences USSR, Moscow.
Kazakov, I.E. and Evlanov, L.G. (1963). On a theory of self-adaptive systems with the gradient method, *Theory Self-Adapt. Contr. Syst. Proc. IFAC 2nd* **1963**.
Kishi, F.H. (1964). On line computer control techniques and their application to re-entry aerospace vehicle control. *In* "Advances in Control Systems" Vol. 1 (C.T. Leondes, ed.). Academic Press, New York.
Knoll, A.L. (1965). Experiment with a pattern classifier on an optimal control problem, *IEEE Trans. Automat. Contr.* **AC-10** (4).
Kogan, B.Ya. (1963). " Electronic Modeling Devices and Their Application in Automatic Control Systems." Fizmatgiz, Moscow. (In Russian.)
Kokotović, P. (1964). The method of sensitivity points in the investigation and optimization of linear control systems, *Automat. Remote Contr.* (*USSR*) **25** (12).
Kostyuk, V.I. (1965). Extremal control without search oscillations using self-adaptive model of the plant, *Avtomatika* **10** (2). (In Ukrainian.)
Kostyuk, V.I. (1966). A relationship between the self-adaptive systems without search, differential systems and the systems based on sensitivity theory, *Avtomatika* **11** (2). (In Ukrainian.)
Krasovskii, A.A. (1963). " Dynamics of Continuous Self-Organizing Systems." Fizmatgiz, Moscow. (In Russian.)
Krutova, I.V., and Rutkovskii, V.Yu. (1964a). A self-adaptive system with a model reference I, *Izv. Akad. Nauk SSSR Tekh. Kibern.* **1964** (1). (Engl. transl.: *Eng. Cybern.* (*USSR*).)
Krutova, I.V., and Rutkovskii, V.Yu. (1964b). A self-adaptive system with a model reference II, *Izv. Akad. Nauk SSSR Tekh. Kibern.* **1964** (2). (Engl. transl.: *Eng. Cybern.* (*USSR*).)
Kubát, L. (1965a). Adaptivni proporcionálni regulace jednoduche soustavy s dopravnim zpožděnim. *In* " Problemy Kibernetiky." Praha.
Kubát, L. (1965b). Regulace jednoduche podle statitckých parametru. *In* " Kubernetika a jejivyužiti." Praha.
Kukhtenko, A.I. (1963). "The Problem of Invariance in Automation." Gostehizdat, Moscow. (In Russian.)
Kulikovski, R. (1963). Optimization of nonlinear random control processes, *Theory Self-Adapt. Contr. Syst. Proc. IFAC 2nd* **1963**.
Kulikovski, R. (1966). " Optimal and Adaptive Processes in Automatic Control Systems." Nauka, Moscow. (In Russian.)
Leonov, Yu.P. (1966). Asymptotically optimal systems as the models of learning control, *Dokl. Akad. Nauk SSSR* **167** (3).
Margolis, M., and Leondes, C.T. (1961). On a theory of adaptive control systems; the method of a learning model, *Theory Self-Adapt. Contr. Syst. Proc. IFAC 1st* **1961**.
Meerov, M.V. (1947). The automatic control systems which are stable with an arbitrary large gain coefficient, *Avtomat. Telemekh.* **8** (4).
Meerov, M.V. (1959) "The Synthesis of High Precision Automatic Control Systems." Fizmatgiz, Moscow.
Meisinger, H.F. (1962). " Parameter Influence Coefficients and Weighting Functions Applied to Perturbation Analysis of Dynamic Systems." Press Academiques, Europeens, Brussels.

Mishkin, E., and Brown, L. (1961). "Adaptive Control." McGraw-Hill, New York.
Nakamura, K., and Oda, M. (1965). Development of learning control, *J. Japan Assoc. Automat. Contr. Eng.* **9** (8).
Narendra, K.S., and McBride, L.E. (1964). Multiparameter self-optimizing systems using correlation techniques, *IEEE Trans. Automat. Contr.* **AC-9** (1).
Narendra, K.S., and Streer, D.N. (1964). Adaptive procedure for controlling on defined linear processes, *IEEE Trans. Automat. Contr.* **AC-9** (4).
Pearson, A.E. (1964). On adaptive optimal control of nonlinear processes, *J. Basic Eng.* **86** (1).
Pearson, A.E., and Sarachik, P.E. (1965). On the formulation of adaptive optimal control problems, *J. Basic. Eng.* **87** (1).
Perlis, H.J. (1965). A self adaptive system with auxilary adjustment prediction, *IEEE Trans. Automat. Contr.* **AC-10** (3).
Petraš, Š. (1965). Učiace sa systemy automaticke go piadenia. *In* "Výskumné Problémy Technickej Kybernetiky a Mechaniky." Slovenska Akademia Vied, Bratislava.
Popkov, Yu.S. (1966). Asymptotical properties of optimal systems of extremal control, *Automat. Remote Contr.* (*USSR*) **27** (6).
Popkov, Yu.S. (1967). Analytical design of statistical optimal systems of extremal control. In "Trudi III Vsesoyuznogo Soveshchania po Avtomaticheskom Upravleniu, Odessa, 1965." Nauka, Moscow. (In Russian.)
Pospelov, G.S. (1955). Vibrational linearization of relay systems of automatic control. Reaction of relay systems on variable perturbations. *In* "Trudi II Vsesoyuznogo Soveshchania po Teorii Avtomaticheskogo Regulirovania," Vol. 1. Academy of Sciences USSR, Moscow.
Propoi, A.I., and Tsypkin, Ya.Z. (1967). On adaptive synthesis of optimal systems, *Dokl. Akad. Nauk SSSR* **175** (6).
Rissanen, J. (1963). On the theory of self-adjusting models, *Automatika* **1** (4).
Rissanen, J. (1966a). On the adjustment of the parameters of linear systema by a functional derivative technique, Sensitivity method in control theory. *In* "Proceedings of International Symposium, Dubrovnik, Sept. 1964." Pergamon Press, New York.
Rissanen, J. (1966b). Drift compensation of linear systems by parameter adjustments, *J. Basic Eng.* **88** (2).
Rozonoer, L.I. (1959). Pontriagin's maximum principle in optimal control theory, *Automat. Remote Contr.* (*USSR*) **20** (10–12).
Šefl, O. (1964). Certain problems of automatic control of processes with unknown characteristics. *In* "Transactions of the Third Prague Conference on Information Theory, Statistical Decision and Random Functions, Prague, 1964."
Šefl, O. (1965a). Principy adaptivity v technicke kybernetice. *In* "Kybernetika a Jeji Veuzity." Praha.
Šefl, O. (1965b). Adaptivni systémy. *In* "Sounru Praci a Automatizace, 1959." Praha.
Shultz, P.R. (1965). Some elements of stochastic approximation theory and its application to a control problem. *In* "Modern Control Systems Theory" (C.T. Leondes, ed.). McGraw-Hill, New York.
Taylor, W.K. (1963). Self-organizing control systems using pattern recognition, *Theory Self-Adapt. Contr. Syst. Proc. IFAC 2nd* **1963**.
Thomas, M.E., and Wilde, D.J. (1963). Feed-forward control by over-determined systems by stochastic relaxation. *In* "Joint Automatic Control Conference, Minneapolis, Minnesota, 1963."
Tomingas, K.V. (1966). On extremal control of flotation processes, *Automat. Remote Contr.* (*USSR*) **27** (3).

Tou, J.T. (1963). "Optimum Design of Digital Control Systems." Academic Press, New York.
Tsypkin, Ya.Z. (1966). "Theory of Relay Systems of Automatic Control." Gostehizdat, Moscow.
Ulrikh, M. (1964). On certain probabilistic methods of automatic control, *Automat. Remote Contr. (USSR)* **25** (1).
Ulrikh, M. (1965). Pravděpodobnorni přistup k určeni optimálnich regulátorů. *In* "Problemy Kiberneticky." Praha.

8

RELIABILITY

8.1 Introduction

The problem of reliability is one of the most important in the design of complex systems. It is at least as important as the characteristics describing the operation of the system. What is reliability? What are we going to discuss in this chapter? Reliability is the capability of a system to function without failures; for instance, the existence of failures in a digital computer which could be caused by improperly operating components produces unsatisfactory results.

An increase in system reliability can be achieved by an increase in the reliabilities of individual components, and also by constructing special structures with components of given reliabilities. In this chapter we shall show that one possible way to solve the problems related to the increase of reliability is the adaptive approach. This approach may prove, in time, to be the only one which provides the solutions to reliability problems.

8.2 Concept of Reliability

It is difficult to give an abstract, and thus a general, definition of reliability. Different people either include various ideas in such abstract concepts, or do not include any ideas at all. But not to give any definition of reliability is also impossible. Reliability is frequently considered to be the capacity of the system to preserve its operating characteristics within given limits which guarantee normal operation of the system under specified conditions of exploitation. If the characteristics of the system fall outside these limits, it is frequently said

8.3 Reliability Indices

that a failure of the system occurred. The phrases "frequently consider" and "frequently say" which are used here should remind us that there is a very large number of other definitions which is commensurable with the number of publications on this subject.

In any case, the concept of reliability is related to the capacity of the system to maintain its correct operation during a given interval of time (faultless operation), to the capabilities to detect and remove the causes of failures (repairability) and finally to the capability for long exploitation (longevity). Faultless operation, repairability and longevity are the binding elements of the reliability theory.

After these definitions are given, we have to specify the performance indices which are to be minimized. The adaptive approach will then be applied to this attractive, but still not completely explored area.

8.3 Reliability Indices

The indices of reliability represent certain characteristics of the system and correspond to those earlier-mentioned performance indices or indices of optimality. Depending upon the problems under consideration, these indices can be different quantities describing the reliability. For our needs, it is convenient to introduce the reliability indices in the following fashion.

Let us assume that the evolution of the system is described by a stochastic process $\mathbf{x}(t)$; $\mathbf{x}(t)$ is the trajectory of the system which describes the state of the system at any instant t. All the possible states form the state space of the system. Let us select a certain region S_{fail} in the state space. If the trajectory of the system, $\mathbf{x}(t)$, falls into the region S_{fail}, i.e., $\mathbf{x}(t') \in S_{\text{fail}}$ in Fig. 8.1, it is considered that a failure of the system occurred.

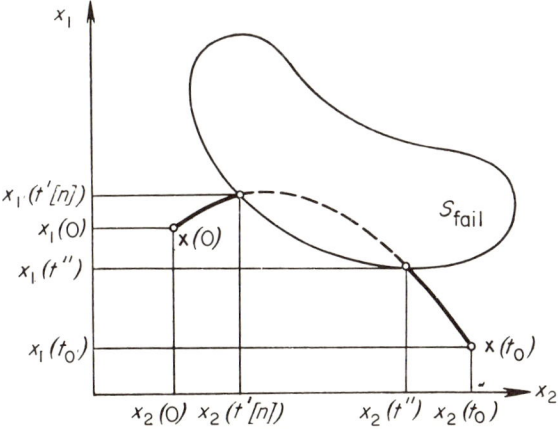

Fig. 8.1

In order to define a quantitative index of reliability, a functional θ defined along the trajectory $\mathbf{x}(t)$ is introduced. The reliability index can then be defined as the mathematical expectation of this functional:

$$J = M\{\theta(\mathbf{x}(t))\} \tag{8.1}$$

The meaning of this index is obvious. Each trajectory is assigned a certain definite weight $\theta(\mathbf{x}(t))$, and the average value of the weight is chosen to be the reliability index.

If the functional $\theta = \theta_1(\mathbf{x}(t))$ is defined as

$$\theta_1(\mathbf{x}(t)) = \begin{cases} 0 & \text{if for at least one } s \le t_0,\, \mathbf{x}(s) \in S_{\text{fail}} \\ 1 & \text{otherwise} \end{cases} \tag{8.2}$$

then, as we already know,

$$J_1 = M\{\theta_1(x(t))\} = P_{t_0} \tag{8.3}$$

i.e., the functional is equal to the probability of faultless operation of the system in the interval $[0, t_0]$.

If the functional $\theta = \theta(\mathbf{x}(t))$ represents the instant when the trajectory $\mathbf{x}(t)$ first enters the region S_{fail}, then the mathematical expectation of this functional

$$J_2 = M\{\theta_2(\mathbf{x}(t))\} = T \tag{8.4}$$

defines the average time of faultless operation. The system is considered to be more reliable if the probability of faultless operation is closer to one, or if the average time of faultless operation is longer. Very often, the problem of designing the systems which are optimal with respect to reliability consists of securing the maximal values of the reliability indices. The problem can now be observed from the standpoint of our interest.

8.4 Definition of Reliability Indices

The definition of the reliability indices (8.3) and (8.4) is reduced to the computation of mathematical expectation, and if the probability density function is unknown, we can use the adaptive algorithms. We obtain the following algorithms for the probability of faultless operation over a fixed interval of time t_0, and the average time of faultless operation T on the basis of discrete adaptive algorithms:

$$P_{t_0}[n] = P_{t_0}[n-1] - \gamma[n][P_{t_0}[n-1] - \theta_1(\mathbf{x}_n(t))] \tag{8.5}$$

8.4 Definition of Reliability Indices

and

$$T[n] = T[n-1] - \gamma[n][T[n-1] - \theta_2(\mathbf{x}_n(t))] \tag{8.6}$$

where $\mathbf{x}_n(t)$ is a realization of the stochastic process $\mathbf{x}(t)$ for $0 \leq t \leq t_0$ at the nth step.

When the modified algorithms are used, we obtain

$$P_{t_0}[n] = P_{t_0}[n-1] - \gamma[n]\left[P_{t_0}[n-1] - \frac{1}{n}\sum_{m=1}^{n}\theta_1(\mathbf{x}_m(t))\right] \tag{8.7}$$

and

$$T[n] = T[n-1] = \gamma[n]\left[T[n-1] - \frac{1}{n}\sum_{m=1}^{n}\theta_2(\mathbf{x}_m(t))\right] \tag{8.8}$$

In the case of ergodic processes, the sequential segments of a single realization, each one of duration t_0, can be used instead of an ensemble of realizations over the interval $[0, t_0]$. This particular case corresponds to the block diagrams for obtaining P_{t_0} and T using standard (Fig. 8.2) and modified (Fig. 8.3) algorithms. It is not difficult to recognize their similarity to the

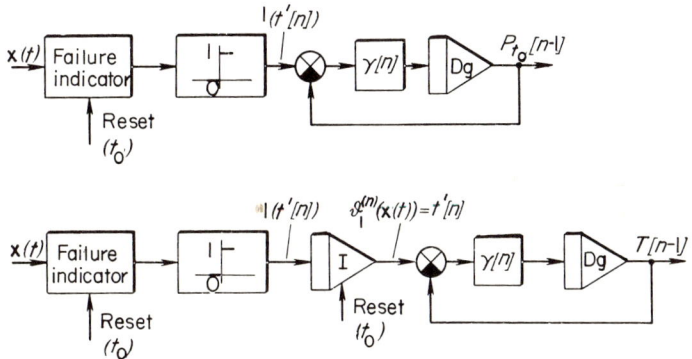

Fig. 8.2

devices which are used for restoration (estimation) of probability density functions.

In the schemes which realize algorithms (8.5)–(8.8), the indicator of failures (FI) operates in the following manner: As soon as the state of the system $\mathbf{x}(t)$ falls for the first time into the region S_{fail} during the time interval $[nt_0, (n+1)t_0]$, the output of FI becomes positive and it preserves its sign to the end of the interval, i.e., to the instant $(n+1)t_0$. At the instants kt_0 ($k = 1, 2, \ldots$), the output of FI is reset to zero.

8 Reliability

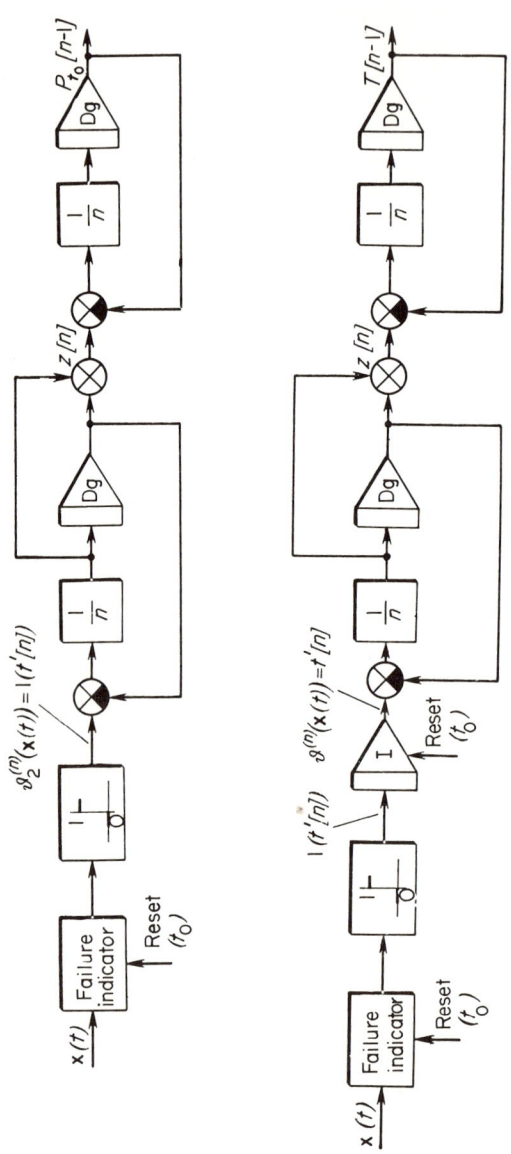

Fig. 8.3

8.5 Minimization of Operating Expenditures

The expenditures of system maintenance include the expenditures of all operations which are necessary to secure a working system. Operating expenditures can be divided into two parts: One part consists of expenditures $W[\theta(\mathbf{x}(t), T)]$ which are necessary to ensure a certain level of reliability $\theta(\mathbf{x}(t), T)$ for the time interval T of system operation (expenditures for preventive repairs, restarting the system after a failure, replacement of the equipment by a better one, etc). The second part includes the expenditures $U(T)$ which do not directly affect the reliability of the system (tendency to improve other characteristics such as work atmosphere, accuracy, etc.).

General expenditures caused by operating the system during the time interval T can be evaluated in an average sense by the functional

$$J = M\{W[\theta(\mathbf{x}(t), T)]\} + \lambda_0 M\{U(T)\} \tag{8.9}$$

where λ is the weighting coefficient. Here, we reserve λ to designate the Lagrange multipliers. This may irritate some readers since in the theory of reliability λ is an established symbol of the coefficient of reliability. We can be excused for doing this because the coefficient of reliability is not used in the sequel.

It is obvious that the functional J will considerably depend upon the "maintenance program" of the system. The system's maintenance program can represent either a time-table of the preventive repairs of the system, the distribution of funds designated for system maintenance, or something else. Therefore, the exact meaning of this concept is determined by the nature of each particular problem. The optimal maintenance program of the system is one for which the general operating expenditures (8.9) become minimal.

8.6 A Special Case

Let us assume that it is necessary to estimate a time interval L between successive preventive repairs of the system for which the losses (8.9) are minimal. For simplicity, we consider $U[T]$ to be independent of L. The criterion (8.9) is then simplified, and

$$J_1 = M_{\mathbf{x}}\{W[\theta_1(\mathbf{x}(t), T, L)]\} \tag{8.10}$$

where $T = \text{const.}$

The expenditures necessary to secure the level of reliability θ_1 can be written as

$$W[\theta_1(\mathbf{x}(t), T, L)] = W_1[\theta_1(\mathbf{x}(t), T, L)] + W_2(L) \tag{8.11}$$

where $W_2(L)$ represents the total expenditures on preventive repairs for the time interval T of system operation; W_1 designates the remaining expenditures necessary to maintain the reliability level θ for a given value of L (expenditures to restart the system after a shutdown, etc.). The expenditures $W_2[L]$ on the preventive repairs are directly proportional to T and inversely proportional to L. With an increase in L, the expenditures on preventive repairs during time interval T are decreased, but the reliability of the system is also reduced at the same time. The system is then frequently out of service and the expenditures of type W_1 increase. The optimal value $L = L^*$ which minimizes the functional J_1 is found using the algorithm

$$L[n] = L[n-1] - \gamma[n][W'_{1L}(x_n(t), T, L[n-1]) + W'_{2L}(L[n-1])] \quad (8.12)$$

8.7 Minimization of Cost, Weight and Volume

Let A indicate a finite set of different components of a certain system which are interconnected in a specific way. Each element α which belongs to A, i.e., $\alpha \in A$, can be in one of two states: either the operating state or the failure state, depending upon different random factors.

The probability $P(\mathbf{x}, \mathbf{c}(\alpha))$ that the system is in the state $\mathbf{x} \in X$ depends upon different parameters of the components $\alpha \in A$, on their quality (and thus cost), perhaps their weight, volume, number of components of the same type, etc. Therefore, $\mathbf{c}(\alpha)$ is a vector function,

$$\mathbf{c}(\alpha) = (c_1(\alpha), c_2(\alpha), c_3(\alpha), \ldots) \quad (8.13)$$

where, for instance, $c_1(\alpha)$ is cost, $c_2(\alpha)$ may be weight, $c_3(\alpha)$ is volume, $c_4(\alpha)$ is the number of reserve components, etc.

Let us designate by $F(\mathbf{x})$ the probability that the system being in state $\mathbf{x}$ is capable of solving a problem which is selected at random (independently of the state $\mathbf{x}$) from a given set of problems according to certain probability law. It is natural to select

$$J = \sum_{\alpha \in X} F(\mathbf{x}) P(\mathbf{x}, \mathbf{c}(\alpha)) = M_{\mathbf{x}}\{F(\mathbf{x})\} \quad (8.14)$$

to represent the reliability index of the system.

In this case the problem of designing a system which is optimal with respect to the chosen efficiency index consists of the following: Define a vector function $\mathbf{c}(\alpha)$ which minimizes the functional (8.14) under the constraints

$$\mathbf{c}(\alpha) \geq 0 \qquad \sum_{\alpha \in A} \mathbf{c}(\alpha) \leq \mathbf{g}_0 \quad (8.15)$$

These inequalities express the fact that the cost, volume, weight, energy consumption, etc., are positive and usually finite.

Let us designate

$$\mathbf{c}(\alpha) = \{c_1(\alpha), \ldots, c_N(\alpha)\} \quad (8.16)$$

We should notice that the vector function $c(\alpha)$, which depends on α, has to be estimated here. In the previous problems we simply had to estimate a vector. The argument α which defines the type of the components in the system, can take a finite number of values.

8.8 Algorithms of Minimization

For every fixed value of α, according to (3.19) and (3.24), one can write the algorithm which minimizes (8.14) under the constraints (8.15). This algorithm has the form

$$c_\alpha[n] = \max\{0, c_\alpha[n-1] - \gamma[n]\{\tilde{\nabla}_{c_\alpha} F(x[n]) + \lambda[n-1]\}\}$$

and (8.17)

$$\lambda[n] = \max\left\{0, \lambda[n-1] + \gamma_1[n]\left[\sum_{\alpha \in A} c_\alpha[n-1] - g_0\right]\right\}$$

where $c_\alpha[n]$ is the value of the vector function $c(\alpha)$ at the nth step. The criterion (8.14) differs from the criteria considered earlier since the function $F(x)$ under the expectation sign does not depend on the parameters $c(\alpha)$.

The probability density function of the random vector x depends on $c(\alpha)$ only: $p(x, c(\alpha))$, i.e., the values of the function $F(x)$ depend on $c(\alpha)$ only through the values of the random function

$$x = x(c(\alpha)) \qquad (8.18)$$

Therefore, the distribution $F(x)$ depends on $c(\alpha)$ in a parametric way,

$$F(x) = Y(c(\alpha)) \qquad (8.19)$$

where $Y(c(\alpha))$ is a certain random function of α. The distribution of $Y(c(\alpha))$ depends on the conditional probability density function of the random vector x, i.e., $p(x, c(\alpha))$. The scheme realizing the algorithms (8.17) is given in Fig. 8.4.

8.9 A Special Case

The reader has probably noticed that the algorithms (8.17) do not include all possible cases of the minimization problems, and particularly one related to the minimization of the functional (8.14) with respect to the number of reserve components. As we shall see further, the problems of minimization of these and other functionals with respect to the parameters which can have a finite number of values are not as rare in the reliability theory as it may seem at first. It is thus important to form the algorithms of minimization which are applicable in such cases. This possibility will now be examined.

For simplicity, we shall consider only the case of a single parameter. Let parameter c take only specified fixed values from a finite set (alphabet),

$$D = \{d_1, d_2, \ldots, d_N\} \qquad (8.20)$$

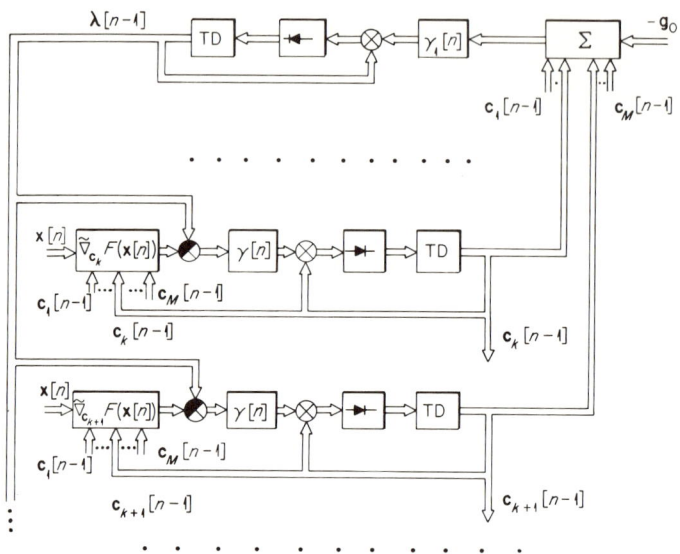

Fig. 8.4

and let it be necessary to determine the extremum of a functional

$$J(c) = M\{Q(\mathbf{x}, c)\} \qquad (8.21)$$

This functional has a physical meaning only for $c \in D$ where D is the set of possible parameters c. Outside of this set we can define $Q(\mathbf{x}, c)$ using linear interpolation. The function $Q(\mathbf{x}, c)$ can then be written for any c as

$$Q^0(\mathbf{x}, c) = Q(\mathbf{x}, d_{v-1}) + \frac{Q(\mathbf{x}, d_v) - Q(\mathbf{x}, d_{v-1})}{d_v - d_{v-1}} (c - d_{v-1}) \qquad (8.22)$$

where

$$d_{v-1} \leq c < d_v \qquad (8.23)$$

It should be obvious that the new functional

$$J^0 = M\{Q^0(\mathbf{x}, c)\} \qquad (8.24)$$

and the functional (8.21) reach their extrema for the same $c^* \in D$.

8.10 Algorithms

Let us apply the search algorithm to (8.24). Considering the piecewise linear nature of $Q(\mathbf{x}, c)$, we obtain

$$c[n] = c[n-1] - \lambda[n]\tilde{Q}^0_{c\pm}(\mathbf{x}[n], c[n-1], a[n]) \qquad (8.25)$$

where

$\tilde{Q}_{c\pm}^0(\mathbf{x}[n], c[n-1], a[n])$

$$= \begin{cases} \dfrac{Q(\mathbf{x}[n], d_v) - Q(\mathbf{x}[n], d_{v-1})}{d_v - d_{v-1}} & \text{when } d_{v-1} < c_k[n-1] < d_v \\ \dfrac{1}{2}\left[\dfrac{Q(\mathbf{x}[n] d_{v-1}) - Q(\mathbf{x}[n], d_v)}{d_{v-1} - d_v} \right. \\ \left. + \dfrac{Q(\mathbf{x}[n], d_v) - Q(\mathbf{x}[n], d_{v-1})}{d_v - d_{v-1}}\right] & \text{when } c[n-1] = d_v \end{cases} \quad (8.26)$$

The meaning of this relation is the following: The search for estimating the gradient of the function $Q^0(\mathbf{x}, c)$ is performed only at those points where the derivative of $Q(\mathbf{x}, c)$ does not exist. If this derivative exists at the point $c[n-1]$, then the nth step is based on the exact value of the gradient. The scheme which realizes algorithm (8.26) (for $d_v = v$; $v = 1, 2, \ldots, M$) is shown in Fig. 8.5. Here we find a new element which has the output signal

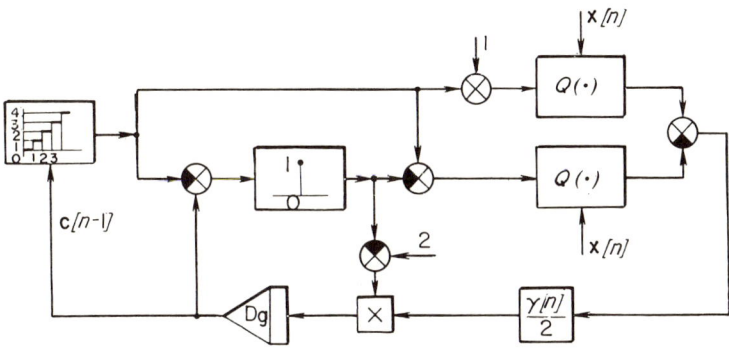

Fig. 8.5

equal to 1 or 0 when the input signal is, respectively, equal to 0 or different from zero.

In order to avoid the difficulties connected with the occurrence of new stationary points during a linear interpolation, a search algorithm can also be used, but under the constraint

$$\prod_{v=1}^{M} (c - d_v) = 0 \quad (8.27)$$

in this case we obtain a new search algorithm,

$$c[n] = c[n-1] - \gamma[n]]\tilde{Q}_{c\pm}(\mathbf{x}[n], c[n-1], a[n])$$
$$+ \lambda[n-1] \sum_{k=1}^{M} \prod_{\substack{v=1 \\ v \neq k}}^{M} (c[n-1] - d_v)]$$

$$\lambda[n] = \lambda[n-1] - \gamma_1[n] \prod_{v=1}^{M} (c[n-1] - d_v) \qquad (8.28)$$

where $\tilde{Q}_{c\pm}(x[n], c[n-1], a[n])$ is defined as before, according to an expression similar to (8.26).

Algorithms (8.25) and (8.28) can be generalized for a more complicated case when **c** is a vector; all or some components of **c** can take certain specified fixed values.

8.11 Increase of Reliability by Reserve Components

One of the methods to increase the reliability of a complex system is to use the additional components of the devices which are connected to the system when the corresponding devices or parts fail, i.e., to use reserve components. The limitations on the cost, weight and also the complexity of the structure do not allow too many additional reserve component devices. Therefore, the operation of the system cannot be absolutely faultless. How can we formulate a problem with realistic constraints and find its best solution?

Let us assume that the system consists of N stages (blocks), $k = 1, \ldots, N$. We shall designate by m_k the number of duplicate devices in the kth stage, and by $P_k(m_k, c_k, w_k)$ the probability that the kth stage is operating correctly (the last function represents the reliability index of the kth stage when m_k devices are used with the weight c_k and the cost w_k in the kth stage).

The reliability index of the whole system—the probability that the whole system (i.e., all N stages) is operating correctly—is obviously defined by

$$R = \prod_{k=1}^{N} P_k(m_k, c_k, w_k) \qquad (8.29)$$

or

$$R = M\{\theta(\mathbf{m}, \mathbf{c}, \mathbf{w})\} \qquad (8.30)$$

where

$$\theta(\mathbf{m}, \mathbf{c}, \mathbf{w}) = \begin{cases} 1 & \text{if all } N \text{ stages are operating correctly} \\ 0 & \text{otherwise} \end{cases} \qquad (8.31)$$

The total weight and cost are defined by the expressions

$$\sum_{k=1}^{N} c_k m_k = \mathbf{c}^T \mathbf{m} \tag{8.32}$$

$$\sum_{k=1}^{N} w_k m_k = \mathbf{w}^T \mathbf{m} \tag{8.33}$$

The problem is to maximize the general reliability index (8.30) over the number of duplicate devices in each stage. The following constraints must also be satisfied:

$$\mathbf{c}^T \mathbf{m} = A \tag{8.34}$$

$$\mathbf{w}^T \mathbf{m} = B \tag{8.35}$$

Here we have a special case, and thus algorithm (8.25) must be applied, but with the equality constraints. The algorithm is

$$\begin{aligned}\mathbf{m}[n] &= \mathbf{m}[n-1] - \gamma[n](\tilde{\nabla}_{m\pm} \theta[\mathbf{m}[n-1], \mathbf{c}, \mathbf{w}] + \lambda_1[n]\mathbf{c} + \lambda_2[n]\mathbf{w}) \\ \lambda_1[n] &= \lambda_1[n-1] + \gamma_1[n](A - \mathbf{c}^T \mathbf{m}[n-1]) \\ \lambda_2[n] &= \lambda_2[n-1] + \gamma_2[n](B - \mathbf{w}^T \mathbf{m}[n-1])\end{aligned} \tag{8.36}$$

where

$$\tilde{\nabla}_{m\pm} \theta = (\tilde{\theta}_{m_1}, \tilde{\theta}_{m_2}, \ldots, \tilde{\theta}_{m_N})$$

$$\tilde{\theta}_{m_k}[\mathbf{m}[n-1], \mathbf{c}, \mathbf{w}] = \begin{cases} \theta[\mathbf{d} + \mathbf{e}, \mathbf{c}, \mathbf{w}] - \theta[\mathbf{d}, \mathbf{c}, \mathbf{w}] & \text{if } m_k[n-1] \text{ is not an integer} \\ \frac{1}{2}[\theta[\mathbf{d} + \mathbf{e}, \mathbf{c}, \mathbf{w}] - \theta[\mathbf{d} - \mathbf{e}, \mathbf{c}, \mathbf{w}]] & \text{if } m_k[n-1] \text{ is an integer} \end{cases} \tag{8.37}$$

Here, $\mathbf{e} = (1, 1, \ldots, 1)$ is an N-dimensional unit vector, $k = 1, \ldots, N$, and

$$\mathbf{d} = (d_1, \ldots, d_N) \tag{8.38}$$

where d_v is an integer part of m_v ($v = 1, \ldots, N$). The problem can be studied in an analogous way even if the inequality constraints are considered, but then more complex expressions for the algorithms are obtained.

8.12 Increase of Reliability by Redundant Components

Let an input signal s be simultaneously applied to $N + 1$ identical communication channels. Output signal x_v of the vth channel differs from the input signal due to the existence of noise ξ_v:

$$x_v = s + \xi_v \tag{8.39}$$

We shall assume that the signal s and the noise ξ_v are uncorrelated, and that ξ_k and ξ_l for $k \neq l$ are also uncorrelated. The problem is to find the best (in a given sense) estimates of the true value of the signal s. The quality of estimation will be measured by the mean square deviation of the output signal x from the true signal s.

The estimate of the true value of the signal will be sought in the form

$$\hat{s} = \sum_{v=0}^{N} c_v x_v \Big/ \sum_{v=0}^{N} c_v \qquad (8.40)$$

or, using the vector notation,

$$\hat{s} = \frac{\mathbf{c}^T \mathbf{x}}{\mathbf{c}^T \mathbf{e}}$$

where $\mathbf{c}$ is still an unknown vector of coefficients. This estimator is indeed the best one according to the principle of maximum likelihood estimation or the least square error estimation when the distribution of the noise is gaussian.

We shall find the vector of coefficients $\mathbf{c} = \mathbf{c}^*$ such that the functional

$$J(\mathbf{c}) = M\{F(s - \hat{s})\} \qquad (8.41)$$

is minimized; F is a convex function.

It is assumed that after each measurement x_v we can measure the noise component in at least one channel (the rth, for instance):

$$\xi_r = x_r \quad \text{for} \quad s = 0 \qquad (8.42)$$

This can be accomplished if we measure the noise during the intervals of time when it is known that the input signal is not present. Considering (8.39) and (8.40), we can write (8.41) as

$$J(\mathbf{c}) = M\left\{F\left(x_r - \xi_r - \frac{\mathbf{c}^T \mathbf{x}}{\mathbf{c}^T \mathbf{e}}\right)\right\} \qquad (8.43)$$

The gradient can be easily computed using the relationship

$$\nabla_{\mathbf{c}} F\left(x_r - \xi_r - \frac{\mathbf{c}^T \mathbf{x}}{\mathbf{c}^T \mathbf{e}}\right) = -F'\left(x_r - \xi_r - \frac{\mathbf{c}^T \mathbf{x}}{\mathbf{c}^T \mathbf{e}}\right)\left(\frac{\mathbf{x}}{\mathbf{c}^T \mathbf{e}} - \mathbf{e}\frac{\mathbf{c}^T \mathbf{x}}{(\mathbf{c}^T \mathbf{e})^2}\right) \qquad (8.44)$$

The gradient could always be determined by the values x and measurements ξ_r in the interval of time when the signal does not exist. Then on the basis of the algorithm of adaptation, we obtain

$$\mathbf{c}[n] = \mathbf{c}[n-1] + \gamma[n] F'\left(x_r[n] - \xi_r[n] - \frac{\mathbf{c}^T[n-1]\mathbf{x}[n]}{\mathbf{c}^T[n-1]\mathbf{e}}\right)$$

$$\times \left(\frac{\mathbf{x}[n]}{\mathbf{c}^T[n-1]\mathbf{e}} - \mathbf{e}\frac{\mathbf{c}^T[n-1]\mathbf{x}[n]}{(\mathbf{c}^T[n-1]\mathbf{e})^2}\right) \qquad (8.45)$$

If the variance σ_2 of the noise ξ is known a priori, then for the special case, when F is a quadratic function, the need for adaptation would cease to exist, and one can find the optimal vector of the coefficients directly from (8.43) by

$$\mathbf{c}^* = k\boldsymbol{\sigma}^{-1} \qquad (8.46)$$

where

$$\boldsymbol{\sigma}^{-1} = \{\sigma_1^{-1}, \ldots, \sigma_M^{-1}\} \qquad (8.47)$$

The adaptive approach is efficient when a priori information about the noise does not exist. Of course, in this simplest case we could have first used adaptive algorithms for computing the noise variance, and then determined the optimal parameters according to (8.46).

8.13 Design of Complex Systems

In designing complex multicomponent systems, we are always confronted with the discrepancies between the nominal and the real values of the parameters. Due to different causes, these parameters are random functions of time. Naturally, the vector of parameters $\mathbf{x}(t) = (x_1(t), \ldots, x_N(t))$ differs from the initial vector of parameters $\mathbf{x}_0 = (x_{01}, \ldots, x_{0N})$. Let us designate by $\psi_k(\mathbf{x}_0, \mathbf{x})$ certain external characteristics of the system. They depend not only on the initial ($\mathbf{x}_0$) but also on the current ($\mathbf{x}$) vector of parameters; these characteristics define the productivity of the system.

The system is operating if these external characteristics satisfy certain specific conditions (for instance, they are within certain limits, or they take the values which are, in a certain sense, close to certain previously specified values).

One of the basic steps in the design of complex systems is to guarantee the productivity of the system. This can be accomplished by selecting an initial parameter vector $\mathbf{x}_0$. Of course, the productivity depends on the current value of the parameter vector $\mathbf{x}$, i.e., it depends on the structure of the system. Since the number of the components is limited, and the structure of the system is frequently determined by its extreme application, changes are impossible. Therefore, the problem is to select an initial parameter vector which gives an optimally productive system.

8.14 Algorithms of Optimal Productivity

If the productivity of the system is characterized by the probability that not a single external characteristic falls outside the permissible limits, i.e.,

$$J(\mathbf{x}_0) = P\{\alpha_k \leq \psi_k(\mathbf{x}_0, \mathbf{x}) < \beta_k\} \qquad (k = 1, 2, \ldots) \qquad (8.48)$$

then we have a problem which is similar to one considered in Section 7.11 and is related to the control by perturbation. Therefore, in order to determine the optimal initial vector $\mathbf{x}_0^*$ we can use the search algorithm

$$\mathbf{x}_0[n] = \mathbf{x}_0[n-1] - \gamma[n]\tilde{\nabla}_{\mathbf{x}_0^+}\theta(\mathbf{x}[n], \mathbf{x}_0[n-1], a[n]) \quad (8.49)$$

where

$$\tilde{\nabla}_{\mathbf{x}_0^+}\theta(\mathbf{x}, \mathbf{x}_0, a) = \frac{\theta(\mathbf{x}, \mathbf{x}_0 + a) - \theta(\mathbf{x}, \mathbf{x}_0)}{a}$$

with

$$\theta(\mathbf{x}, \mathbf{x}_0, a) = \begin{cases} 0 & \text{if } \alpha_k \le \psi_k(\mathbf{x}, \mathbf{x}_0 + a) < \beta_k \\ 1 & \text{otherwise} \end{cases} \quad (8.50)$$

In a number of cases it can be appropriate to use another criterion of optimality:

$$J(\mathbf{x}_0) = M\{F(\psi(\mathbf{x}, \mathbf{x}_0) - A)\} \quad (8.51)$$

and the constraint

$$M\{L_k(\psi_k(\mathbf{x}, \mathbf{x}_0)) - B_k\} \le 0 \quad (k = 1, \ldots, N) \quad (8.52)$$

where A and B_k are constants.

The functional (8.51) can be minimized under the constraints (8.52) using algorithm (3.24):

$$\begin{aligned}\mathbf{x}_0[n] &= \mathbf{x}_0[n-1] - \gamma[n](\nabla_{\mathbf{x}_0} F(\psi(\mathbf{x}[n], \mathbf{x}_0[n-1]) - A) \\ &\quad + H_c(\mathbf{x}[n], \mathbf{x}_0[n-1]\lambda[n-1] \\ \lambda[n] &= \max\{0, \lambda[n-1] + \gamma_1[n](L(\mathbf{x}[n], \mathbf{x}_0[n-1]) - B)\}\end{aligned} \quad (8.53)$$

where

$$\lambda[0] \ge 0 \quad (8.54)$$

and

$$L(\mathbf{x}, \mathbf{x}_0) = (L_1(\psi_1(\mathbf{x}, \mathbf{x}_0)), \ldots, L_N(\psi_N(\mathbf{x}, \mathbf{x}_0)))$$
$$B = (B_1, \ldots, B_N)$$
$$H_c(\mathbf{x}, \mathbf{x}_0) = \left\|\frac{\partial L_k(\mathbf{x}, \mathbf{x}_0)}{\partial x_{0\mu}}\right\| \quad (\nu = 1, \ldots, N; \mu = 1, \ldots, M) \quad (8.55)$$

In these cases, when the components of the parameter vector can take only discrete values (these values could be nominal values of the parameters in different components of the scheme–resistors, condensers), the algorithms described in Section 8.10 should be applied. The scheme realizing algorithm (8.55) is shown in Fig. 8.6.

8.15 The min-max Criterion of Optimization

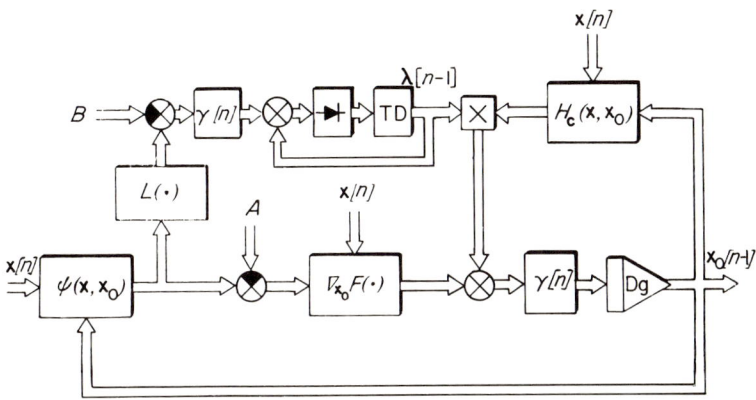

Fig. 8.6

8.15 The min-max Criterion of Optimization

It is sometimes necessary to satisfy stricter criteria than those considered in Section 8.14 in order to guarantee the productivity of the system. In particular, we could use the criterion of minimizing the least upper bound of a family of average characteristics of the system with respect to the parameter c:

$$J(c) = \max_k M\{F_k(\psi(\mathbf{x}, c))\} \tag{8.56}$$

In this case we have the so-called min-max problem.

For the special types of the function $F_k(\psi_k(\mathbf{x}, c))$, there exist iterative procedures which allow the solution of the problem using only the values of F_k evaluated at fixed values of c. For instance, if $F_k(\psi_k(\mathbf{x}, c))$ represents the probability that the kth characteristic falls outside the permissible limits $[\alpha_k, \beta_k]$ for $c = c^0$, i.e.,

$$F_k(\psi_k(\mathbf{x}, c^0)) = P\{\psi_k(\mathbf{x}, c^0) \notin [\alpha_k, \beta_k]\} \tag{8.57}$$

the problem can be solved with the help of algorithm (7.51):

$$c[n] = c[n-1] - \gamma[n](c[n-1] - c_m^0) \tag{8.58}$$

Here, m is the index of the characteristic of the system which at the nth stage falls outside the permissible limits (for more details see Sections 7.11–7.13). As it can be seen, by selecting a broader class of functions $F_k(\cdot)$, we can construct an iterative procedure for estimating the points of the min-max (8.56) for the regression function using the values $F_k(\psi_k(\mathbf{x}, c))$ for certain fixed values of c. It is possible that such a solution can also be obtained using the game theoretic approach to this problem (see Chapter 10).

8.16 More about the Design of Complex Systems

In the problems considered earlier, it was possible to optimize the system productivity only by varying the initial values of the parameters of the system. This influences the statistical characteristics of the current parameters of the system. Another more general approach to problems of this type is possible.

Let the structure of the system be given. The values of the component parameters are the random variables with unknown distributions. We shall determine the statistical characteristics of these random variables or the values of the parameters specifying their distributions for which a certain criterion of optimality for this system will be satisfied. For instance, if the distribution of the parameter vector $\mathbf{x}[n] = (x_1[n], \ldots, x_N[n])$ of the system components depends on the parameters $\mathbf{u} = (u_1, \ldots, u_N)$, then the output signal can be written as

$$\mathbf{y}[n] = T(\mathbf{x}[n]) \tag{8.59}$$

where T is a known operator defined by the equation of the system. In the statistical sense, the signal $\mathbf{y}[n]$ is defined completely by the vector $\mathbf{u}$. The problem is then to determine the optimal values of the parameters in the probability density function using the values of a certain performance index $J(\mathbf{y}, \mathbf{u})$, i.e., to find the value of a vector $\mathbf{u}^*$ for which the mathematical expectation

$$\bar{J} = M_\mathbf{x}\{J(\mathbf{y}, \mathbf{u})\} \tag{8.60}$$

is minimal. The criterion of optimality $J(\mathbf{y}, \mathbf{u})$, like the cost functions in other problems, is selected on the basis of physical requirements imposed on the system under consideration. In particular, $J(\mathbf{y}, \mathbf{u})$ should be defined as

$$J(\mathbf{y}, \mathbf{u}) = J_1(\mathbf{y}) + J_2(\mathbf{u}) \tag{8.61}$$

where J_1 describes the error in the output variable of the system, and J_2 describes the expenses related to the improvement of tolerances in the system components. Some other additional constraints are possible.

If the variations in the statistical characteristics of the system parameters are allowed, this problem can be solved using probabilistic algorithms with or without search. Moreover, it is possible to use adaptation for improving the index of optimality alone.

8.17 Comment

Many of the problems in queuing theory which are related to the quantitative investigations of the processes and the qualities of the serviced systems, are reduced to models studied in connection with the problems of reliability. Therefore, it is possible to apply many algorithms described in the previous sections of this chapter to the solution of problems in queuing theory.

It is useful then to keep in mind the correspondence between the concepts and terms of the theory of reliability and the queuing theory. Usually, the "failure" and "repair" correspond to "demand" and "maintenance." In this manner the adaptive approach is extended even to the problems of queuing theory.

8.18 Certain Problems

The adaptive approach is obviously the most valuable in the accelerated reliability tests. Development of the adaptation method applicable to this situation can greatly simplify the evaluation of the reliability of the tested system. Development of the synthesis methods for highly reliable systems in which the adaptation is used both to introduce the additional essential components, and to remove less active parts of the system, is very much needed.

As it was mentioned in Section 8.14, we can select the maximum of deviation from the average characteristics of the system to be the criterion of optimality. Is it possible then to minimize this criterion using only the performance index evaluations without computing the maximum of the deviation from the average characteristics?

8.19 Conclusion

In this chapter we have demonstrated the applicability of the adaptive approach to solving the problems of estimating and optimizing reliability indices. Of course, the problems which were considered here are relatively simple. They may also be academic to the extent that specialists interested in the problems of system reliability increase could study them with a deserved seriousness and appreciation. Let us remember that many important things have simple beginnings. We hope that the simplicity of these problems does not cause laughter, but that it will be a source of serious thinking toward the application of the adaptive approach to more complicated realistic problems which need urgent solutions.

COMMENTS

8.2–8.4   On the basic definition, see the book by Gnedenko *et al.* (1965), which contains a bibliography on the mathematical reliability theory.

8.11   Bellman looked at this problem from the "heights" of the dynamic programming method.

8.12   This problem was solved by Pierce (1962) in the case of sufficient a priori information.

8.13   A solution of this min-max problem on the basis of the Monte Carlo method was described in the paper by Evseev *et al.* (1965).

8.15 A number of optimization problems based on the reliability criterion were considered in the book by Iiudu (1966).

The problems of optimizing information networks were considered by Kelmans (1964) and Kelmans and Mamikonov (1964). Interesting results in the area of search for failures were obtained by Kletsky (1961) and Ruderman (1963).

8.16 The statement of this problem and its solution on the basis of a search algorithm is due to Gray (1964).

BIBLIOGRAPHY

Evseev, V.F., Korablina, T.D., and Rozhkov, S.I. (1965). An iterative method of designing systems with stationary random parameters, *Izv. Akad. Nauk SSSR Tekh. Kibern.* **1965** (6). (Engl. transl.: *Eng. Cybern.* (*USSR*).)

Gnedenko, B.V., Belyaev, Yu.K., and Solovev, A.D. (1965). "Mathematical Methods in Reliability Theory." Nauka, Moscow.

Gray, K. (1964). Application of stochastic approximation to the optimization of random circuits, *Proc. Symp. Appl. Math.* **16**.

Iiudu, K.A. (1966). "Optimization of Automatic Systems Using Reliability Criteria." Energia, Moscow-Leningrad. (In Russian.)

Kelmans, A.K. (1964). On certain optimal problems in the reliability theory of information networks, *Automat. Remote Contr.* (*USSR*) **25** (5).

Kelmans, A.K., and Mamikonov, A.G. (1964). On the design of systems for information transmission which are optimal in reliability, *Automat. Remote Contr.* (*USSR*) **25** (2).

Kletsky, E. (1961). Application of information theory in the detection of failures in engineering systems. *In* "Zarubezhnaia Radioelektronika," No. 9.

Moskowitz, F., and McLean, T.P. (1956). Some reliability aspects of systems design, *IRE Trans. Reliability Quality Contr.* **PGRQC-8**.

Pierce, W.H. (1962). Adaptive design elements to improve the reliability of redundant systems, *IRE Inter. Conv. Rec.* 4.

Ruderman, S.Yu. (1963). The questions of reliability and of the search for the failures in the systems, *Izv. Akad. Nauk SSSR Tekh. Kibern.* **1963** (6). (Engl. transl.: *Eng. Cybern.* (*USSR*).)

9

OPERATIONS RESEARCH

9.1 Introduction

We shall now enter another area which is related to the optimal methods of organizing the goal-directed processes and the results of human activities. These problems are studied in a special branch of science called operations research. Everything related to the organization of activities directed toward the fulfillment of certain goals belongs to operations research. The subject of operations research is so broad that it is doubtful that a sufficiently complete definition can be given.

It is frequently said (and this is not without grounds) that operations research represents a "quantitative expression of common sense" or "art of giving bad answers" to those practical problems for which other methods give even worse answers. It is naive and purposeless to discuss operations research in general, but it is courageous and impossible to attempt to embrace all of its concrete applications. Therefore, we can only select several typical problems of operations research and show that it is not only possible to look at them from some unusual points of view which stem from the adaptive approach developed above, but that their solution can also be obtained when other approaches are neither appropriate nor suitable.

Among typical operations research problems are the problems of planning, distribution of reserves and resources, construction of servicing systems, etc. These problems are extremely important. The solution of similar problems frequently provides an answer to the question of what is better: fast but expensive or slow but inexpensive. These problems are usually considered under the conditions of sufficient a priori information regarding the probability density functions appearing in the problems; frequently, these are unknown. In such cases, the problem is replaced by a min-max problem. But are we always justified to replace one problem by another? Perhaps it is better to observe these problems from the standpoint of adaptive approach which does not require a priori knowledge about probability distributions. We shall attempt to show the fruitfulness of such an approach.

9.2 Planning of Reserves

The problem of optimally planning reserves consists of finding the volume of production or purchases, the sequence of supply or the level of reserves necessary to satisfy future demands under minimal losses and expenses. It is clear that extremely large reserves lead to redundancy of material values and demand large expenses for storage facilities. Insufficient reserves may cause interruptions in production. We shall now begin by formulating the problems.

Let us assume that there is a single base designated for both the reserve and the delivery of goods (commodities, wares). The goods arrive and leave the base at discrete instants of time. An order arrives at the base L units of time after it was made. The demand for the goods, i.e., the number of goods which are needed in a unit of time, is determined by the conditions outside the base, and they do not depend on the reserves in the base. In the operation of a base, the expenses are related to the preparatory closing operations, to the maintenance of reserves and to the losses due to deficit (insufficient goods in the base to satisfy demands). The effectiveness of operating the base depends upon the operating conditions (in particular, on the demands, the length of the interval L, characteristics of the expenses during a unit of time, etc.), and on the policy of ordering the goods. The policy of ordering can, for instance, consist of specifying the quantity of ordered goods and the instant when the order is made.

Let us assume that an order for quantity q is made every time the reserves in the base become equal to a certain level p. The problem of planning the reserves in the given case consists of defining a policy of ordering (i.e., in defining such values p^* and q^*) for which a certain criterion of optimality in planning the orders is satisfied. The demand $x(t)$ is usually not known a priori. Therefore, it is reasonable to assume that $x(t)$ is a certain random process with statistical characteristics determined by external conditions (by activity of other enterprises, commodity market, etc.). Considering this, we can choose the mathematical expectation of the losses during a unit of time to be a criterion of optimality in planning the reserves. Averaging is performed over the interval $[t_0, t_0 + T]$ between two arrivals of supplies.

We define a vector

$$\mathbf{c} = (p, q) \tag{9.1}$$

which characterizes an order. The losses in a time interval $[t_0, t_0 + T]$, where t_0 is the instant when an order arrives at the base, can be written as

$$F(x, \mathbf{c}, T(x, \mathbf{c})) \tag{9.2}$$

Here, T is the interval between the arrivals of two successive orders; this is a random variable. Its distribution can depend on the vector $\mathbf{c}$ and the distribution of $x(t)$. Therefore, the criterion of optimality in planning can be written in a general case as

$$J(\mathbf{c}) = M_x\{F(x, \mathbf{c}, T(x, \mathbf{c}))\} \tag{9.3}$$

The problem is now to select a vector $\mathbf{c} = \mathbf{c}^*$ for which the criterion $J(\mathbf{c})$ reaches a minimal value.

9.3 The Criterion of Optimality in Planning

Since examples are more educational than rules, we shall examine a specific form of the functional (9.3). Let us assume that the demand $x(t)$ is a certain stationary random process with a constant mathematical expectation equal to r. We shall also assume that the arrival of an order of any quantity q is instantaneous, and that the delay time is L; the losses d on the unit of deficit, and the expenses h on the storage of goods are constant and a priori known to us. For simplicity, we shall also assume that the conditions of operating the base and the policy of ordering are such that the total quantity of stored goods and the total quantity of the scarce goods (critical commodities) during a time interval $[t_0, t_0 + T]$ are such that the demand during the same interval is constant and equal to

$$x(t) = \frac{I(t_0) - I(t_0 + T)}{T} \tag{9.4}$$

for any t_0 where $I(t)$ is the level of reserves in the base at the instant t. The critical commodities are considered to be goods of which the reserves on the base are less than the demands.

Then, in estimating the expenses for a unit of time over the interval $[t_0, t_0 + T]$, we can assume that the level of reserves is a piecewise linear function of time (Fig. 9.1).

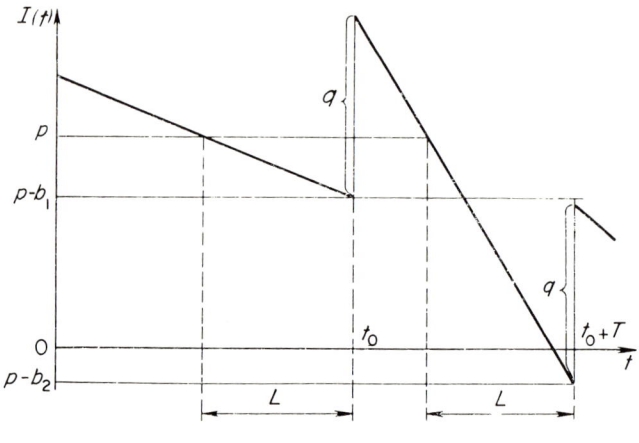

Fig. 9.1

For this case, the expenses for storing the reserves per unit of time are equal to

$$\mathscr{F}_1 = \begin{cases} \dfrac{1}{2}(2p - b_1 + q - b_2) & \text{when } b_2 \leq p \\ \dfrac{1}{2}\dfrac{(p + b_1 + q)^2}{q - b_1 + b_2} & \text{when } b_2 > p \end{cases} \quad (9.5)$$

The losses due to deficit per unit of time are

$$\mathscr{F}_2 = \begin{cases} (b_2 - p)\dfrac{d}{T} & \text{when } b_2 > p \\ 0 & \text{when } b_2 \leq p \end{cases} \quad (9.6)$$

Finally, the losses of preparatory closing operations per unit of time are equal to

$$\mathscr{F}_3 = \frac{s}{T} \quad (9.7)$$

Using vector notation (9.1) for an order, we can write the criterion of optimality in planning as

$$J(\mathbf{c}) = M\{F(x, \mathbf{c}, T(x, \mathbf{c}))\} \quad (9.8)$$

9.4 Algorithm of Optimal Planning

where the expenses $F(\cdot)$ per unit of time over the interval $[t_0, t_0 + T]$ are defined by the expression

$$F(x, \mathbf{c}, T(x, \mathbf{c})) = \frac{1}{2}(2p + q - b_1 - b_2) \operatorname{sgn}(p - b_2)$$

$$+ \frac{1}{2}\frac{(p + q - b_1)^2}{q - b_1 + b_2} \operatorname{sgn}(b_2 - p)$$

$$+ \frac{d}{T(q)}(b_2 - p)\operatorname{sgn}(b_2 - p) + \frac{s}{T(q)} \quad (9.9)$$

where

$$b_1 = \int_{t_0 - L}^{t_0} x(t)\, dt \qquad b_2 = \int_{t_0 + T - L}^{t_0 + T} x(t)\, dt$$

and

$$\operatorname{sgn} z = \begin{cases} 1 & \text{when } z \geq 0 \\ 0 & \text{when } z < 0 \end{cases}$$

It is not difficult to show that in the model described above, T is only a function of q and does not depend on p. This simplifies the procedure of finding the optimal solution.

9.4 Algorithm of Optimal Planning

If the probability distribution of demands $x(t)$ is known, we can define $J(\mathbf{c})$ in explicit form and minimize it with respect to $\mathbf{c}$. This is a usual procedure, but frequently we do not have any information regarding the probability distribution of $x(t)$. Then, instead of guessing and choosing with uncertainty a probability distribution, we can use the firm basis of the adaptive approach and avoid such difficulties. Let us apply the algorithms of minimization in minimizing (9.9). We then obtain

$$\mathbf{c}[n] = \mathbf{c}[n-1] - \gamma[n]\nabla_{\mathbf{c}} F(x[n], \mathbf{c}[n-1], T(x[n], \mathbf{c}[n-1])) \quad (9.10)$$

if the function $F(\cdot)$ is differentiable, and

$$\mathbf{c}[n] = \mathbf{c}[n-1] - \gamma[n]\tilde{\nabla}_{\mathbf{c} \pm} F(x[n], \mathbf{c}[n-1], a[n], T(x[n], \mathbf{c}[n-1], a[n]))$$

$$(9.11)$$

if it is not convenient to find the gradient $\nabla_{\mathbf{c}} F(\cdot)$.

9.5 More about the Planning of Reserves

Let us consider a slightly different model of the production and sale of goods (commodities). Assume that the goods arrive at the base in equal intervals of time T (planned intervals). Arrival of goods is instantaneous. After each arrival, during time interval T there will be a sale of goods determined by the initial reserve of goods at the beginning of the interval T and the demand for goods during that interval. Furthermore, let the demand r for goods during the planned interval of time T be a random variable. Probability density function $p(r)$ exists, but it is not known to us.

The process of obtaining, storing and selling the goods during time interval T is necessarily connected with certain losses $F(\cdot)$. These losses depend on the demand r and on the received supplies s at the beginning of the interval T, i.e.,

$$F = F(r, s) \tag{9.12}$$

We shall find the optimal value of the supply s for which the average losses, (expenses) over the time interval

$$J(s) = M_r\{F(r, s)\} \tag{9.13}$$

are minimal. The function $F(r, s)$ very much depends upon the type of losses which have to be taken into consideration. If it is possible to determine the values of the function $F(r, s_k)$ for different but specific values s_k, then the optimal value s^* can be obtained by applying corresponding algorithms of adaptation. In the following sections we shall examine several particular cases of special interest.

9.6 Optimal Supply

Let the losses $F(\cdot)$ during a time interval T be determined by the losses due to an excess of goods $R_1(T)$ and a lack of goods (deficit) $R_2(T)$:

$$R_1(T) = \frac{h}{2}[|s - r| + (s - r)]$$

$$R_2(T) = \frac{d}{2}[(r - s) + |s - r|] \tag{9.14}$$

and

$$F(r, s) = R_1(T) + R_2(T) \tag{9.15}$$

where h and d are the corresponding losses per unit of goods. Then the criterion (9.13) has the form

9.7 Optimal Reserves

$$J(s) = h \int_0^s (s-r)p(r)\,dr + d \int_s^\infty (r-s)p(r)\,dr \qquad (9.16)$$

and the problem of finding optimal supply is reduced to one of solving the equation,

$$\frac{dJ(s)}{ds} = (h+d)\int_0^s p(r)\,dr - d = 0 \qquad (9.17)$$

which can be written as

$$M_r\{\text{sgn}\,(s-r)\} = \frac{d}{h+d} \qquad (9.18)$$

or, since sgn $z = \frac{1}{2}(\text{sign } z + 1)$,

$$M_r\{\text{sign}\,(r-s)\} = \frac{h-d}{h+d} \qquad (9.19)$$

and the algorithm for finding optimal supply has the following form:

$$s[n] = s[n-1] + \gamma[n]\left[\text{sign}\,(r[n] - s[n-1]) - \frac{h-d}{h+d}\right] \qquad (9.20)$$

This simple algorithm is realized by the scheme shown in Fig. 9.2.

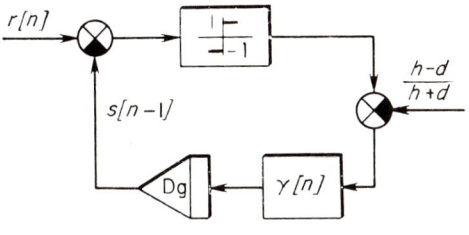

Fig. 9.2

9.7 Optimal Reserves

Let the total losses during the time interval T be defined as a sum of the losses on upkeep of reserves $R_1(T)$ and the losses due to the lack of goods $R_2(T)$:

$$R_1(T) = \frac{h}{2}\int_0^T (|I(t)| + I(t))\,dt$$

$$R_2(T) = \frac{d}{2}\int_0^T (|I(t)| - I(t))\,dt \qquad (9.21)$$

where $I(t)$ is the level of reserves at the instant t.

If we assume, as was done in the problem of optimal planning of reserves (Sections 9.2–9.4), that $I(t)$ is a linear function over the time interval t (see Fig. 9.1), then we obtain the loss function

$$\mathscr{F}(s, r) = \begin{cases} h\left(s - \dfrac{r}{2}\right)T & \text{when } r \le s \\ \dfrac{h}{2} st_1 + \dfrac{d}{2}(r - s)t_2 & \text{when } r > s \end{cases} \quad (9.22)$$

where s is the level of reserves at the beginning of the time interval T; t_1 is the time after the arrival of goods within which all reserves are exhausted; $t = t_1(r)$; $t_2 = T - t_1$; and r is the demand for goods over the time interval T.

The problem of finding the optimal initial level of reserves $s = s^*$ in the beginning of the planned interval T, for which the functional $J(s) = M\{\mathscr{F}(s, r)\}$ is minimized, becomes a problem of solving the equation

$$\frac{dJ(s)}{ds} = (h + d)\left[\int_0^s p(r)\,dr + s\int_s^\infty \frac{p(r)}{r}\,dr\right] - d = 0 \quad (9.23)$$

which, as before, can be transformed into a convenient form,

$$M_r[(s - 1)\,\text{sign}\,(r - s) + s] = \frac{d - h}{d + h} \quad (9.24)$$

It is now obvious that the algorithm for obtaining s^* has the form

$$s[n] = s[n - 1] - \gamma[n]$$
$$\times \left[(s[n-1] - 1)\,\text{sign}\,(r[n] - s[n-1]) + s[n-1] - \frac{d - h}{d + h}\right] \quad (9.25)$$

A slightly more complicated scheme, shown in Fig. 9.3, corresponds to this algorithm.

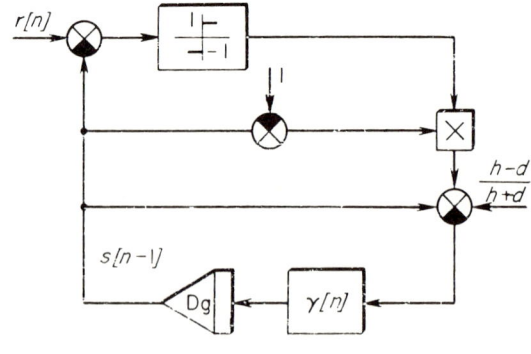

Fig. 9.3

9.8 A Comment

The problems of optimal supply (Section 9.6) and optimal level of reserves (Section 9.7) are special cases in the problem of optimal planning (Sections 9.2–9.5). However, these problems differ from the general problem of planning. They are characterized by a finite flow of reserves, and by a constant and a priori chosen time interval T, during which the losses (expenses) are observed. Finally, the supply of goods is also evaluated in these special problems.

9.9 Distribution of Production Resources

A production process usually consists of a sequence of interrelated processes (for instance, production of components of a final product). Each such process is characterized by a definite production rate. The production system has a number of storage facilities with volumes $v_1, \ldots, v_N$ for the storage of intermediate and final products. The rate of supply (raw material) and the rate of demand in the system can vary in a random manner. The system can be considered to work in a satisfactory fashion if each warehouse is neither full nor empty. The problem is to distribute the rate of completion in different production processes for which the probability of unsatisfactory operation of the enterprise, i.e., the probability that at least one warehouse is either full or empty, is minimized.

Let us designate by y_k the production volume which has to be stored in the kth warehouse, and by Ω_k the region defined by

$$\Omega_k = \{y: 0 < y_k \leq v_k\} \tag{9.26}$$

Then, by introducing the vector of production volume,

$$\mathbf{y} = (y_1, \ldots, y_N) \tag{9.27}$$

we can write the performance index of productivity in the enterprise:

$$J = P\{\mathbf{y} \notin \Omega\} \tag{9.28}$$

By introducing the characteristic function

$$\theta(\mathbf{y}) = \begin{cases} 0 & \text{if } \mathbf{y} \in \Omega_k \\ 1 & \text{if } \mathbf{y} \notin \Omega_k \end{cases} \tag{9.29}$$

we can write (9.28) as

$$J = M\{\theta(\mathbf{y})\} \tag{9.30}$$

The quantity of products (both **y** and θ) depends on the vector of completion rates in different processes $\mathbf{c}=(c_1, \ldots, c_N)$. Minimum $J(\mathbf{c})$ is obtained by using a search algorithm,

$$\mathbf{c}[n] = \mathbf{c}[n-1] - \gamma[n]\tilde{\nabla}_{\mathbf{c}^+}\theta(\mathbf{y}(\mathbf{c}[n-1], a[n])) \tag{9.31}$$

For other criteria of optimality in the operation of a system, the solution of the problem can be obtained either by using analogous search algorithms of adaptation, or by using relaxation algorithms (see Section 7.13).

9.10 An Example

As an illustration, we shall briefly consider a problem of planning the operation of a plant for producing sulfuric acid. The scheme describing the connections between the production units and the storage facilities is shown in Fig. 9.4. Blocks 1–5 correspond to the facilities for storing initial and inter-

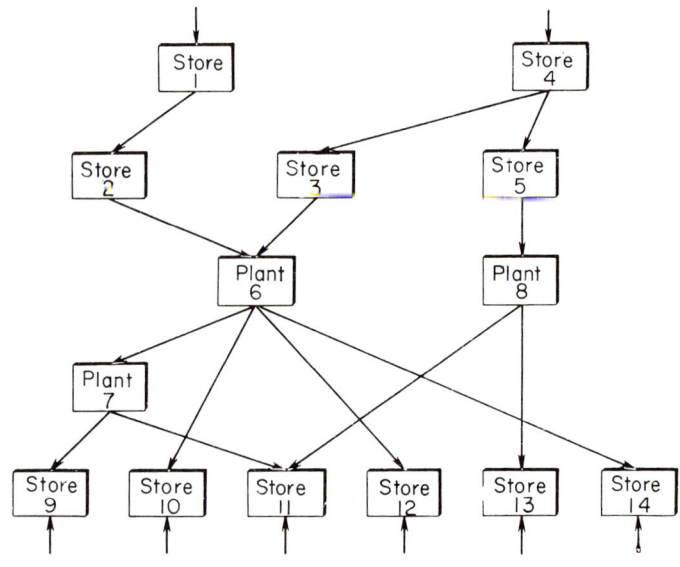

Fig. 9.4

mediate products. Blocks 6–8 are the devices which manufacture the products at the rates v_6, v_7 and v_8. Blocks 9–14 are the storage facilities for the final product. It is assumed that the demand y_0 for the sulfuric acid of different concentrations, and the rate **c** at which the raw material arrives, are determined by external factors, and that they are random variables distributed with certain unknown probability distributions p_1 and p_2. The volume of storage facilities 1–5 and 9–14, and therefore the rates of flow of the products between the corresponding blocks, are limited by the quantities $v_1, \ldots, v_5$ and $v_9, \ldots, v_{14}$. It could be stated that the plant operates in a satisfactory fashion if a single storage facility is neither empty nor overfilled. Optimal planning for a plant operation consists of establishing production rates $c_1^*, \ldots, c_N^*$ for which the probability of unsatisfactory plant operation is minimized. This is accomplished by algorithm (9.31).

9.11 Distribution of Detection Facilities

In radars for target detection, it is very important to find effective means for observing a given zone, i.e., the distribution of energy in that zone, the choice of sequential stages of observation in a certain fixed interval of time. We can select the functionals which characterize either the efficiency of observations under constant expenditures of energy, or the expenditures of energy under the conditions of constant probabilities of detection, to serve as the criteria of efficiency in observation.

Average efficiency of probing, which characterizes the efficiency of observing the desired zone, can be written in the following form:

$$J = \int_X F(f(\mathbf{x}), \mathbf{x}) p(\mathbf{x}) \, d\mathbf{x} \qquad (9.32)$$

or, briefly,

$$J = M\{F(f(\mathbf{x}), \mathbf{x})\} \qquad (9.33)$$

where $F(\cdot)$ is a reward function. For instance, $F(\cdot)$ can be the probability of detection under constant $\mathbf{x}$. The reward function is assumed to be given and to depend upon the distribution of available detection facilities, $f(\mathbf{x})$. For instance, the detection facilities are the strength of the probing signals, time of the search, etc. Since there are limitations in energy and time, $f(\mathbf{x})$ has to satisfy the condition

$$\int_X \psi(\mathbf{x}) f(\mathbf{x}) \, d\mathbf{x} = A \qquad (9.34)$$

which characterizes the limitations imposed on the resources (energy, number of alarms, time). In (9.34), $\psi(\mathbf{x})$ is a certain weighting function; frequently, $\psi(\mathbf{x}) = 1$.

The problem then is to find a function $f(\mathbf{x})$ for which the effectiveness of observation or the average reward (9.32) is maximized under the constraints (9.34). If the probability density function $p(\mathbf{x})$ is known, then under certain conditions imposed upon the reward F, the problem of finding the optimal function $f(\mathbf{x})$ can be solved analytically. However, $p(\mathbf{x})$ is usually unknown, and this problem is sometimes substituted by a min-max problem. An a priori distribution which is most unfavorable from the standpoint of average efficiency of observation (for which J is maximal) is first found. Then the distribution of the observation facilities, $f(\mathbf{x})$, for which this maximal value J is minimized, is determined. This substitution is not always justified and is often inconvenient. The adaptive approach offers an algorithm for finding the optimal function, and its method of realization.

9.12 Algorithm for Optimal Distribution

It is natural to try to find the optimal function $f(\mathbf{x})$ in the form

$$f(\mathbf{x}) = \mathbf{c}^T \boldsymbol{\phi}(\mathbf{x}) \tag{9.35}$$

Then the functional (9.33) and the constraint (9.34) can be written as

$$J(\mathbf{c}) = M\{F(\mathbf{c}^T \boldsymbol{\phi}(\mathbf{x}), \mathbf{x})\} \tag{9.36}$$

and

$$\mathbf{c}^T \mathbf{b} = A \tag{9.37}$$

where

$$\mathbf{b} = \int_X \psi(\mathbf{x}) \boldsymbol{\phi}(\mathbf{x}) \, d\mathbf{x} \tag{9.38}$$

It is now necessary to find a vector $\mathbf{c} = \mathbf{c}^*$ which maximizes the functional (9.36) under the constraint (9.37). This is a problem of conditional extremum under the equality constraints. An analogous problem, but with inequality constraints, was already considered in Section 8.8. We shall form the functional

$$J(\mathbf{c}, \lambda) = M\{F(\mathbf{c}^T \boldsymbol{\phi}(\mathbf{x}), \mathbf{x}) + \lambda(\mathbf{c}^T \mathbf{b} - A)\} \tag{9.39}$$

and apply to it algorithm (3.19), which in this case has a very simple form, since λ is a scalar (Lagrange multiplier). We then obtain

$$\mathbf{c}[n] = \mathbf{c}[n-1] - \gamma[n][F'(\mathbf{c}^T[n-1]\boldsymbol{\phi}(\mathbf{x}[n]), \mathbf{x}[n])\boldsymbol{\phi}(\mathbf{x}[n]) + \lambda[n]\mathbf{b}]$$

and $\tag{9.40}$

$$\lambda[n] = \lambda[n-1] - \gamma_1[n](\mathbf{c}^T[n-1]\mathbf{b} - A)$$

These algorithms are realized by a discrete system shown in Fig. 9.5. The basic contour realizes the first algorithm of (9.40), and the additional one realizes the second algorithm of (9.40).

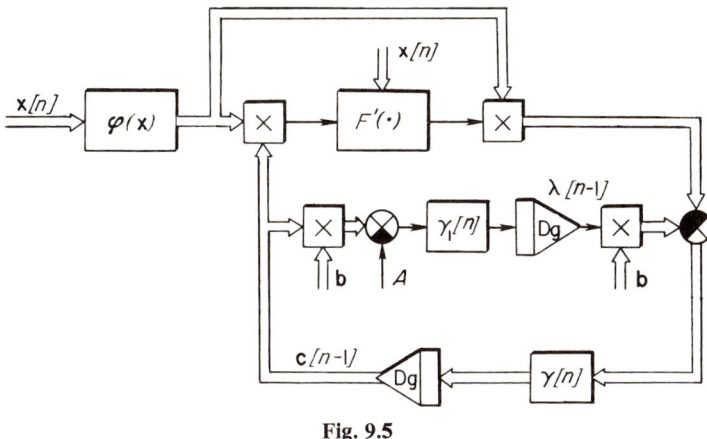

Fig. 9.5

If the data are observed continuously, we can also use the continuous algorithms in the problem considered above. Then, instead of (9.40), we obtain

$$\frac{d\mathbf{c}(t)}{dt} = -\gamma(t)[F'(\mathbf{c}^T(t)\boldsymbol{\phi}(\mathbf{x}(t), \mathbf{x}(t))\boldsymbol{\phi}(\mathbf{x}(t)) - \lambda(t)\mathbf{b}]$$

$$\frac{d\lambda(t)}{dt} = -\gamma_1(t)(\mathbf{c}^T(t)\mathbf{b} - A)$$

(9.41)

For the realization of this algorithm, we could use the previous scheme in which the discrete integrators are replaced by continuous integrators. After a period of training, the discrete continuous system determines the sought optimal function $f(\mathbf{x})$.

9.13 Distribution of Quantization Regions

Quantization is considered to be any transformation of a continuous set of values of a function or its arguments into a discrete set. Quantization is used in various communication systems for storage and processing of information, and it is one of the irreplaceable operations when digital computers are employed. For instance, photographs (functions of two variables) are decomposed into discrete lines before telegraphic transmissions, and the pictures (functions of three variables) are decomposed into nonoverlapping regions before television transmission. Transmission of audio signals (functions of a single variable) by pulse-code modulation is related to quantization of continuous signals and subsequent coding.

A geometrical description of quantization is shown in Fig. 9.6; the space of the continuous signal is partitioned into disjoint (nonoverlapping) regions Λ_k. Then, for instance, a certain discrete value equal to the index of a region can correspond to all the values of the continuous signal which belong to that region.

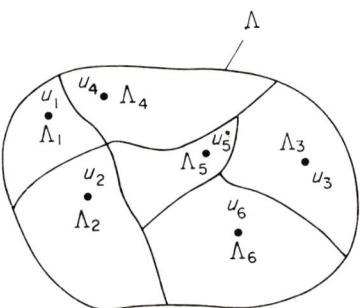

Fig. 9.6

Quantization of the arguments of a function always reduces the quantity of information which is contained in the continuous signal. Quantization of the values of a function introduces an error, since an uncountable set of values which belong to the same region is substituted by a single discrete value. In communication theory, this error, which is analogous if not identical to the truncation error, is called the quantization noise. The quantization error depends on the configuration and the dimensions of the regions Λ_k.

After transmitting discrete signals across a communication channel (the number of these discrete values is limited by the channel capacity), we have to determine the true signal. How should the signal space Λ be partitioned into the regions Λ_k so that the quantization noise is minimized? This problem is very closely related to the problems which were encountered in the statistical theory of detection and pattern recognition since it is again necessary to find the boundaries of the regions. But we shall consider it in this chapter in order to emphasize that the problem of quantizing the signal space can be studied with some success in parallel with the problems of distribution of resources and producing powers.

9.14 Criterion of Optimal Distribution

Let us designate by $\mathbf{x}$ a continuous signal which is to be quantized; its probability density function $p(\mathbf{x})$ is not known a priori. Furthermore, let u be an estimate which becomes constant for a given region such that

$$u = u_k \quad \text{when} \quad \mathbf{x} \in \Lambda_k \quad (k = 1, \ldots, N) \quad (9.42)$$

The criterion of optimal distribution and thus the accuracy criterion of an estimate u, can be average losses given in the form of a functional,

$$J(\mathbf{u}) = \sum_{k=1}^{N} \int_{\Lambda_k} F(\mathbf{x}, u_k) p(\mathbf{x}) \, d\mathbf{x} \tag{9.43}$$

Here, $\mathbf{u} = (u_1, \ldots, u_N)$. The optimal estimate $\mathbf{u} = \mathbf{u}^*$ minimizes average losses.

For simplicity we shall consider an interesting one-dimensional case where the regions Λ_k are the segments of the real line

$$\Lambda_k = \lambda_k - \lambda_{k-1} \qquad (k = 1, \ldots, N) \tag{9.44}$$

and u_k are real numbers. The average losses (9.43) are then equal to

$$J(\mathbf{u}) = \sum_{k=1}^{N} \int_{\lambda_{k-1}}^{\lambda_k} F(x, u_k) p(x) \, dx \tag{9.45}$$

For a quadratic loss function,

$$F(x, u) = (x - u)^2 \tag{9.46}$$

we have

$$J(\mathbf{u}) = \sum_{k=1}^{N} \int_{\lambda_{k-1}}^{\lambda_k} (x - u_k)^2 p(x) \, dx \tag{9.47}$$

The remaining problem is to find optimal estimates u_k ($k = 1, \ldots, N$) and the boundaries λ_k ($k = 1, \ldots, N$) such that the average losses are minimized.

9.15 Algorithms of Optimal Evaluation

The results of Section 4.20 could be used in defining an algorithm of optimal evaluation, but it is simpler for this case to differentiate (9.47) with respect to λ_k and u_k. The conditions which define the minimum of the functional can be written as

$$\lambda_k = \tfrac{1}{2}(u_k + u_{k+1}) \tag{9.48}$$

$$\int_{\lambda_{k-1}}^{\lambda_k} (x - u_k) p(x) \, dx = 0 \tag{9.49}$$

From (9.48) and (9.49), it follows that

$$\int_{\frac{1}{2}(u_k + u_{k-1})}^{\frac{1}{2}(u_{k+1} + u_k)} (x - u_k) p(x) \, dx = 0 \tag{9.50}$$

An analytic solution of this system of equations with respect to u_k is only possible for a simple case of $p(x) = \text{const}$. We consider the case where $p(x)$ is unknown.

Let us introduce the characteristic function,

$$\varepsilon(u_{k-1}, u_k, u_{k+1}) = \begin{cases} 1 & \text{if } \tfrac{1}{2}(u_k + u_{k-1}) < x < \tfrac{1}{2}(u_{k+1} + u_k) \\ 0 & \text{otherwise} \end{cases} \quad (9.51)$$

Then the condition (9.50) can be written as the mathematical expectation,

$$M\{(x - u_k)\varepsilon(u_{k-1}, u_k, u_{k+1})\} = 0 \quad (9.52)$$

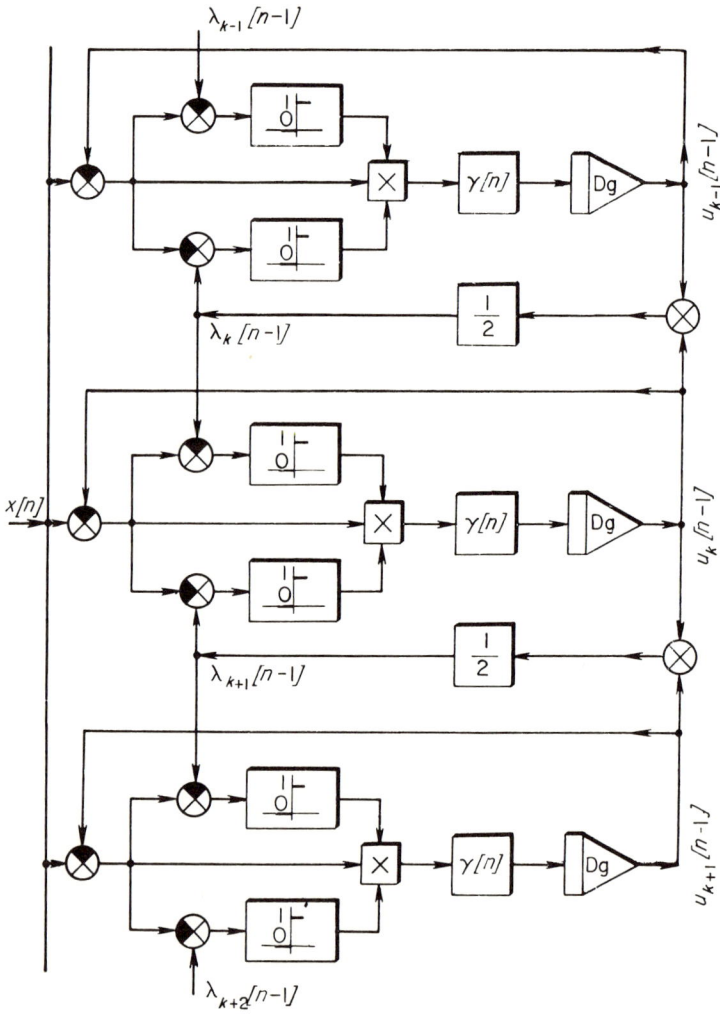

Fig. 9.7

Now it is not difficult to write an algorithm which determines optimal estimates

$$u_k[n] = u_k[n-1] + \gamma[n][(x[n] - u_k[n-1])$$
$$\times \varepsilon(u_{k-1}[n-1], u_k[n-1], u_{k+1}[n-1])] \qquad (k = 1, \ldots, N) \quad (9.53)$$

After obtaining the optimal estimates $u_k{}^*$ using this algorithm, we obtain the boundaries λ_k of the region Λ according to (9.48). A block diagram of the discrete system which determines these estimates is shown in Fig. 9.7; it is a multivariable closed-loop system. The reader can attempt to find the best values $\gamma_{\text{opt}}[n]$ for the algorithms (9.53).

9.16 Certain Problems

The problems of operations research are a fertile field for the application of probabilistic iterative algorithms which take into consideration different constraints. However, the difficulties which may arise here are not so much related to the development of the methods of solution, but rather to the statement of the problems concerning planning and control under the conditions of uncertainty and thus, risk. Therefore, it is of great interest to formulate various problems of planning under insufficient a priori information, and to extend the adaptive approach to these new problems.

9.17 Conclusion

This chapter differs from the preceding chapters by an abundance of problems. In spite of the variety encountered, it is possible to notice a general theme which is characteristic for all the problems considered here. In all the problems mentioned, we have to deal with the distribution and planning of limited resources. To such resources belong matter or energy, time or space. This variety is characteristic for the modern theory of operations research, and we have not tried to change its aspect.

The basic problem of this chapter was to show the usefulness of the adaptive approach even in this vast area embraced by an ambiguous but intriguing term—"operations research."

COMMENTS

9.1 The book by Kaufmann and Faure (1968) can serve as a good introduction to the theory of operations research. If the reader is not inclined to learn the details, he should start with the book by Saaty (1959).

9.2–9.8 Numerous similar problems were considered in the book by Hannsmenn (1966). However, it was assumed that sufficient a priori information exists.

9.9 and 9.10   The statement of these problems and the example are due to Wilde (1964).

9.11 and 9.12   On these questions, see the works by de Guenin (1961) and Kuznetsov (1962, 1964, 1965).

9.13–9.15   A problem of this kind was considered in relation to the synthesis of discrete systems with minimal distortion (see the works by Max (1960), Koshelev (1963) and Larson (1966)). The problems of quantization were also studied by Ignatiev and Blokh (1961) and Stratonovich and Grishanin (1966). The problem under consideration is closely related to the problem of self-learning discussed in Chapter 5. Max (1960) has solved this problem for the case of a uniform probability density function by using an approximate method.

BIBLIOGRAPHY

Basharinov, A.E., and Flaishman, B.S. (1962). "Methods of Statistical Analysis and Their Application in Radio Engineering." Sovyetskoe Radio, Moscow. (In Russian.)

Bellman, R., Glicksberg, I., and Gross, O. (1958). "Some Aspects of the Mathematical Theory of Control Processes." Rep. R-313. The Rand Corp.

Chuev, Yu.V. (ed.) (1965). "Foundations of Operations Research in Military Operations." Sovyetskoe Radio, Moscow. (In Russian.)

de Guenin (1961). Optimum distribution of effort: an extension of the Koopman basic theory, *Oper. Res.* **9** (1).

De Toro, M.T. (1956). Reliability criterion for constrained systems, *IRE Trans. Reliability Quality Contr.* **PGRQC-8**.

Fleisher, P.E. (1964). Sufficient conditions for achieving minimum distortion in a quantizer, *IEEE Inter. Conv. Rec.* **1**.

Hanssmenn, F. (1966). "Application of Mathematical Methods in Process and Storage Control." Nauka, Moscow. (In Russian.)

Ignatiev, N.K., and Blokh, E.L. (1961). Optimal discretization of multicomponent messages, *Radiotekhnika* (*Moscow*) **16** (6). (In Russian.)

Kaufmann, A., and Faure, R. (1968). "Introduction to Operations Research." Academic Press, New York.

Koopman, B.O. (1957). The theory of search, *Oper. Res.* **5** (5).

Koshelev, V.N. (1963). Quantization with minimal entropy. *In* "Problemi Peredachi Informatsii," Vol. 14. Academy of Sciences USSR, Moscow. (In Russian.)

Kuznetsov, I.N. (1962). The problems of radar search, *Radiotekh. Elektron.* **9** (2). (In Russian.)

Kuznetsov, I.N. (1964). Optimal distribution of limited resources for a convex-concave cost function, *Izv. Akad. Nauk SSSR Tekh. Kibern.* **1964** (5). (Engl. transl.: *Eng. Cybern.* (*USSR*).)

Kuznetsov, I.N. (1965). Minimization problem in the theory of search for a convex-concave cost function, *Izv. Akad. Nauk SSSR Tekh. Kibern.* **1965** (5). (Engl. transl.: *Eng. Cybern.* (*USSR*).)

Larson, R. (1966). Optimum quantization in dynamic systems. *In* "Proceedings of the Joint Automatic Control Conference, Seattle, Washington, 1966."

Max, J. (1960). Quantizing for minimum distortion, *IRE Trans. Inform. Theory.* **IT-6** (1).

Saaty, T.L. (1959). "Mathematical Methods of Operations Research." McGraw-Hill, New York.

Stratonovich, R.L., and Grishanin, B.A. (1966). The value of information when it is impossible to observe the estimated parameter directly, *Izv. Akad. Nauk SSSR Tekh. Kybern.* **1966** (3). (Engl. transl.: *Eng. Cybern. (USSR).*)

Tartakovskii, G.P. (ed.) (1962). "The Questions of Statistical Theory in Radar Systems." Sovyetskoe Radio, Moscow.

Wilde, D.J. (1964). "Optimum Seeking Methods." Prentice-Hall, Englewood Cliffs, N.J.

Wilde, D.J., and Acrivos, A. (1960). Short-range prediction planning of continuous operations, *Chem. Eng. Sci.* **12** (3).

10

GAMES AND AUTOMATA

10.1 Introduction

Until now we have considered the problems of finding absolute or relative extrema of various functionals under the conditions of uncertainty. But these are relatively peaceful situations. In life, we often encounter situations which have a conflicting character. Conflicting situations occur when the interests of the interacting sides do not coincide with each other. Of course, the effect called conflict of interests depends upon the actions of all sides participating in the conflict. Such conflicting situations are studied in the relatively new branch of science which has an appealing name—game theory.

A large number of works is devoted to game theory but we do not intend to present different results of this theory here. Our task is completely different. We want to observe a number of problems in game theory from the standpoint of the adaptive approach, and to show the suitability of this approach to typical game-theoretic problems and to such problems which are reduced to game-theoretic problems.

We shall consider the methods of learning the solutions of games and find the corresponding algorithms of learning. These algorithms will be used in the solution of linear programming and control problems which are conveniently observed as games, and also in defining the conditions of realizability of logical functions by a threshold element. We shall briefly illuminate possible methods of training perceptrons and the role of threshold elements in their realization. We shall also consider the behavior and the games of stochastic automata which are not designed using a priori information about transition probabilities from one state to another, and about the environments in which these automata interact.

10.2 The Concept of Games

Several persons—players—who have different interests usually participate in a game. The actions of the players—moves—consist of selecting a certain specific variant from a set of possible variants. The game is characterized by a system of rules which define the sequence of moves and the gains and losses of the players depending upon their moves. In the course of a game, different situations occur in which the players must make a choice. A complete system of instructions which define these choices in all possible situations is actually the strategy of each of the players. A player tries to maximize his gains at the end of a game. Of course, not all the players can achieve this.

The presented description of the game, although very similar to many games of chance which we confront in life, is sufficiently general to be of help in the popularization of the game theory, but it is not suitable for obtaining any concrete results. Therefore, we shall consider a particular form of games for which the theory is well-developed. Zero-sum matrix games between two persons, i.e., games in which the interests of the players are conflicting and the gain of one player is the loss of another, will also be discussed.

Zero-sum matrix games between two players are characterized by the cost matrix,

$$A = \|a_{\nu\mu}\| \quad (\nu = 1, \ldots, N; \mu = 1, \ldots, M) \quad (10.1)$$

In the game, the first player selects a strategy ν from N possible strategies, and the second player selects a strategy μ from M possible strategies. The gain of the first player, and thus the loss of the second player, is equal to $a_{\nu\mu}$. The choice of the strategies ν and μ can be made in a random fashion according to the probabilities,

$$\begin{aligned} \mathbf{p} &= (p_1, \ldots, p_N) \\ \mathbf{q} &= (q_1, \ldots, q_M) \end{aligned} \quad (10.2)$$

In this case, the outcome of the game is random, and it has to be evaluated by a mathematical expectation of the cost function:

$$V(\mathbf{p}, \mathbf{q}) = \sum_{\nu=1}^{N} \sum_{\mu=1}^{M} a_{\nu\mu} p_\nu q_\mu = \mathbf{p}^T A \mathbf{q} \quad (10.3)$$

This game is called the game with mixed strategies.

The probabilities (10.2) have to satisfy the following usual conditions:

$$\begin{aligned} p_\nu \geq 0 &\quad \sum_{\nu=1}^{N} p_\nu = 1 \\ q_\mu \geq 0 &\quad \sum_{\mu=1}^{M} q_\mu = 1 \end{aligned} \quad (10.4)$$

which indicate that **p** and **q** belong to a simplex. The mixed strategies of the players are defined by the vectors **p** and **q** where the kth components of each vector are equal to the probability of employing the kth simple strategy. If this probability is equal to one, we again arrive at a game with simple strategies.

We shall designate the simple strategies of the first and second player in the form of unit vectors

$$\boldsymbol{\alpha}_v = (0, \ldots, \underbrace{0, 1}_{v}, \ldots, 0)^{\overbrace{}^{N}}$$

$$\boldsymbol{\beta}_\mu = (0, \ldots, \underbrace{0, 1}_{\mu}, \ldots, 0)^{\overbrace{}^{M}}$$

(10.5)

It should be mentioned that for given probabilities **p** and **q**, according to (10.2),

$$\begin{aligned} p_v &= \mathbf{p}^T \boldsymbol{\alpha}_v \equiv M\{\boldsymbol{\alpha}_v\} & (v = 1, \ldots, N) \\ q_\mu &= \mathbf{q}^T \boldsymbol{\beta}_\mu \equiv M\{\boldsymbol{\beta}_\mu\} & (\mu = 1, \ldots, M) \end{aligned}$$

(10.6)

Then it follows that the components of the mixed strategies are the mathematical expectations of the corresponding employed simple strategies. The conditions (10.6) can be written in vector form as

$$\begin{aligned} \mathbf{p} &= M\{\boldsymbol{\alpha}\} \\ \mathbf{q} &= M\{\boldsymbol{\beta}\} \end{aligned}$$

(10.7)

where

$$\begin{aligned} \boldsymbol{\alpha} &= \{\boldsymbol{\alpha}_1, \ldots, \boldsymbol{\alpha}_N\} \\ \boldsymbol{\beta} &= \{\boldsymbol{\beta}_1, \ldots, \boldsymbol{\beta}_M\} \end{aligned}$$

(10.8)

are the collections of the unit vectors (10.5).

10.3 The min-max Theorem

Let us consider a game with the cost matrix (10.1). The natural desire of the first player to increase his gain leads to the selection of a strategy from all the strategies $\boldsymbol{\alpha}_v$ which guarantees maximal gain of all minimal gains, $\max_v \min_\mu a_{v\mu}$. The second player tends to select a strategy $\boldsymbol{\beta}_\mu$ which guarantees the smallest of all the maximal losses, $\min_\mu \max_v a_{v\mu}$. It is always

$$\max_v \min_\mu a_{v\mu} \le \min_\mu \max_v a_{v\mu}$$

(10.9)

It could be shown that this inequality becomes an equality,

$$\max_v \min_\mu a_{v\mu} = \min_\mu \max_v a_{v\mu} = a_{v^*\mu^*}$$

(10.10)

10.4 Equations of Optimal Strategies

It is then said that such a game of simple strategies has a saddle point $(\alpha_{v*}, \beta_{\mu*})$. For this strategy, $\hat{\alpha}_v = \alpha_{v*}$ and $\beta_\mu = \beta_{\mu*}$ are optimal. A deviation of any one of the players from the optimal strategy leads to an increase in the opponent's gain. If that is not enough, he can gain even more by selecting a proper strategy.

If equation (10.10) is satisfied by any cost matrix A, then the search for optimal forms of the game is ended. But frequently the cost matrix cannot have a saddle point. In this case, the player is forced to select mixed strategies in each game, and then the opponent cannot exactly determine the result of such a choice. Therefore, we arrive at the game with mixed strategies **p** and **q**. The basis of the game theory is von Neumann's theorem, as follows: The cost function (10.3) satisfies the relationship

$$\max_{\mathbf{p}} \min_{\mathbf{q}} V(\mathbf{p}, \mathbf{q}) = \min_{\mathbf{q}} \max_{\mathbf{p}} V(\mathbf{p}, \mathbf{q}) = v \qquad (10.11)$$

In other words, there are optimal strategies **p*** and **q*** for which

$$V(\mathbf{p}, \mathbf{q}^*) \le v = V(\mathbf{p}^*, \mathbf{q}^*) \le V(\mathbf{p}^*, \mathbf{q}) \qquad (10.12)$$

The quantity v is called the value of the game. This theorem states that a matrix game with mixed strategies always has a saddle point $(\mathbf{p}^*, \mathbf{q}^*)$. It is interesting that neither the first nor the second player have any advantage if they each know the probabilities of their opponent's moves.

10.4 Equations of Optimal Strategies

The solution of matrix games is reduced to finding the saddle point, i.e., we have to obtain the optimal strategies **p*** and **q*** for which the condition (10.12) is satisfied. There exist direct methods based on a theorem by Shappley-Snow for obtaining optimal strategies. These methods provide an exact solution of the game in a finite number of operations, but they are only practical for a game with a small number of strategies. Due to a relationship which exists between the game theory and linear programming, different methods of linear programming can be used in the solution of game-theoretic problems. There are also many indirect methods. Our interest lies in the methods where the player is taught during the course of the game and improves his mastery of the game.

Let us designate by $\hat{\mathbf{p}}(\mathbf{q})$ and $\hat{\mathbf{q}}(\mathbf{p})$ the optimal answers on the corresponding mixed strategies **q** and **p**. The optimal mixed strategies, determined by von Neumann's theorem, satisfy the equations

$$\begin{aligned} \mathbf{p}^* &= \hat{\mathbf{p}}(\mathbf{q}^*) \\ \mathbf{q}^* &= \hat{\mathbf{q}}(\mathbf{p}^*) \end{aligned} \qquad (10.13)$$

which express the obvious fact that the optimal responses on mixed strategies are optimal mixed strategies.

It follows from (10.13) that the optimal mixed strategies **p*** and **q*** are the solutions of the equations

$$\mathbf{p} = \hat{\mathbf{p}}(\mathbf{q})$$
$$\mathbf{q} = \hat{\mathbf{q}}(\mathbf{p}) \tag{10.14}$$

Since the mixed strategies can be given as mathematical expectations of the simple strategies (10.7), equations (10.14) can be written as

$$\mathbf{p} = M\{\hat{\boldsymbol{\alpha}}(\mathbf{q})\}$$
$$\mathbf{q} = M\{\hat{\boldsymbol{\beta}}(\mathbf{p})\} \tag{10.15}$$

where $\hat{\boldsymbol{\alpha}}$ and $\hat{\boldsymbol{\beta}}$ are the optimal responses (simple strategies) on the simple strategies, which appear with probabilities **p** and **q**, respectively.

Let us now assume that we cannot find exact optimal responses $\hat{\mathbf{p}}(\mathbf{q})$ and $\hat{\mathbf{q}}(\mathbf{p})$ but that we can define them with a certain error ξ, i.e., instead of $\hat{\mathbf{p}}(\mathbf{q})$ and $\hat{\mathbf{q}}(\mathbf{p})$, we actually define

$$\hat{\mathbf{p}}_\xi(\mathbf{q}) = \hat{\mathbf{p}}(\mathbf{q}) + \xi_1$$
$$\hat{\mathbf{q}}_\xi(\mathbf{p}) = \hat{\mathbf{q}}(\mathbf{p}) + \xi_2 \tag{10.16}$$

If the mean value of the errors is zero, then

$$\hat{\mathbf{p}}(q) = M_\xi\{\hat{\mathbf{p}}_\xi(\mathbf{q})\}$$
$$\hat{\mathbf{q}}(p) = M_\xi\{\hat{\mathbf{q}}_\xi(\mathbf{p})\} \tag{10.17}$$

and instead of (10.12), we obtain

$$\mathbf{p} = M_\xi\{\hat{\mathbf{p}}_\xi(\mathbf{q})\}$$
$$\mathbf{q} = M_\xi\{\hat{\mathbf{q}}_\xi(\mathbf{p})\} \tag{10.18}$$

The mathematical expectation is taken here with respect to the probability distributions of the errors. Equations (10.15) and (10.18) are actually the equations of optimal strategies.

10.5 Algorithms of Learning the Solution of Games

Let us write the equations (10.15) in the form

$$M\{\mathbf{p} - \hat{\boldsymbol{\alpha}}(\mathbf{q})\} = 0$$
$$M\{\mathbf{q} - \hat{\boldsymbol{\beta}}(\mathbf{p})\} = 0 \tag{10.19}$$

and apply to them the probabilistic iterative algorithm of a simple type. We then obtain an algorithm of learning for the solution of a game,

$$\mathbf{p}[n] = \mathbf{p}[n-1] - \gamma_1[n]\{\mathbf{p}[n-1] - \hat{\boldsymbol{\alpha}}_\nu(\mathbf{q}[n-1])\}$$
$$\mathbf{q}[n] = \mathbf{q}[n-1] - \gamma_2[n]\{\mathbf{q}[n-1] - \hat{\boldsymbol{\beta}}_\mu(\mathbf{p}[n-1])\} \tag{10.20}$$

10.5 Algorithms of Learning the Solution of Games

Here, $\hat{\alpha}_\nu(\mathbf{q}[n-1])$ and $\hat{\beta}_\mu(\mathbf{p}[n-1])$ are optimal simple strategies at the $(n-1)$st step; $\hat{\alpha}_\nu(\mathbf{q}[n-1])$ is defined by the index of the maximal component of the vector $A\mathbf{q}[n-1]$ and $\hat{\beta}_\mu(\mathbf{p}[n-1])$ is defined by the index of the minimal component of the vector $\mathbf{p}^T[n-1]A$. For convergence, it is sufficient that the coefficients $\gamma_1[n]$ and $\gamma_2[n]$ satisfy the usual conditions (3.34a). The algorithms (10.20) correspond to the process of sequential improvement in the strategies of the players, i.e., to learning a game by experience.

Similarly, by writing the equations (10.18) in the form

$$M_\xi\{\mathbf{p} - \hat{\mathbf{p}}_\xi(\mathbf{q})\} = 0$$
$$M_\xi\{\mathbf{q} - \hat{\mathbf{q}}_\xi(\mathbf{p})\} = 0 \qquad (10.21)$$

we obtain the algorithms for learning the solution of a game under uncertainty:

$$\mathbf{p}[n] = \mathbf{p}[n-1] - \gamma_1[n]\{\mathbf{p}[n-1] - \hat{\mathbf{p}}_\xi(\mathbf{q}[n-1])\}$$
$$\mathbf{q}[n] = \mathbf{q}[n-1] - \gamma_2[n]\{\mathbf{q}[n-1] - \hat{\mathbf{q}}_\xi(\mathbf{p}[n-1])\} \qquad (10.22)$$

The algorithms (10.22) correspond to the process of sequential improvement in the strategies of the players through experience in a perturbing, noisy environment.

It is interesting to notice that the existence of errors with zero mean values is not a barrier to the optimal strategies, but that it only prolongs the learning time.

At each step, according to $\mathbf{p}[n-1]$ and $\mathbf{q}[n-1]$, the function

$$v[n-1] = V(\mathbf{p}[n-1], \mathbf{q}[n-1]) = \mathbf{p}^T[n-1]A\mathbf{q}[n-1] \qquad (10.23)$$

is determined. This function, as $n \to \infty$, tends to a quantity called the value of the game. The algorithms of learning for the solution of the games (10.20) and (10.22) are realized by the systems shown in Fig. 10.1 and 10.2. The

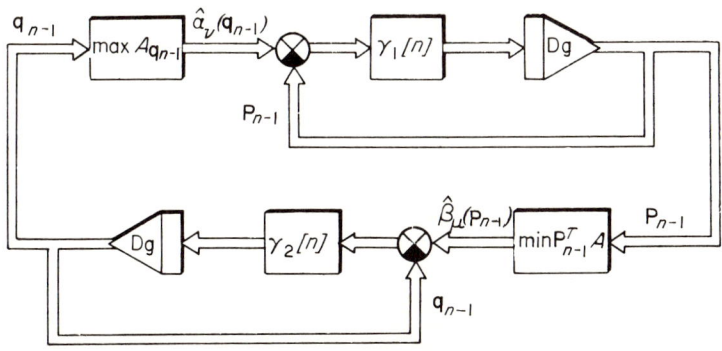

Fig. 10.1

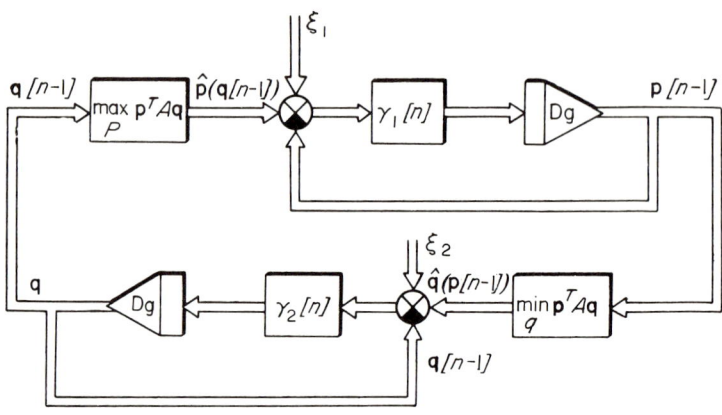

Fig. 10.2

obtained general algorithms of learning (10.20) and (10.22) lead in special cases to the known iterative algorithms presented in Table 10.1.

Generally speaking, the algorithms of learning for the solution of the games converge slowly. Different methods, such as those discussed in Section 3.15, can be used to increase the rate of convergence of the algorithms. All these methods are related to the modification of the coefficients $\gamma_1[n]$ and $\gamma_2[n]$ in the course of learning the solution of the game. For instance, the convergence of the algorithms can be accelerated if the iterations in which the same simple strategy is repeated are replaced by a simple iteration. This corresponds to

$$\gamma[n] = \gamma_1[n] = \gamma_2[n] = \left(r_n \bigg/ \sum_{v=1}^{n} r_v \right) \quad (10.24)$$

where r_v is the number of iterations at the vth step in which the same simple strategies are repeated. We can now apply the algorithms for learning the solution of games to various problems.

10.6 Games and Linear Programming

The solution of games is closely related to the solution of linear programming problems. Actually, if in (10.12) we replace the arbitrary mixed strategies **p** and **q** by the numbers α_v and β_μ, respectively, we obtain inequalities,

$$V(\alpha_v, \mathbf{q}) = A\mathbf{q} \leq v$$
$$V(\mathbf{p}, \beta_\mu) = \mathbf{p}^T A \geq v \quad (10.25)$$

The solutions of these inequalities satisfy (10.3) and thus define the optimal strategies and the value of the game. From here it follows that a game with

10.6 Games and Linear Programming

Table 10.1

ITERATIVE ALGORITHMS

Number	Equations	Algorithms	Comments	
1	$M\{\mathbf{p} - \hat{\boldsymbol{\alpha}}_\nu(\mathbf{q})\} = 0$	$\mathbf{p}[n] = \mathbf{p}[n-1]$ $-\gamma[n][\mathbf{p}[n-1] - \hat{\boldsymbol{\alpha}}_\nu(\mathbf{q}[n-1])]$	$\sum_{n=1}^{\infty} \gamma[n] = \infty,\ \sum_{n=1}^{\infty} \gamma^2[n] < \infty$	Brown
	$M\{\mathbf{q} - \hat{\boldsymbol{\beta}}_\mu(\mathbf{p})\} = 0$	$\mathbf{q}[n] = \mathbf{q}[n-1]$ $-\gamma[n][\mathbf{q}[n-1] - \hat{\boldsymbol{\beta}}_\mu(\mathbf{p}[n-1])]$	$\gamma[n] = 1/n$ $\gamma[n] = r_n / \sum_{s=1}^{n} r_s$ where r_s is the number of combined iterations of the sth stage	Amvrosienko
2	$M\{\mathbf{p} - \hat{\boldsymbol{\alpha}}_\nu(\mathbf{q})\} = 0$	$\mathbf{p}[n] = \mathbf{p}[n-1]$ $-\gamma_1[n][\mathbf{p}[n-1] - \hat{\boldsymbol{\alpha}}_\nu(\mathbf{q}[n-1])]$	Successive steps with $\gamma_1[n] = \gamma_1(\min_j (\mathbf{p}^T[n]A)_j)$ $\gamma_2[n] = \gamma_2(\max_i (A\mathbf{q}[n])_i)$ which satisfy the condition $\sum_{n=1}^{\infty} \gamma_l[n] = \infty \quad (l = 1, 2)$	Borisova, Magarik
	$M\{\mathbf{q} - \hat{\boldsymbol{\beta}}_\mu(\mathbf{p})\} = 0$	$\mathbf{q}[n] = \mathbf{q}[n-1]$ $-\gamma_2[n][\mathbf{q}[n-1] - \hat{\boldsymbol{\beta}}_\mu(\mathbf{p}[n-1])]$		
3	$M_\xi\{\mathbf{p} - \hat{\mathbf{p}}_\xi(\mathbf{q})\} = 0$	$\mathbf{p}[n] = \mathbf{p}[n-1]$ $-\gamma[n][\mathbf{p}[n-1] - \mathbf{p}_\xi(\mathbf{q}[n-1])]$	$\sum_{n=1}^{\infty} \gamma[n] = \infty,\ \sum_{n=1}^{\infty} \gamma^2[n] < \infty$ $\hat{\mathbf{p}}_\xi(\mathbf{q}) = \hat{\mathbf{p}}(\mathbf{q}),\ \hat{\mathbf{q}}_\xi(\mathbf{p}) = \hat{\mathbf{q}}(\mathbf{p})$ $\xi = 0;$	Volkonskii
	$M_\xi\{\mathbf{q} - \hat{\mathbf{q}}_\xi(\mathbf{p})\} = 0$	$\mathbf{q}[n] = \mathbf{q}[n-1]$ $-\gamma[n][\mathbf{q}[n-1] - \mathbf{q}_\xi(\mathbf{p}[n-1])]$	$\sum_{n=1}^{\infty} \gamma[n] = \infty,\ \lim_{n \to \infty} \gamma[n] = 0$	

matrix A is equivalent to a pair of coupled problems of linear programming:

$$v \to \min \qquad\qquad v \to \max$$

$$A\mathbf{q} \le v \qquad\qquad \mathbf{p}^T A \ge v \qquad (10.26)$$

$$q_\mu \ge 0 \quad \sum_{\mu=1}^{M} q_\mu = 1 \qquad p_\nu \ge 0 \quad \sum_{\nu=1}^{N} p_\nu = 1$$

Let us introduce new variables $\tilde{p}_\nu = p_\nu/v$, $\tilde{q}_\mu = q_\mu/v$ and designate $\tilde{v} = 1/v$. Then, from the pair of coupled linear programming problems (10.26) we obtain a pair of dual problems in linear programming:

$$\sum_{\mu=1}^{M} \tilde{q}_\mu = \tilde{v} \to \max \qquad\qquad \sum_{\nu=1}^{N} \tilde{p}_\nu = \tilde{v} \to \min$$

$$A\hat{\mathbf{q}} \le 1 \qquad\qquad \hat{\mathbf{p}}^T A \ge 1 \qquad (10.27)$$

$$\tilde{q}_\mu \ge 0 \qquad\qquad \tilde{p}_\nu \ge 0$$

Therefore, the algorithms of learning the solution of games can be used in solving typical problems of linear programming. But we shall leave such a possibility since we do not intend to anger the "fanatics" of linear programming who prefer to apply the methods of linear programming in the solution of games. We shall only mention that the algorithm of learning for the solution of games can be useful in the solution of different problems in optimal planning of grandiose dimensions.

10.7 Control as a Game

In many cases it is useful to consider a process of control as a certain game. We illustrate this in a simple example. Let the controlled system be described by a differential equation,

$$\frac{dx(t)}{dt} + x(t) = u(t) \qquad (x(0) = x_0) \quad (10.28)$$

We shall try to find a control function $u(t)$ for which the functional

$$J(u(t), x_0) = \int_0^T |1 - x(t)| \, dt \qquad (10.29)$$

is minimized for a given initial condition x_0. The control function must also satisfy the following constraints:

$$0 \le u(t) \le 1$$
$$\int_0^T u(t) \, dt = T_1 \qquad (10.30)$$

In order to reduce this problem of optimal control to a game-theoretic problem, we use an obvious identity,

10.8 Algorithms of Control

$$|1 - x| = \max_{|w| \leq 1} w(1 - x) \tag{10.31}$$

Then

$$\min_u \int_0^T |1 - x(t)| \, dt = \min_u \max_{|w| \leq 1} \int_0^T w(1 - x(t)) \, dt \tag{10.32}$$

On the basis of the min-max theorem (10.11), we can interchange the symbols max and min. Therefore,

$$\min_u \max_{|w| \leq 1} \int_0^T w(1 - x(t)) \, dt = \max_{|w| \leq 1} \min_u \int_0^T w(1 - x(t)) \, dt \tag{10.33}$$

Equation (10.33) guarantees that u^* and w^* form a saddle point.

Let us designate by U a set of functions u which satisfy the condition (10.30), and by W the set of functions w which satisfy the condition $|w| \leq 1$. Then

$$\min_{u \in U} \int_0^T |1 - x(t)| \, dt = \min_{u \in U} \max_{w \in W} \int_0^T w(1 - x(t)) \, dt \tag{10.34}$$

and thus, according to the min-max theorem,

$$\min_{u \in U} \max_{w \in W} \int_0^T w(1 - x(t)) \, dt = \max_{w \in W} \min_{u \in U} \int_0^T w(1 - x(t)) \, dt \tag{10.35}$$

10.8 Algorithms of Control

It follows from (10.35) that a problem of optimal control is reduced to a problem of solving a continuous game. We introduce a cost function of a continuous game,

$$V(u, w) = \int_0^T w(1 - x(t)) \, dt \tag{10.36}$$

Then, it follows from the min-max theorem (10.11) that the problem of optimal control considered in Section 10.7 is reduced to one of solving a continuous game with the cost function (10.36). In order to apply the algorithms of learning to the solution of games like (10.20), it is necessary to approximate the continuous game by a discrete one. On the other hand, one can first obtain the continuous algorithm from a discrete algorithm which is then used in the continuous game. For instance, the continuous algorithms

$$\frac{du(t)}{dt} = -\gamma_1(t)[u(t) - \hat{u}_\xi(w(t))]$$

$$\frac{dw(t)}{dt} = -\gamma_2(t)[w(t) - \hat{w}_\xi(u(t))] \tag{10.37}$$

correspond to the discrete algorithms (10.22). Now $\hat{u}_\xi(w(t))$ and $\hat{w}_\xi(u(t))$ are the optimal answers under noisy conditions. The scheme of a control which computes optimal control actions is similar to the one in Fig. 10.2.

10.9 A Generalization

It is possible that the control problem considered in Sections 10.7 and 10.8 will not be received with approval, since the minimized functional is very sensitive on initial conditions. And if we do not know the initial conditions, then we cannot determine optimal control. In order to circumvent this difficulty, we consider averaging the functional (10.29) over the initial conditions:

$$J(u(t)) = M\{J^0(u(t), x_0)\} \qquad (10.38)$$

The cost function corresponding to the continuous game is then

$$\bar{V}(u, w) = M\left\{\int_0^T w(1 - x(t))\, dt\right\} \qquad (10.39)$$

This "average" cost function is not known to us since we do not know the probability distribution of the initial conditions. But, as we know, this is not a barrier for the adaptive approach. We can employ algorithms of learning (10.37) which use the estimates obtained by the algorithms for estimating average values instead of the unknown "average" cost function for obtaining $\hat{u}_\xi(w(t))$ and $\hat{w}_\xi(u(t))$. Of course, in this case the learning time increases; lack of knowledge is frequently paid for by time.

We shall not go deeper into this interesting, but not yet well-studied problem. Instead, we consider other problems to which the algorithms of learning the solution of games can be applied.

10.10 Threshold Elements

Threshold elements have a broad application in the perceptrons and models of neural nets. They permit realizations of various logical functions. A threshold element consists of an adder and a relay (Fig. 10.3). It can be realized on different physical devices (for example, an amplifier with saturation, a ferrite core with a rectangular hysteresis cycle and constant magnetization, etc.).

The equation of a threshold element can be written as

$$y = \text{sgn}\left(\sum_{v=1}^N c_v x_v - c_0\right) \qquad (10.40)$$

10.11 Logical Functions on a Single Threshold Element

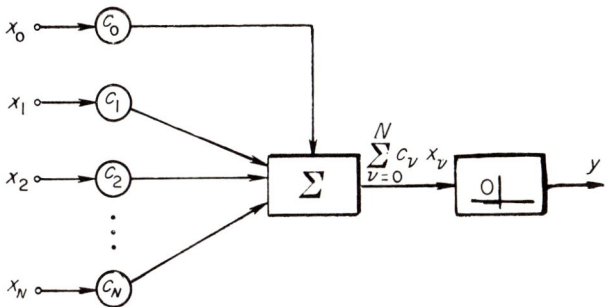

Fig. 10.3

where x_ν is a variable which characterizes the state of the νth input, $\nu = 1, \ldots, N$ (input variable), y is a variable characterizing the state of the output (output variable), c_ν is the weighting coefficient of the νth input, $\nu = 1, \ldots, N$, and c_0 is the threshold. We recall that

$$\text{sgn } z = \begin{cases} 1 & \text{when } z > 0 \\ 0 & \text{when } z \leq 0 \end{cases} \tag{10.41}$$

Input and output variables take only two values, 0 and 1.

The equation of a threshold element can be written in vector form,

$$y = \text{sgn } (\mathbf{c}^T \mathbf{x} - c_0) \tag{10.42}$$

where $\mathbf{x} = (x_1, \ldots, x_N)$ is the vector of input variables, and $\mathbf{c} = (c_1, \ldots, c_N)$ is the vector of weighting coefficients.

10.11 Logical Functions and Their Realizations on a Single Threshold Element

The logical or Boolean function,

$$y = y(x_1, \ldots, x_N) \tag{10.43}$$

briefly written as

$$y = y(\mathbf{x}) \tag{10.44}$$

has independent variables which can take only two values, 0 or 1. It is easy to see that the Boolean function (10.43) is completely defined by a table of 2^N entries. Since the input and output variables of a threshold element also take the values 0 or 1, it is interesting to examine the problem of realization of a Boolean function on a single threshold element.

Let us assume that

$$y = \begin{cases} 1 & \text{for } \mathbf{x}_{(1)} \in X_1 \\ 0 & \text{for } \mathbf{x}_{(0)} \in X_0 \end{cases} \tag{10.45}$$

Obviously, the following conditions must be satisfied by a threshold element:

$$\begin{aligned} \mathbf{c}^T \mathbf{x}_{(1)} &> c_0 \\ \mathbf{c}^T \mathbf{x}_{(0)} &< c_0 \end{aligned} \qquad (10.46)$$

We shall now use the property of the threshold element, that for

$$c_0 = \frac{1}{2} \sum_{v=1}^{N} c_v \qquad (10.47)$$

the inversion of the coordinates of the input signal (i.e., one is substituted by zero, and zero by one) causes the inversion of the output signal. Then, from (10.46), we obtain

$$\begin{aligned} \mathbf{c}^T \mathbf{x}_{(1)} &> \frac{1}{2} \sum_{v=1}^{N} c_v \\ \mathbf{c}^T \bar{\mathbf{x}}_{(0)} &> \frac{1}{2} \sum_{v=1}^{N} c_v \end{aligned} \qquad (10.48)$$

where $\bar{\mathbf{x}}_{(0)}$ indicates the inversion of $\mathbf{x}_0$. This system of inequalities can be written in a more compact form,

$$A\mathbf{c} > \frac{1}{2} \sum_{v=1}^{N} c_v \qquad (10.49)$$

where

$$A = \begin{Vmatrix} \mathbf{x}_{(1)}^T \\ \bar{\mathbf{x}}_{(0)}^T \end{Vmatrix}$$

We can conclude now that the necessary and sufficient condition for realization of a Boolean function (10.43) on a single threshold element is the existence of a vector $\mathbf{c} = \mathbf{c}^*$ which satisfies the system of inequalities (10.49).

10.12 Criterion of Realizability

The realizability of a Boolean function on a single threshold element can be determined by the matrix A. Let us first add the coordinates 1 and 0 to both vectors $\mathbf{x}$ and $\bar{\mathbf{x}}$, and then form the extended matrix

$$\tilde{A} = \begin{Vmatrix} \mathbf{x}_{(1)}^T & 1 & 0 \\ \mathbf{x}_{(0)}^T & 0 & 1 \end{Vmatrix} \qquad (10.50)$$

10.13 Algorithm of Realizability

This extended matrix corresponds to a new threshold element which differs from the original one by two additional inputs. The weights of these new inputs, c_{N+1} and c_{N+2}, can be chosen to satisfy the condition of normalization

$$\sum_{v=1}^{N+2} c_v = 1 \tag{10.51}$$

In that case, as it follows from (10.47),

$$c_0 = 1/2 \tag{10.52}$$

and the condition (10.49) is replaced by the condition

$$\tilde{A}\mathbf{c} > 1/2 \tag{10.53}$$

By its construction the matrix $\tilde{A}$ is such that there is a vector $\mathbf{d}$ with the property

$$\mathbf{d}^T\tilde{A} > 1/2 \tag{10.54}$$

when $\sum_{\mu=1}^{M} d_\mu = 1$ for $M < 2^N$. If we now consider matrix $\tilde{A}$ to be a cost matrix of a certain game, and the vectors $\mathbf{c}$ and $\mathbf{d}$ to be the mixed strategies of the players, the conditions (10.53) and (10.54) are then satisfied when the value of the game, v, is greater than $1/2$. Therefore we obtain the following formulation of the realizability criterion. For realizability of the Boolean functions on a single threshold element, it is necessary and sufficient that the value of the game, v, determined by the cost matrix $\tilde{A}$, be greater than $1/2$.

10.13 Algorithm of Realizability

The realizability on a single threshold element can be determined by using the algorithms for learning the solution of the games. For instance by applying (10.20), we obtain

$$\begin{aligned}\mathbf{c}[n] &= \mathbf{c}[n-1] - \gamma_1[n]\{\mathbf{c}[n-1] - \hat{\alpha}_v(\mathbf{d}[n-1])\} \\ \mathbf{d}[n] &= \mathbf{d}[n-1] - \gamma_2[n]\{\mathbf{d}[n-1] - \hat{\beta}_\mu(\mathbf{c}[n-1])\}\end{aligned} \tag{10.55}$$

and

$$v[n-1] = \mathbf{c}^T[n-1]A\mathbf{d}[n-1] \tag{10.56}$$

If $v[n-1] \to v > 1/2$ for $n \to \infty$, we can realize a given logical function on a threshold element. Vector $\mathbf{c}[n]$ converges (as $n \to \infty$) to the optimal strategy $\mathbf{c}^*$ which defines the vector of the sought weighting coefficients in the threshold element. A test of realizability can be accomplished by a scheme similar to one presented in Fig. 10.1.

It is sometimes convenient that the input and output variables of a threshold element are -1 and $+1$ instead of 0 and 1, respectively. In this last case the behavior of the threshold element is described by the equation

$$y = \text{sign}\left(\sum_{v=1}^{N} c_v x_v - c_0\right) \qquad (10.57)$$

instead of by (10.49). The variables x_v and y are either -1 or $+1$, and

$$\text{sign } z = \begin{cases} +1 & \text{when } z > 0 \\ -1 & \text{when } z \leq 0 \end{cases} \qquad (10.58)$$

The transition from one type of definition to another is based on the relationship

$$\text{sign } z = 2 \text{ sgn } z - 1 \qquad (10.59)$$

In that case, x_v must also be replaced by $2x_v - 1$. Using this substitution in the extended matrix (10.50), the realizability of a Boolean function of a similar type can also be tested by the algorithm (10.55) and (10.56).

10.14 Rosenblatt's Perceptron

The algorithms discussed in Section 10.13 can be used to train the threshold elements so that they can realize given Boolean functions. However, it would be very unjust to delegate only this role to the threshold elements. Learning threshold elements are very important parts of Rosenblatt's perceptron. A simplified scheme for such a perceptron is shown in Fig. 10.4. The system of

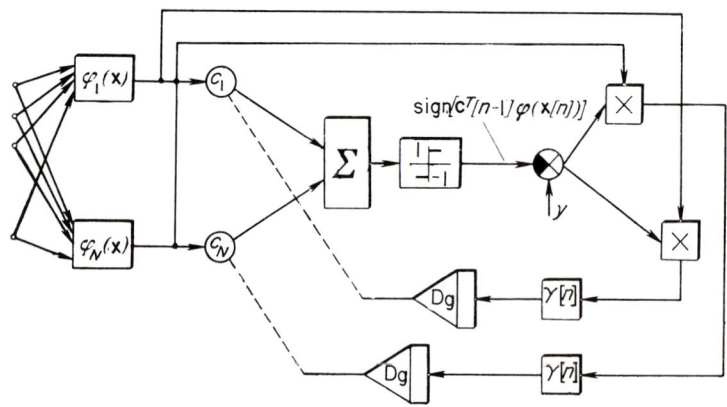

Fig. 10.4

linearly independent functions $\phi_\nu(\mathbf{x})$, $\nu = 1, \ldots, N$, appearing in the general scheme of the perceptron presented in Fig. 10.4, is here a system of threshold functions,

$$\phi_\nu(\mathbf{x}) = \text{sign}\,(\mathbf{c}^{(\nu)T}\mathbf{x} - c_0^{(\nu)}) \tag{10.60}$$

with a priori given thresholds and weights.

The algorithms used to train a perceptron are analogous to algorithm 1 of Tables 4.1 or 4.2:

$$\mathbf{c}[n] = \mathbf{c}[n-1] + \gamma_0[y[n] - \text{sgn}\,\mathbf{c}^T[n-1]\phi(\mathbf{x}[n-1])]\phi(\mathbf{x}[n-1]) \tag{10.61}$$

where the components of $\phi(\mathbf{x})$ are defined by the expressions (10.60). The process of training a perceptron converges under the conditions stated in Section 3.13.

10.15 Widrow's Adaline

The beautiful name "adaline" is a shortened form of "adaptive linear threshold element." Widrow's adaline is an extremely simplified perceptron (Fig. 10.5). The input signals form a vector

$$\mathbf{x}[n] = (1, x_1[n], \ldots, x_N[n]) \tag{10.62}$$

which defines all possible input situations (corners of a hypercube) for $n = 1, 2, \ldots, 2^N$. These situations have to be partitioned into two classes which correspond to two values of y:

$$y = \pm 1 \tag{10.63}$$

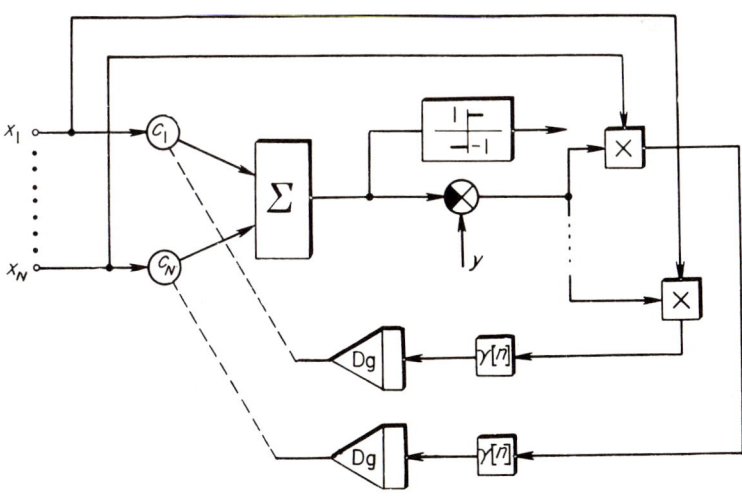

Fig. 10.5

Let us now select the functions $\phi_\nu(x)$ such that

$$\phi_0(\mathbf{x}) = 1$$
$$\phi_\nu(\mathbf{x}) = x_\nu \qquad (\nu = 1, \ldots, N) \qquad (10.64)$$

It is difficult to think of anything simpler. Thanks to this choice of functions, there is no need for the functional transformers of many variables and, in particular, for the threshold elements which were used in Rosenblatt's perceptron.

Using the quadratic functionals and their corresponding algorithm 2 (Table 4.1) with (10.64) we obtain

$$\mathbf{c}[n] = \mathbf{c}[n-1] - \gamma[n](y[n] - \mathbf{c}^T[n-1]\mathbf{x}[n])\mathbf{x}[n] \qquad (10.65)$$

where

$$\mathbf{c} = (c_0, c_1, \ldots, c_N) \qquad (10.66)$$

is an $(N+1)$-dimensional vector of coefficients. The coefficient $\gamma[n]$ is usually chosen to be a constant,

$$\gamma[n] = \frac{\gamma_0}{N+1} \qquad (10.67)$$

Algorithm (10.65), with γ defined by (10.67), is used to train adalines. An adaline is therefore an example of a trainable threshold element. The training is based not upon the binary values of the output variable, but upon a linear combination of the input stiuations, which is briefly called a composite.

10.16 Training of a Threshold Element

A slightly different method of training a threshold element which does not use a linear combination of the input situations directly, is also possible. In the general case, the output signal of the threshold elements is equal to

$$\hat{y}[n, m] = \text{sign } \mathbf{c}^T[m]\mathbf{x}[n] \qquad (10.68)$$

where n is the index of one of the input situations, and m is the index of the instant when the measurements of the vector of weighting coefficients are performed. In particular, when $n = m + 1$,

$$\hat{y}[n] = \text{sign } \mathbf{c}^T[n-1]\mathbf{x}[n] \qquad (10.69)$$

Until the learning process is completed, the output signal differs from the desired $y[n]$. Learning is completed when, by varying the weighting coefficients,

$$y[n] = \hat{y}[n] = 0 \qquad (10.70)$$

or, equivalently,

$$y[n]\mathbf{c}^T[n-1]\mathbf{x}[n] > 0 \qquad (10.71)$$

10.16 Training of a Threshold Element

A similar condition was already encountered in Section 4.11. We should recall that now the components of the vector $\mathbf{x}[n]$ do not have arbitrary values but only $+1$ and -1. Although this does not prevent us from using algorithm (4.41), we suggest that the reader write them down for this special case. Here we give a simple algorithm which takes into consideration the specific nature of threshold elements.

Using the algorithm for training the perceptrons (10.61), but where $\phi(\mathbf{x}[n])$ is substituted by $\mathbf{x}[n]$, we obtain

$$\mathbf{c}[n] = \mathbf{c}[n-1] + \gamma[n](y[n] - \operatorname{sign} \mathbf{c}^T[n-1]\mathbf{x}[n])\mathbf{x}[n] \qquad (10.72)$$

In the special case when $\gamma[n] = 1/2$ we have

$$\mathbf{c}[n] = \mathbf{c}[n-1] + \tfrac{1}{2}(y[n] - \operatorname{sign} \mathbf{c}^T[n-1]\mathbf{x}[n])\mathbf{x}[n] \qquad (10.73)$$

If the initial vector $\mathbf{c}[0]$ has integer-valued components, then the vectors $\mathbf{c}[n]$ will also have integer-valued components. A block diagram of the system for realization of algorithm (10.73) is shown in Fig. 10.6.

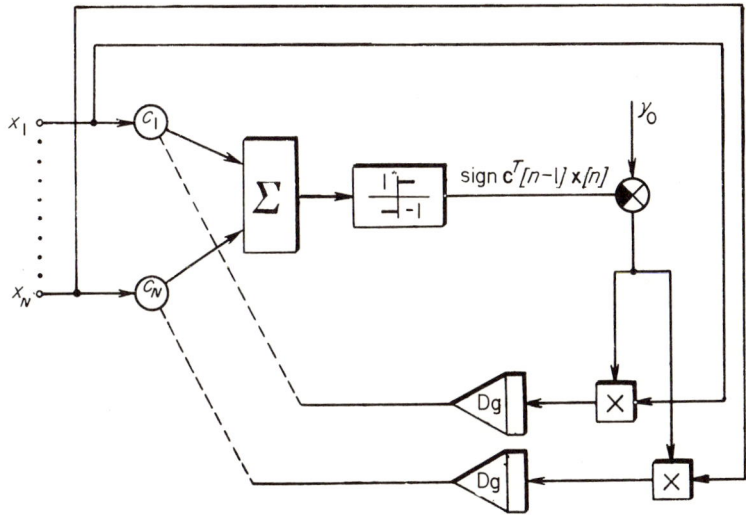

Fig. 10.6

When the situations, i.e., corners of a hypercube, can be at least in principle separated by a hyperplane, then nonconvexity of the functional which generates this algorithm (we have discussed this in Section 4.10) will not have a significant meaning.

Generally speaking, with the passage of time, $\mathbf{c}[n]$ will converge in probability to the optimal vector $\mathbf{c}^*$. This convergence is not simple, but is such that the most probable value of the limit $c[n]$, sometimes called the mode of $c[n]$, converges to c^*. When there are no disturbances, the optimal value $c = c^*$ may be reached in a finite number of steps.

Let us now be more rigorous. We shall apply at the input of the threshold element a vector $x[n]$, which corresponds to the same situation, until the learning process for this situation is finished. After that, we consider the next input situation, and repeat the procedure without paying any attention to the fact that we may modify previously obtained results. In this case, we must use the condition (10.68) instead of (10.69), and then, by repeating with a certain small modification our previous reasoning, we obtain a "rigorous" algorithm for training a threshold element,

$$c[m] = c[m-1] + \tfrac{1}{2}(y[n] - \operatorname{sign} c^T[m-1]x[n])x[n] \qquad (10.74)$$

This rigorous algorithm for training can be written in a slightly different form. Let us designate by r_n the number of iterations when the vector $x[n]$ is shown, and by $c[m_0 - 1]$ the vector with which the iterations for the input situation $x[n]$ begin. Then, from (10.74), we obtain

$$c[m_0 - 1 + r_n] = c[m_0 - 1] + r_n y[n]x[n] \qquad (10.75)$$

It should be noticed that $y[n] = \pm 1$ and $x_v[n] = \pm 1$, since $y[n]x_v[n] = \pm 1$. The algorithm modifies the weights until the learning of the threshold element is not completed.

The capability for learning can be improved if, instead of a single threshold element, a network of threshold elements is observed. This was done in the case of Rosenblatt's perceptron. However, such a generalization is not new in principle, and thus we shall not study trainable threshold networks.

10.17 Automata

The dynamic systems which have been considered until now were characterized by generalized coordinates defined on a continuum, i.e., the coordinates could take any real value. The independent variable time is continuous for continuous systems, and discrete for discrete systems.

The processes in the continuous systems are described by differential or integral equations, and the processes in the discrete systems by difference or sum equations. Now, we shall consider a special case of dynamic systems in which the generalized coordinates can take either a finite or a countable number of a priori defined values, and where time is also discretized.

An automaton is considered to be a dynamic system in which a change in its internal state $a[n]$, and a response defined by the vector $x[n]$ of output variables are created under the influence of a vector $u[n]$ of input actions. In a finite automaton, the sets correspond to the components of the input actions; states and output variables are finite. The components of the vector of input actions, or the input vector, can take the values from the input alphabet:

$$\mathbf{u} \in U = \{u^0, u^1, \ldots, u^N\} \qquad (10.76)$$

The components of the vector of output variables, or the output vector, can take the values from the output alphabet:

$$\mathbf{x} \in X = \{x^0, x^1, \ldots, x^M\} \qquad (10.77)$$

Here, u^0 and x^0 are the symbols which correspond to the case when the input and output signals do not exist, i.e., u^0 and x^0 are empty symbols.

The signals or the symbols forming an alphabet are called letters, and their combinations are called words. Therefore, a finite automaton transforms the letters of an input alphabet into the letters of an output alphabet. The alphabet of internal states defines a finite number of internal states which correspond to the number of components in the state vector:

$$\mathbf{a} \in A = \{a^1, \ldots, a^k\} \qquad (10.78)$$

The number of internal states, k, corresponds to the memory capacity of the automaton. These terms, which are specific for the theory of finite automata, have other meanings and names in the theory of discrete systems. Input signals, states, and output signals (called "words") correspond to certain discontinuous functions of time. The values of such functions are quantized into levels which correspond to the letters of corresponding alphabets.

10.18 Description of Finite Automata

The behavior of finite automata, or the processes in the finite automata, can be described by two equations: the state (or transition) equation and the output equation:

$$\begin{aligned} \mathbf{a}[n] &= \mathbf{f}_\mathbf{I}(\mathbf{a}[n-1], \mathbf{u}[n]) \qquad \mathbf{a}[0] = \mathbf{a}_0 \\ \mathbf{x}[n] &= \mathbf{\psi}_\mathbf{I}(\mathbf{a}[n]) \end{aligned} \qquad (10.79)$$

In these equations, $\mathbf{f}_\mathbf{I}(\cdot)$ is an amplitude-quantized function of two variables, $\mathbf{\psi}_\mathbf{I}(\cdot)$ is an amplitude-quantized function of a single variable, and $\mathbf{a}[0] = \mathbf{a}_0$ is the initial state vector. The special index in the functions $\mathbf{f}_\mathbf{I}(\cdot)$ and $\mathbf{\psi}_\mathbf{I}(\cdot)$ should remind us that these functions are amplitude-quantized, and thus they can take the "values" which correspond to the letters of their alphabet.

The intermediate variables—states $\mathbf{a}[n]$—do not appear in the equations of continuous and impulse systems. Their importance in the construction of a model for a finite automaton will not be investigated here. A block diagram

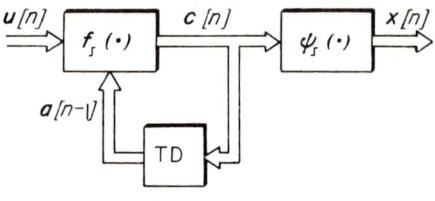

Fig. 10.7

of a finite automaton which corresponds to the equations (10.79) is shown in Fig. 10.7. This finite automaton has l_1 inputs and l outputs. For an automaton with a single output and a single input (Fig. 10.8), instead of equation (10.79) we have

$$a[n] = f_{\mathbf{s}}(a[n-1], u[n]) \qquad a[0] = a_0$$
$$x[n] = \psi_{\mathbf{s}}(a[n]) \qquad (10.80)$$

However, there is little difference between the finite automata described by vector equations (10.79) and finite automata described by scalar equations (10.80). Let all the possible combinations of l_1 input signals be coded by corresponding letters, and the same also be done with the states and output signals. Then, an automaton with l inputs, and a set of output signals coded in an r-letter alphabet, is equivalent to an automaton with a single input signal and with a single output coded in an l^r-letter alphabet.

The amplitude-quantized functions $f_{\mathbf{s}}(\cdot)$ and $\psi_{\mathbf{s}}(\cdot)$, and thus the equations of finite automata, are given in different ways (for instance, in tabular form or in the form of graphs). The functions $f_{\mathbf{s}}(\cdot)$ and $\psi_{\mathbf{s}}(\cdot)$ are completely defined either by a transition table of inputs and outputs (Table 10.2), or by a transition diagram (Fig. 10.9). In Table 10.2, the symbols of the states are written above the output values. On the graphs, they are written around the directed

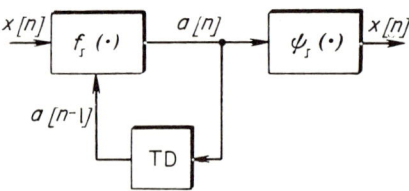

Fig. 10.8

10.18 Description of Finite Automata

Table 10.2

TRANSITION TABLE

u \ a	a^1	a^2	a^3	a^4	a^5
u^1	a^2 / x^1	a^3 / x^2	a^4 / x^3	a^4 / x^3	a^1 / x^1
u^2	a^1 / x^1	a^4 / x^3	a^2 / x^1	a^5 / x^3	a^2 / x^1

lines indicating the corresponding state transitions. The function $f_{\mathbf{J}}(\cdot)$ can be defined by a state matrix

$$D(u^r) = \|d_{\nu\mu}(u^r)\| \qquad (\nu = 1, \ldots, N; \mu = 1, \ldots, M) \quad (10.81)$$

Each row of the matrix contains only one element which is equal to one, and the remaining elements are equal to zero. When at an instant $n - 1$ the automaton is in the state $a[n - 1] = a^\nu$, and the input signal is $u[n] = u^r$, then the next state is $a[n] = a^\mu$ only if $d_{\nu\mu} = 1$.

We shall next consider stochastic automata with a single input and a single output. As we mentioned earlier, this is not a limitation.

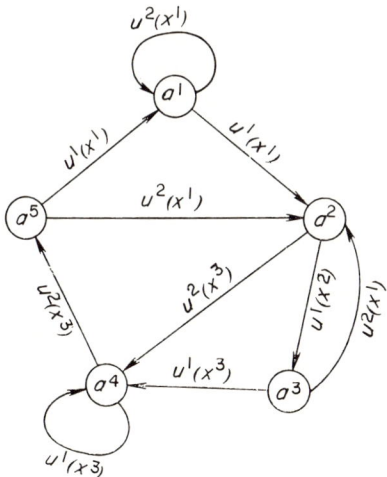

Fig. 10.9

10.19 Stochastic Finite Automata

Stochastic automata are generalizations of the deterministic automata discussed in the preceding section. In stochastic automata we can discuss only the probabilities of transition from one state to another. The equations of such stochastic automata may be written in the following form:

$$a[n] = f_\mathbf{s}(a[n-1], u[n], \zeta[n])$$
$$x[n] = \psi_\mathbf{s}(a[n]) \tag{10.82}$$

In the first equation of (10.82), $\zeta[n]$ is a random discontinuous function where the probability of an occurrence of a certain value of the function does not depend on the occurrences of the other values. Such a random discontinuous function is called the Bernoulli process. Therefore, the states of a stochastic automaton depend on the random discontinuous function $\zeta[n]$ which can, for instance, vary as the "parameters" of a finite automaton, or simply be an additional random component of the input signal (Fig. 10.10). In this last

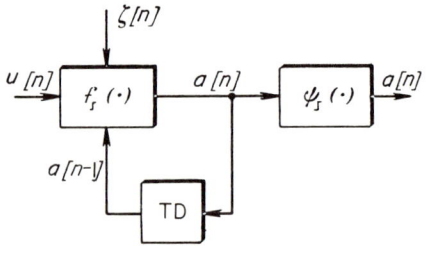

Fig. 10.10

case, the first equation of (10.82) has a more specific form,

$$a[n] = f_\mathbf{s}(a[n-1], u[n] \oplus \zeta[n]) \tag{10.83}$$

where the symbol of addition $\oplus$ indicates that the sum $u \oplus \zeta$ always belongs in the input alphabet.

A stochastic automaton is usually defined by its transition matrix,

$$P(r) = \|p_{\nu\mu}(r)\| \quad (\nu = 1, \ldots, N; \mu = 1, \ldots, M) \tag{10.84}$$

This matrix differs from the state matrix (10.81) since the elements $d_{\nu\mu}(u^r)$ are replaced by the transition probabilities $p_{\nu\mu}(r)$ which define the probability of transition from the μth state into the νth state. Of course, $p_{\nu\mu}(r)$ must satisfy the conditions

$$p_{\nu\mu}(r) \geq 0 \qquad \sum_{\mu=1}^{M} p_{\nu\mu}(r) = 1 \quad (\nu = 1, \ldots, N) \tag{10.85}$$

Deterministic automata are a special case of stochastic automata when $\zeta[n] = 0$.

10.20 Automaton–Environment Interaction

An automaton interacts with its environment in the following way: The automaton's action, $x[n]$, causes a response in the environment, $u[n]$, which in turn acts as an input signal for the automaton. Here the input signal $u[n]$ depends upon the output signals of the automaton and on the properties of the environment.

The equation of the environment can be written as

$$u[n] = \theta_J(x[n], \xi[n]) \tag{10.86}$$

where $\theta_J(\cdot)$ is a certain amplitude-quantized function, and $\xi[n]$ is a random discontinuous function. The automaton–environment interaction is evident from the feedback loop existing around the automaton (Fig. 10.11). As

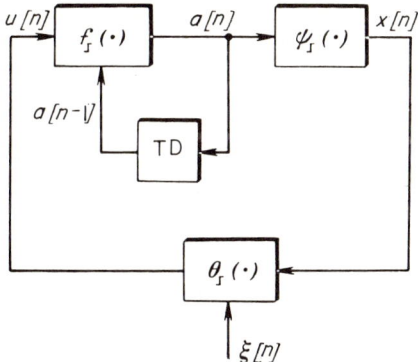

Fig. 10.11

should be expected, the behavior of the automaton is then changed, and there is a problem of investigating the influence of the environment on the behavior of the automaton, and a problem of improving such behavior. This circle of problems was posed and considerably solved by Tsetlin. Here we shall consider this problem from a slightly different point of view which leads to the synthesis of optimal automata. The criterion of optimality is based upon the expediency of behavior.

10.21 The Expediency of Behavior

In the automaton–environment interaction, each satisfactory behavior of the automaton causes $u[n] = 0$ (a reward), and each unsatisfactory one generates $u[n] = 1$ (a penalty). The mathematical expectation of the penalty, i.e.,

$$\rho = M\{u\} \tag{10.87}$$

can serve as a performance index of the automaton's expediency. Here ρ is the probability of penalizing the automaton.

If the automaton generates the responses $x[n]$ independently of the environment's reactions $u[n]$, its output variable is then a Bernoulli sequence $b = b[n]$. To this indifferent behavior of the automaton corresponds the following conditional expectation of the penalty:

$$\rho_0 = M\{u \,|\, x = b\} \tag{10.88}$$

An automaton is considered to be expedient if

$$\rho < \rho_0 \tag{10.89}$$

Inequality (10.89) is very important in the analysis of behavior of finite automata under sufficiently complete a priori information. We cannot resist the temptation to describe an example given by Tsetlin which illustrates well the concept of expediency. Taken from a stack of books on a table, a book can be used in different ways. For instance, we can select and take a needed book, and then (after use) put it back in its original place in the stack. Or, we can put it back on top of the stack. This second approach is definitely expedient. A more needed book is more often at the top of the stack.

We should notice that the quantities ρ and ρ_0 can be found by using the algorithms for estimating the mathematical expectations (5.8) and (5.9), if we can observe the automaton's actions.

10.22 Training of Automata

In the synthesis and training of automata, it is reasonable to request that the automata are not simply expedient, but that they are optimal with respect to expedient behavior. In other words, the behavior must be such that the mathematical expectation of the penalty is minimized. This can be achieved during a training period by modifying the automaton (i.e., $f_\mathbf{I}(\cdot)$) or the characteristic of the transformer $\psi_\mathbf{I}(\cdot)$. For simplicity, we shall consider the last case. Due to the equations (10.80) and the second equation in (10.82), we can write ρ as

$$\rho = M\{\theta_\mathbf{I}(\psi_\mathbf{I}(a), \xi)\} \tag{10.90}$$

Let us now search for a characteristic of the transformer $\psi_\mathbf{I}(a)$ for which ρ is minimal. We shall seek $\psi_\mathbf{I}(a)$ in the familiar form

$$\psi_\mathbf{I}(a) = \sum_{\kappa=1}^{k} c_\kappa \phi_\kappa(a) = \mathbf{c}^T \boldsymbol{\phi}(a) \tag{10.91}$$

10.22 Training of Automata

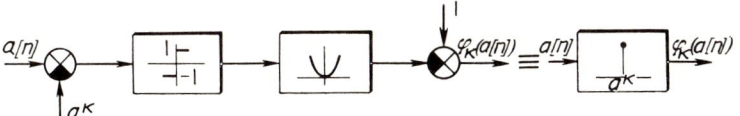

Fig. 10.12

This is possible if we define a system of linearly independent functions. For instance, we can select

$$\phi_\kappa(a) = \begin{cases} 1 & \text{when } a = a^\kappa \\ 0 & \text{when } a \neq a^\kappa \end{cases} \quad (10.92)$$

or, analytically,

$$\phi_\kappa(a) = 1 - \text{sg}^2(a - a^\kappa) \quad (10.93)$$

where

$$\text{sg } z = \begin{cases} +1 & \text{when } z > 0 \\ 0 & \text{when } z = 0 \\ -1 & \text{when } z < 0 \end{cases} \quad (10.94)$$

The scheme of such transformers and their symbolic representations is shown in Fig. 10.12. The functions $\phi_\kappa(a)$ are unit impulse functions, and the coefficients c_κ in (10.94) are members of the alphabet X. Therefore, $\psi_\mathbf{I}(a)$ is not approximately but exactly equal to the linear combination $\mathbf{c}^T\boldsymbol{\phi}(a)$ (Fig. 10.13).

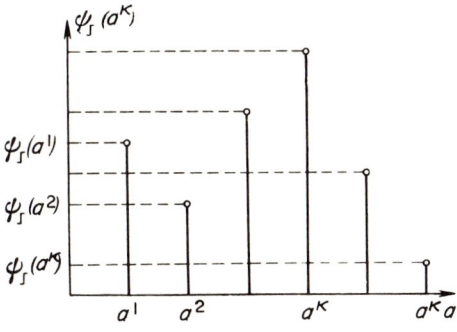

Fig. 10.13

The condition of optimality of the automaton can now be written as

$$\rho = \rho_{\min} = \min_c M\{\theta_\mathbf{I}(\mathbf{c}^T\boldsymbol{\phi}(a), \xi)\} \quad (10.95)$$

and the problem of training is to reach this condition by observing the results of the automaton's behavior.

Since it is not possible to define the gradient of the realization $\theta_f(\cdot)$, we compute an estimate

$$\tilde{\nabla}_{\mathbf{c}\pm}\theta_\mathbf{I}(\mathbf{c}, 1) = \theta_\mathbf{I}{}^+(\mathbf{c}, 1) - \theta_\mathbf{I}{}^-(\mathbf{c}, 1) \quad (10.96)$$

where

$$\theta_\mathbf{I}{}^\pm(\mathbf{c}, 1) = (\theta_\mathbf{I}(\mathbf{c} \pm \mathbf{e}_1), \ldots, \theta_\mathbf{I}(\mathbf{c} \pm \mathbf{e}_k)) \quad (10.97)$$

and use the search algorithms of the type (3.15) with $\gamma[n] = 1$. Then the algorithm of training is

$$\begin{aligned}\mathbf{c}[n] &= \mathbf{c}[n-1] - \tilde{\nabla}_{\mathbf{c}\pm}\theta_\mathbf{I}(\mathbf{c}[n-1], 1) && \text{if} \quad \mathbf{c}[n] \in A \\ \mathbf{c}[n] &= \mathbf{c}[n-1] && \text{if} \quad \mathbf{c}[n] \notin A\end{aligned} \quad (10.98)$$

We can now construct a block diagram of a trainable automaton; it is shown in Fig. 10.14. As a result of training in a certain environment, there will be

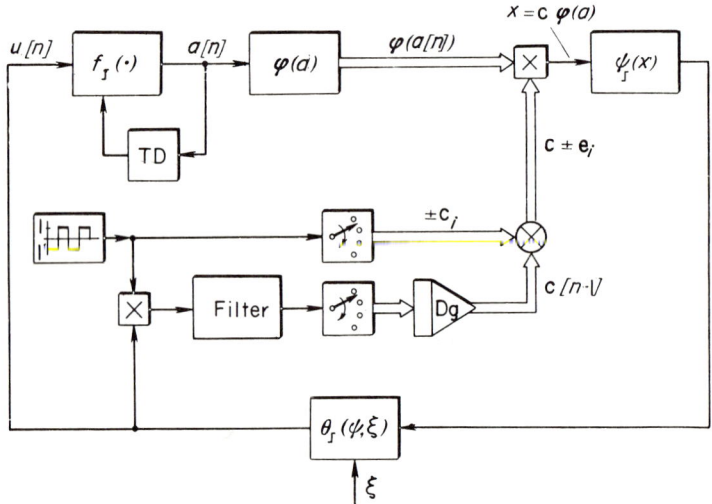

Fig. 10.14

an "adjustment" of the transformer's characteristic, and the automaton will have optimal behavior. Training can also be accomplished when the transformer's characteristic is constant but the input signal ξ is varying; ξ is generated by a special generator. The transition probabilities are reduced for the transitions which were penalized.

Such learning automata have flexible behavior and are adaptable to relatively slow changes in the environment. They pay a minimal penalty "in the case when for the sins of yesterday one is rewarded today, and also in the case when sins remain sins."

Before we start discussing another topic, we shall ask the reader the following question: "Do automata learn by correspondence or otherwise?" We have talked about this in Section 4.4.

10.23 About Markov Chains

A stochastic automaton corresponds to a Markov chain if it is assumed that the initial condition $a[0]$ is random, and the output variable $x[n]$ is identified with the state $a[0]$. In Markov chains, the probability of an outcome depends only on the outcome of the immediately preceding experiment. The stochastic matrix of transition probabilities (10.84) describes a Markov chain. As an example of a Markov chain, we give a problem of an observer who has to detect a useful signal which is masked by noise. This problem is closely related to one considered in Section 6.13. The observer can make mistakes. If the observer decides that there is a signal when such a signal actually does not exist, we have an error of the first kind or the error of false alarm. If the observer decides that the signal does not exist when the signal is actually present, we have an error of the second kind or the error of missing the signal. A Markov chain which corresponds to this situation can be given in the form of a graph (Fig. 10.15). The nodes of this graph represent the outcomes. In

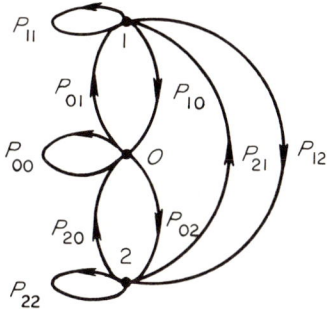

Fig. 10.15

the given case we have three outcomes: 1 and 2, which correspond to the errors of the first and the second kind, respectively, and 0, if there are no errors. The branches represent the corresponding conditional probabilities $p_{\nu\mu}$ (the probability that the νth outcome is followed immediately by the μth outcome).

Of course, all the outcomes are not equally expensive. Outcomes 1 and 2 are not desired, and outcome 0 is desired. Learning is then reduced to a modification of conditional probabilities $p_{\nu\mu}$ so that the desired output becomes more probable. This can be accomplished by a sensible use of punishment and reward. Using the equivalence which exists between the automata and the Markov chains, we shall consider Markov learning by studying the synthesis of a stochastic automaton which has to solve the problem of Markov learning.

10.24 Markov Learning

Let a certain information source emit the signals. These useful signals $s[n]$ which belong to a two-letter alphabet, $\{0, 1\}$, are mixed with noise $\xi[n]$. The observer receives the signal masked by disturbances, $y[n] = s[n] + \xi[n]$, and compares it with the threshold $c[n]$.

The problem of learning is then to adjust the threshold $c = c^*$ sequentially so that the conditional probabilities of the desired outcomes are increased and the undesired outcomes are decreased. In contrast to the results of Section 6.13, we shall assume here that the threshold varies in a discrete fashion throughout a finite alphabet. Discrete variations and the finiteness of the alphabet characterize Markov learning. Let us assume that the alphabet corresponding to the threshold is actually the alphabet of an automaton's states without an output transformer. In other words, let us assume that $c[n] = a[n]$ and $x[n] = a[n]$. Then

$$a[n] = f_{\mathbf{I}}(a[n-1], v[n]) \qquad (10.99)$$

Let this automaton operate in an environment described by

$$u[n] = \theta_{\mathbf{I}}(y[n] - a[n-1]) \qquad (10.100)$$

This environment is stochastic since the signal $y[n]$ is random. Finally, in contrast to the reinforcement schemes described earlier, the punishments and rewards here are defined by the difference between the teacher's answer $y_0[n] = s[n]$ and the reaction of the environment, $u[n]$:

$$v[n] = y_0[n] - u[n] \qquad (10.101)$$

where $v[n]$ is defined by the alphabet $\{+1, -1\}$. In the simplest case we can assume

$$\begin{aligned} f_{\mathbf{I}}(a[n-1], v[n]) &= a[n-1] - v[n] \\ \theta_{\mathbf{I}}(y_0[n] - a[n-1]) &= \operatorname{sgn}(y_0[n] - a[n-1]) \end{aligned} \qquad (10.102)$$

and then from (10.99)–(10.101) we easily obtain the algorithms of Markov learning:

$$\begin{aligned} a[n] &= a[n-1] - (y_0[n] - \operatorname{sgn}(y_0[n] - a[n-1])) && \text{if} \quad a[n] \in A \\ a[n] &= a[n-1] && \text{if} \quad a[n] \notin A \end{aligned} \qquad (10.103)$$

This algorithm differs from the algorithm of learning in the adaptive receiver (6.65). In (10.103), $\gamma = 1$. Algorithm (10.103) is very similar to the algorithm of learning in threshold elements (10.74).

The constant value $\gamma = 1$ in such Markov algorithms indicates that the state $a[n]$ does not converge to a single optimal state $a = a^*$ either in probability or with probability one, but that only the mode of $a[n]$ converges to a^* as $n \to \infty$. This becomes obvious if we remember that in considering stochastic automata, which are equivalent to Markov chains, we can talk only about the probabilities that certain states are reached.

There is a close relationship between the algorithms of Markov learning and sequential decoding which has received great interest lately. With an understanding that a desire to accomplish everything is never satisfied, we shall not discuss this topic.

10.25 Games of Automata

An automaton's learning of optimal behavior can be considered as an automaton's game with nature. But automata can play among themselves if they are in the same environment (Fig. 10.16). The strategies of the player-

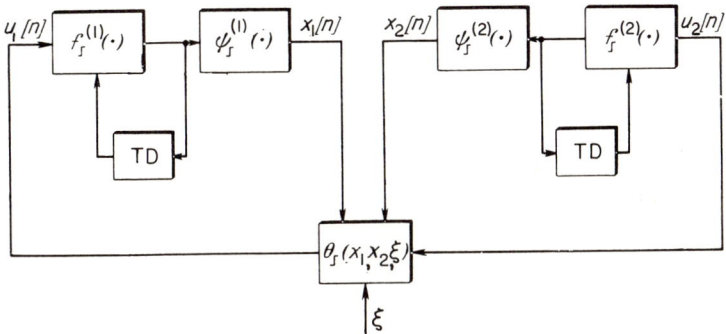

Fig. 10.16

automata are the states. The number of strategies is determined by the memory of the automata. Now, the penalty (or the reward) corresponds to the loss (or gain) of the automata. The games of the automata determine collective behavior of learning automata. For the zero-sum games of automata, the basic theorem of min-max is valid, and we can supply the algorithms of learning the solution of the games and the algorithms of learning in automata. We hope that for the simplest problems related to the games of automata, the reader will note the algorithms which determine the strategies of the playing automata.

10.26 Certain Problems

Among the problems related to games, we shall indicate the problem of defining the algorithms of learning the solution of the games in those cases when the elements of the cost matrix are not constant but are either random variables or unknown. The last case includes the so-called "blind" games.

In all the algorithms of learning and adaptation, it is important to accelerate the convergence of the algorithms using more complete information about the environment which is obtained during the operation of the automaton. This is usually accomplished by specific variations of $\gamma_1[n]$ and $\gamma_2[n]$. Are there any other possibilities of using the results of the preceding experiments in order to accelerate convergence?

It would be desirable to evaluate the effectiveness of various algorithms for training a threshold element, and also of training threshold sets which are constructed of threshold elements. The problems of identification and learning are also existent in finite automata. A systematic study of these possibilities would be of definite interest.

Very interesting problems appear in connection with the games of automata. For instance, the behavior of automata which possess not only the information regarding the outcome of the game, but also the information about the strategies used in the game. The "blind" games of automata (where the cost matrix is unknown), the zero-sum games of automata, and the games of a larger number of automata are also of special interest and importance. In the adaptive approach, sufficiently small a priori information is assumed, and a direct application of the well-developed theory of Markov chains is not possible.

10.27 Conclusion

In this last chapter we have briefly discussed the zero-sum game between two persons, and have developed the algorithms for learning the solutions of the games. These algorithms were useful in the solutions of a number of problems related to the realizability and control by threshold elements. The threshold elements which realize Boolean functions form the basis for the design of Rosenblatt's perceptron and Widrow's adaline. Their "capability" for learning is used here. We have also given considerable attention to finite automata which form a special class of dynamic systems.

The adaptive approach, broadly used in the preceding chapters for continuous and impulse systems, was also applied in studies of finite automata. Thus we were able to consider not only the expediency of an automaton's behavior, but also the training of automata to behave in an optimal fashion. Optimal behavior of automata can consist of pattern recognition, signal detection, identification of other automata, etc.

A characteristic property of the events and the objects studied in this chapter was that the sought variables must either satisfy the conditions of a simplex or that they can take the fixed values from a finite alphabet. We have tried to show that the adaptive approach even in this special situation can provide solutions of optimization problems under uncertainty.

COMMENTS

10.2 An exposition of the foundations of game theory can be found in the books by Saaty (1959) and Goldstein and Yudin (1966). The applications to the particular problem in the various areas of human activity are described by Luce and Raiffa (1957).

10.3 Von Neumann's min-max theorem is the basis of the game theory. A simple proof of this theorem is described in the book by Luce and Raiffa (1957).

10.4 Volkonskii indicated to us the importance of studying the errors in defining optimal strategies.

10.5 The algorithms of learning the solution of games (Table 10.1) include the algorithm proposed by Brown (1951). A modification of this algorithm was proposed by Amvrosienko (1965). Convergence of Brown's algorithm was proved by Robinson (1961). This proof is presented in the book by Bellman et al. (1958). Shapiro has estimated the rate of convergence and Braithwaite has applied Brown's algorithm, or as it is now called, Brown-Robinson's algorithm, to the solution of a system of linear equations, and to the solution of certain games in a finite number of steps. The algorithms (10.22) become the algorithms of Volkonskii (1962) when there are no errors in defining the optimal response (i.e., when $\xi_1 = \xi_2 = 0$).

The convergence of these algorithms is guaranteed by the conditions of convergence of the iterative methods. Therefore, this also establishes another relationship between various approaches to the game-theoretic problems. The theory of one class of the solutions of matrix games which also contains Brown's method was developed by Likhterov (1965). In addition, he has proposed computational algorithms with accelerated convergence.

The modifications of Brown's algorithm with accelerated convergence were proposed by Amvrosienko (1965) and Borisova and Magarik (1966).

10.6 The classical result due to Dantzig (1951) is presented here. See also Gale (1963) and Goldstein and Yudin (1966).

Some interesting applications of the game-theoretic approach to the optimal planning of grandiose dimensions were considered by Volkonskii (1962). Galperin (1966) has applied Volkonskii's algorithm in the solution of the transportation problem of grandiose dimension, and Meditskii (1965) has considered the problem of distributing the planned assignments.

10.7 This example was borrowed from the book by Bellman (1957). Certain systems of equations describing games can be found in the paper by Lebedkin and Mikhlin (1966).

10.8 The convergence of these continuous algorithms can be proven using the conditions of convergence for the continuous probabilistic methods. Another approach to the iterative method of solving games was considered by Danskin (1963).

10.10 and 10.11 A survey of literature describing the threshold elements and the synthesis of systems using threshold elements was written by Bogolyubov et al. (1963). See also the paper by Varshavskii (1962).

10.12   The relationship between the realizability of systems using a single threshold element and the two-person zero-sum games which is used here was established by Akers (1961).

10.14   The description of Rosenblatt's perceptrons can be found in the papers by Hay *et al.* (1960) and Keller (1961).

10.15   The description of Widrow's adaline is given in the papers by Widrow (1963, 1964). Very interesting applications of adaline as a trainable functional transformer can be found in the work by Smith (1966).

10.16   The "rigorous" algorithm of the form (10.75) was proposed by Ishida and Steward (1965). It is now clear that this "new" algorithm is only a small modification of the algorithm for the perceptron (10.61). For training of threshold elements and threshold networks, see the paper by Sagalov *et al.* (1966).

10.17   The basic concepts of the theory of finite automata are described in the books by Aizerman *et al.* (1963) and Pospelov (1966). See also the collection of papers edited by Shannon and McCarthy (1956). We shall consider the finite automata as a finite dynamic system.

10.18 and 10.19   The finite automata were not described here by transition matrices in order to emphasize the generality of finite impulsive and continuous systems. They are described as discrete dynamic systems. This system point of view of "linear" automata, or more exactly, of linear sequential machines was developed by Huffman (1956). See also the work of Tsypkin and Faradzhev (1966). The point of view presented here explains why we had not considered numerous excellent works in the theory of finite automata presented in the language which is characteristic for the theory.

10.20   The problems of interaction between the automata and the environment were posed and considerably solved by Tsetlin (1961a–c). An excellent survey of results in this area up to 1961 can be found in the work by Tsetlin (1961c). See also the book by Pospelov (1966). There is a vast literature devoted to the problems of analyzing the behavior of various automata in the environment. We shall mention the works by Blank (1964), Bardzin (1965), Brizgalov *et al.* (1965), Ponomarev (1964) and Krilov (1963). A slightly different approach was developed by McLaren (1966) and McMurtry and Fu (1965).

The behavior of stochastic automata in random media was studied by Varshavskii and Vorontsova (1963), Kandelaki and Tsertsvadze (1966) and Sragovich and Flerov (1965).

The expediency of the behavior of automata was determined in all of these works. The theory of Markov chains was used for that purpose.

10.22   A relatively small number of works is devoted to the problems of learning automata which are related to the problems of synthesis. The paper by Varshavskii *et al.* (1962) is concerned with learning automata. Milyutin (1965) has studied the choice of transition matrices in the automata which guarantee an optimal behavior. Dobrovidov and Stratonovich (1964) have described the synthesis of optimal automata which takes into consideration a posteriori probability of penalties during the process of operation. McMurtry and Fu (1965) have applied the automata with variable structure for the search of extrema.

As the reader has perhaps noticed, we have considered learning in finite automata exactly as in ordinary discrete or impulsive systems. The characteristic of learning automata presented at the end of this section was taken from the paper by Varshavskii *et al.* (1962).

10.23 The relationship between automata and Markov chains was considered in the book by Glushkov (1966). The example presented here was considered many times in the works by Sklansky (1965, 1966). Thanks to Feldbaum (1959, 1960), Markov chains are broadly used in the studies of transition processes in discrete extremal systems. These questions have been considered in the book by Pervozvanskii (1965), and also in the papers by Tovstukha (1960, 1961) and Kirschbaum (1963). They have also been applied in the work by Pchelintsev (1964) for solving the problems in search of the failures.

10.24 The theory of Markov learning was developed by Sklansky (1965) with respect to trainable threshold elements.

If the threshold γ is not assumed to be fixed, but to be a variable which satisfies the usual conditions $\gamma[n] > 0$, $\sum_{n=1}^{\infty} \gamma[n] = \infty$ and $\sum_{n=1}^{\infty} \gamma^2[n] < \infty$, as Sklansky (1966) has done, we arrive at the scheme of a threshold receiver which we have already considered in Section 6.13. This adaptive receiver corresponds to a trainable threshold element. Markov learning was studied from a slightly different viewpoint by McLaren (1964, 1966). The relationship between Markov learning and stochastic approximation was considered by Nikolic and Fu (1966).

The stochastic models of learning proposed by Bush and Mosteller (1955) are closely related to Markov models of learning. In such models, learning is accomplished in accordance with a mechanism postulated in advance, and not on the basis of optimization.

These problems can also be observed from the adaptive point of view. However, this would carry us slightly away from our main goal.

On sequential decoding, see the book by Wozencraft and Reiffen (1961).

10.25 The games of automata and their collective behavior are discussed in the works by Tsetlin (1961c, 1963), Tsetlin and Krilov (1963), Krinskii (1964), Varshavskii and Vorontsova (1963), Ginzburg and Tsetlin (1965), Stefanyuk (1963) and Volkonskii (1962). Very interesting results in modeling the games between automata on digital computers were described by Brizgalov *et al.* (1966) and Tsetlin *et al.* (1966).

The blind games of automata were examined in the works by Krinskii and Ponomarov (1964, 1966).

BIBLIOGRAPHY

Aizerman, M.A., Gusev, L.A., Rozonoer, L.I., Smirnova, I.M., and Tali, A.A. (1963). "Logic, Automata, and Algorithms." Fizmatgiz, Moscow. (In Russian.)

Akers, S.B. (1961). Threshold logic and two-person zero-sum games, *Proc. IRE* **49**.

Akers, S.B., and Rutter, B.H. (1963). The use of threshold logic in pattern recognition, *WESCON Conv. Rec.* **7** (4).

Amvrosienko, V.V. (1965). The condition of convergence for Brown's method of solving matrix games, *Ekon. Mat. Met.* **1965** (4). (In Russian.)

Bardzin, Ya.M. (1965). Capacity of the environment and behavior of automata, *Dokl. Akad. Nauk SSSR* **160** (2). (In Russian.)

Bellman, R. (1957). "Dynamic Programming." Princeton University Press, Princeton, N.J.

Bellman, R., Glicksberg, I., and Gross, O. (1958). "Some Aspects of the Mathematical Theory of Control Processes," Rep. R-313. The Rand Corp.

Blank, A.M. (1964). Expediency of automata, *Automat. Remote Contr.* (*USSR*) **25** (6).

Blazhkin, K.A. and Friedman, V.M. (1966). On an algorithm of training a linear perceptron, *Izv. Akad. Nauk SSSR Tekh. Kibern.* **1966** (6). (Engl. transl.: *Eng. Cybern.* (*USSR*)).

Bogolyubov, I.N., Ovsievich, B.L., and Rozenblum, L.Ya. (1963). Synthesis of networks with threshold elements. *In* "Seti Peredachi. Informatsii i ikh Avtomatizacia." Nauka, Moscow. (In Russian.)

Borisova, E.P., and Magarik, I.V. (1966). On two modifications of Brown's method of solving matrix games, *Ekon. Mat. Met.* **1966** (5).

Braithwaite, R. B. (1963). Finite iterative algorithm for solving certain games and corresponding systems of linear equations. *In* "Matrix Games." Fizmatgiz, Moscow. (In Russian.)

Brizgalov, V.I., Pyatetskii-Shapiro, I.I., and Shik, M.L. (1965). On a two-level model of interaction between automata, *Dokl. Akad. Nauk SSSR* **160** (5).

Brizgalov, V.I., Gelfand, I.M., Pyatetskii-Shapiro, I.I., and Tsetlin, M.L. (1966). Homogeneous games of automata and their modeling on digital computers. *In* "Samoobuchayushchiesia Avtomaticheskie Sistemi." Nauka, Moscow. (In Russian.)

Brown, G.W. (1951) Iterative solution of games by fictions play. *In* "Activity Analysis of Production and Allocation (T.S. Koopman, ed.). Wiley, New York.

Bush, R.R., and Mosteller, F.L. (1955). "Stochastic Models for Learning." Wiley, New York.

Daly, T.A., Joseph, R.D., and Ramsey, D.M. (1963). An iterative design technique for pattern classification logic, *WESCON Conv. Rec.* **7** (4).

Danskin, J.M. (1963). An iterative method of solving games. *In* "Beskonachnie Antagonisticheskie Igri." Fizmatgiz, Moscow. (In Russian.)

Dantzig, G.B. (1951). A proof of the equivalence of programming problem and the game problem. *In* "Activity Analysis of Production and Allocation." (R.C. Koopman, ed.). Wiley, New York.

Dobrovidov, A.V., and Stratonovich, R.L. (1964). On synthesis of optimal automata operating in random media, *Automat. Remote Contr. (USSR)* **25** (10).

Feldbaum, A.A. (1959). Transient processes in simple discrete extremal systems under random disturbances, *Automat. Remote Contr. (USSR)* **20** (8).

Feldbaum, A.A. (1960). Statistical theory of the gradient systems of automatic optimization for a quadratic plant characteristic, *Automat. Remote Contr. (USSR)* **21** (2).

Gale, D. (1963). "Theory of Linear Economic Models." IL, Moscow. (In Russian.)

Galperin, Z.Ya. (1966). Iterative method and the principle of decomposition in the solution of transportation problems, *Ekon. Mat. Met.* **2** (4). (In Russian.)

Gelfand, I.M., Pyatetskii-Shapiro, I.I., and Tsetlin, M.L. (1963). On certain classes of games and games between automata, *Dokl. Akad. Nauk SSSR* **152** (4). (In Russian.)

Gill, A. (1962). "Introduction to the Theory of Finite-State Machines." McGraw-Hill, New York.

Ginzburg, S.L., and Tsetlin, M.L. (1965). On some examples of modeling collective behavior of automata, *Probl. Peredachi Informa.* **1** (2). (In Russian.)

Glushkov, V.M. (1966). "Introduction to Cybernetics." Academic Press, New York.

Goldstein, E.G., and Yudin, D.B. (1966). "New Directions in Linear Programming." Sovetskoe Radio, Moscow. (In Russian.)

Hay, J.S., Martin, F.S., and Wightman, S.W., (1960). The MARK I perceptron—design and performance, *IRE Inter. Conv. Rec.* **8** (2).

Huffman, D.A. (1956). The synthesis of linear sequential coding machines. *In* "Information Theory" (C. Cherry, ed.). Academic Press, New York.

Ishida, H., and Stewart, R.H. (1965). A learning network using threshold element, *IEEE Trans. Electron. Comp.* **EC-14** (3).

Ito, T., and Fukanaga, K. (1966). An iterative realization of pattern recognition networks, *Proc. Nat. Electron. Conf.* **22**.

Kandelaki, N.P., and Tsertsvadze, G.N. (1966). On behavior of a certain class of stochastic automata in random media, *Automat. Remote Contr. (USSR)* **27** (6).

Kaszerman, P. (1963). A geometric test-synthesis procedure for a threshold device, *Inform. Contr.* **6** (4).

Keller, H. (1961). Finite automata, pattern recognition and perceptrons, *J. Assoc. Comp. Mach.* **8** (1).

Kirschbaum, H.S. (1963). The application of Markov chains to discrete extremum seeking adaptive systems. *In* "Adaptive Control Systems" (E.P. Caruthers and H. Levenstein, (eds.).) Pergamon Press, New York.

Krilov, V.Yu. (1963). On an automaton asymptotically optimal in random media, *Automat. Remote Contr. (USSR)* **24** (9).

Krinskii, V.I. (1964). On a construction of a cascade of automata and their behavior in the games, *Dokl. Akad. Nauk SSSR* **156** (6). (In Russian.)

Krinskii, V.I., and Ponomarev, V.A. (1964). On blind games, *Biofizika* **9** (3). (In Russian.)

Krinskii, V.I., and Ponomarev, V.A. (1966). On blind games. *In* "Samoobuchayuchiesia Avtomaticheskie Sistemi." Nauka, Moscow. (In Russian.)

Lebedkin, V.F., and Mikhlin, I.S. (1966). On design of game systems of control for separation of mixtures, *Izv. Akad. Nauk SSSR Tekh. Kibern.* **1966** (6). (Engl. transl.: *Eng. Cybern. (USSR)*.)

Likhterov, Ya.M. (1965). On a class of processes for solving matrix games, *Izv. Akad. Nauk SSSR Tekh. Kibern.* **1965** (5). (Engl. transl.: *Eng. Cybern. (USSR)*.)

Luce, R. D., and Raiffa, H. (1957). "Games and Decisions." Wiley, New York.

McLaren, R.W. (1964). A Markov model for learning system operating in an unknown environment, *Proc. Nat. Electron. Conf.* **20**.

McLaren, R.W. (1966). A stochastic automaton model for the synthesis of learning systems, *IEEE Trans. Sys. Sci. Cybern.* **2** (2).

McMurtry, G.J., and Fu, K.S. (1965). A variable structure automaton used as a multi-modal searching technique, *Proc. Nat. Electron. Conf.* **21**.

Meditskii, V.G. (1965). On a method of solving the problems of optimal distributions, *Ekon. Mat. Met.* **1** (6). (In Russian.)

Milyutin, A.A. (1965). On automata with optimal expedient behavior in random media, *Automat. Remote Contr. (USSR)* **26** (1).

Nikolic, Z.J., and Fu, K.S. (1966). A mathematical model of learning in an unknown random environment. *Proc. Nat. Electron. Conf.* **22**.

Pashkovskii, S. (1963). Self-organizing system of control for plants with a finite number of states, *Theory Self-Adapt. Contr. Syst. Proc. IFAC 2nd* **1963**.

Pchelintsev, L.A. (1964). Search for the failures as an absorbing Markov chain, *Izv Akad. Nauk SSSR Tekh. Kibern.* **1964** (6). (Engl. transl.: *Eng. Cybern. (USSR)*.)

Pervozvanskii, A.A. (1965). "Random Processes in Nonlinear Control Systems." Academic Press, New York.

Ponomarov, V.A. (1964). On a construction of automata asymptotically optimal in stationary random media, *Biofizika* **9** (1). (In Russian.)

Pospelov, D.A. (1966). "Games and Automata." Energia, Moscow-Leningrad. (In Russian.)

Robinson, J. (1961). An iterative method for solving games. *In* "Matrix Games." Fizmatgiz, Moscow. (In Russian.)

Saaty, T.L. (1959). "Mathematical Methods of Operations Research." McGraw-Hill, New York.

Sagalov, Yu.E., Frolov, V.I., and Shubin, A.B. (1966). Automatic training of threshold elements and threshold networks. *In* "Samoobuchayushchiesia Avtomaticheskie Sistemi." Nauka, Moscow. (In Russian.)

Shannon, C.E., and McCarthy, J. (eds.) (1956). Automata studies. "Annals of Mathematics Studies," Vol. 34. Princeton University Press, Princeton.

Shapiro, G.N. (1961). Comments on computational methods in game theory. In "Matrix Games." Fizmatgiz, Moscow. (In Russian.)

Sklansky, J. (1965). Threshold training of two mode signal detection, *IEEE Trans. Inform. Theory* **IT-11** (2).

Sklansky, J. (1966). Time-varying threshold learning. In "Proceedings of the Joint Automatic Control Conference, Seattle, Washington, 1966."

Smith, F.W., (1966). A trainable nonlinear function generator, *IEEE Trans. Automat. Contr.* **AC-11** (3).

Sragovich, V.G., and Flerov, Yu.A. (1965). On a class of stochastic automata, *Izv. Akad. Nauk SSSR Tekh. Kibern.* **1965** (2). (Engl. transl.: *Eng. Cybern. (USSR)*.)

Stefanyuk, V.L. (1963). An example of collective behavior of two automata, *Automat. Remote Contr. (USSR)* **24** (6).

Tovstukha, T.I. (1960). Influence of random noise on transient behavior of extremal systems for parabolic characteristic of the plant, *Automat. Remote Contr. (USSR)* **21** (5).

Tovstukha, T.I. (1961). On a question of selecting the parameters of the controllers for the gradient system of automatic optimization, *Automat. Remote Contr. (USSR)* **22** (8).

Tsertsvadze, G.N. (1964). Stochastic automata and the problem of designing reliable automata from unreliable elements, *Automat. Remote Contr. (USSR)* **25** (2) and (4).

Tsetlin, M.L. (1961a) On the behavior of finite automata in random media, *Automat. Remote Contr. (USSR)* **22** (10).

Tsetlin, M.L. (1961b). Certain problems of the behavior of finite automata, *Dokl. Akad. Nauk SSSR* **139** (4). (In Russian.)

Tsetlin, M.L. (1961c). Finite automata and modeling of simple forms of behavior, *Usp. Mat. Nauk* **18** (4).

Tsetlin, M.L. (1963). A comment on a game of a finite automaton with a partner, *Dokl. Akad. Nauk SSSR* **149** (1). (In Russian.)

Tsetlin, M.L., and Krilov, V.Yu. (1963). Examples of games of automata, *Dokl. Akad. Nauk SSSR* **149** (2).

Tsetlin, M.L., and Krilov, V.Yu. (1965). Games of automata. In "Theoria Konechnikh i Veroyatnostnikh Avtomatov." Nauka, Moscow. (In Russian.)

Tsetlin, M.L., Ginzburg, S.L., and Krilov, V.Yu. (1966). An example of collective behavior of automata. In "Self-Learning Automatic Systems." Nauka, Moscow. (In Russian.)

Tsypkin, Ya.Z., and Faradzhev, R.G. (1966). Laplace-Galoi transformation in the theory of sequential machines, *Dokl. Akad. Nauk SSSR* **166** (3). (In Russian.)

Varshavskii, V.I. (1962). Certain problems of the theory of logical networks based on threshold elements. In "Voprosi Teorii Matematicheskikh Mashin," Part 2. Fizmatgiz, Moscow. (In Russian.)

Varshavskii, V.I., and Vorontsova, I.P. (1963). On behavior of stochastic automata with variable structure, *Automat. Remote Contr. (USSR)* **24** (3). (In Russian.)

Varshavskii, V.I., and Vorontsova, I.P. (1965). Stochastic automata with variable structure. In "Teoria Konechnikh i Veroyatnostnikg Avtomatov." Nauka, Moscow. (In Russian.)

Varshavskii, V.I., and Vorontsova, I.P. (1966). Application of stochastic automata with variable structure in the solution of certain problems of behavior. In "Samoobuchayushchie Avtomaticheskie Systemi." Nauka, Moscow. (In Russian.)

Varshavskii, V.I., Vorontsova, I.P., and Tsetlin, M.L. (1962). Teaching stochastic automata. In "Biologicheskie Aspecti Kibernetiki." Nauka, Moscow.

Volkonskii, V.A. (1962). Optimal planning in the conditions of large dimensions (iterative methods and the principle of decomposition), *Ekon. Mat. Met.* **1** (2). (In Russian.)

Volkonskii, V.A. (1965). Asymptotic behavior of simple automata in the games, *Probl. Peredachi Inform.* **1** (2). (In Russian.)

Widrow, B. (1963). A statistical theory of adaptation. *In* "Adaptive Control Systems" (F.P. Caruthers and H. Levenstein, eds.). Pergamon Press, New York.

Widrow, B. (1964). Pattern recognition and adaptive control, *IEEE Trans. Appl. Ind.* **83** (74).

Wong, E. (1964). Iterative realization of threshold functions. *In* "International Conference on Microwave Circuit Theory and Information Theory, Tokyo, 1964," p. 3.

Wozencraft, J.M., and Reiffen, B. (1961). "Sequential Decoding." Wiley, New York.

EPILOGUE

The problems of modern theory and the practice of automatic control are characterized by an absence of a priori information about the processes and plants, and by an initial uncertainty. It appears now that adaptation and learning can serve as a common foundation for the solution of optimization problems under these conditions.

The concepts of adaptation and learning originally were always associated with the behavior of either separate living organisms or their aggregates. It is now customary to apply these concepts to those automatic systems which are capable of performing their functions in the conditions of initial uncertainty. This does not mean, however, that one should identify adaptation and learning in living organisms with that in automatic systems. But apparently, the theory of adaptation in automatic systems could become useful in explaining the amazing behavior of living organisms.

The reader may have noticed that we have purposely not used the analogy between the behavior of engineering and biological systems. We did not consider such presently fashionable questions related to an understanding of artificial intelligence, and which are also closely interlaced with adaptation and learning. This decision was made not only by the author's fear of competing with an enormous number of popular articles and books on cybernetics, but also by the fact that the present concept of "intelligence" has to contain something that is in principle unknown, unrecognizable and not subject to a formalization. Therefore, regardless of which algorithm is used for adaptation and learning in an automatic system, it is not advisable to attribute to it intelligence even if it is labeled artificial.

Epilogue

By considering adaptation and learning to be probabilistic iterative processes, it was possible to cover a number of different problems in the modern theory of automatic control, to recognize their common features, and finally to develop efficient methods for their solution. Such an approach is valuable not only because it enables us to discover the essence of the problems and to find ways for their solution, but also because it gives new importance to the "old" and neglected problems of classical control theory (for instance, the problems of stability and quality). We have seen that every algorithm of adaptation and learning which converges can be realized. The convergence of the algorithms is nothing else but an expression of the stability of the corresponding stochastic nonlinear closed-loop systems.

Estimation of the rate of convergence and the optimal algorithms are closely related to the problems of quality in the stochastic nonlinear systems. Therefore, the problems of stability and quality play an important and sometimes main role in the new problems of adaptation and learning. Is this not evidence that the theory of automatic control " grows, but does not age," that the new directions appear in it, but the problems of stability and quality remain eternally young?

At the present time, it is difficult to discuss the completion of the theory of adaptation and learning. We are only at the beginning of a road which is perhaps long but has a promising future. Along this road will be discovered new and perhaps unexpected relationships between various branches in control theory and the related sciences. New problems will also be created, and they will most probably broaden the application of the theory of adaptation and learning.

More complex plants in the control systems and an absence of a priori information regarding the operating conditions have brought the adaptive control systems to life. But the role of adaptive and learning processes is not limited only to an elimination of uncertainty and to control under insufficient information. The elimination of uncertainty and the gathering of useful information represent the elements of a creative process. Who knows? The adaptive systems may help us in the very near future to accomplish better control under given conditions, and to develop new and more general methods, theories, and concepts.

SUBJECT INDEX

A

Acceleration of convergence, 35, 38, 62
Adaline, Widrow's, 263
Adaptation, 44, 69
Adaptive approach, 11, 15
Adaptive equalizer, 152
Algorithm, 19, 39
  of adaptation, 48
    continuous, 54
    multistage, 52
    search, 48
    the best, 61
  of dual control, 178
  of extremal control, 192
  of Markov learning, 276
  of optimal evaluation, 243
  of optimal productivity, 223
  of optimization, 19
    continuous, 31
    feasible directions, 28
    multistage, 29
    relaxation, 22
  of restoration, 96
  of self-learning, 102
  of signal extraction, 163
  of training, 79
    continuous, 87
    discrete, 83
    search, 108
Alphabet, input, 267
  output, 267
Approximation, stochastic, 46, 70
A priori information, 9
Automaton, 266
  finite, 267
  stochastic, 270
Average risk, 6, 100

B

Barrier of stability, 177
Bayesian approach, 65
Bayesian criterion, 6
Box, "black," 120
  "grey," 120

C

Classification, 77
Coefficient of statistical linearization, 128
Constraints, 24, 50, 51
Convergence, 33
  conditions, 56
  probabilistic, 54
Correlation function, 125
Cost matrix, 6, 249
Criterion of optimality, 6
  Bayesian, 6
  Kotelnikov's, 159
  Neyman-Pearson, 159

D

Decoding, sequential, 277, 281
Deterministic process, 10
Dichotomy, 78
Digrator, 20, 39
Distribution of detection facilities, 239
 of production resources, 237
Dual control, 179

E

Equalizer, adaptive, 153
Ergodicity, 8, 15
Expectation, 6
 of the penalty, 271
Expediency of behavior, 271
Extrapolation, 78
Extremal control, 191

F

Feasible directions, 28
Filter, adaptive, 149
 based on search, 154
 predictor, 156
 Kolmogorov-Wiener, 157
Filtering, 147
Function, amplitude-quantized, 242, 267
 characteristic, 6
 logical, 259
 sensitivity, 184
Functional, 6

G

Games, 248
 between automata, 277
 blind, 278
 with optimal strategies, 252
 zero sum, 249
Graph, 268

H

Hypothesis of compactness, 77, 110
 of representation, 81, 110

I

Identification, 119
 of linear plants, 136
 of nonlinear plants, 130
Information, a priori, 9
 current, 9, 10
Intelligence, artificial, 286
Interaction, automaton-environment, 271

L

Learning, 45
 Markov, 276
 pattern recognition, 78
 the solution of games, 252
Least-square method, 65
Linear programming, 254
Linearization, statistical, 128

M

Markov chains, 275
Maximum likelihood, 66
Method, least square, 65
 probabilistic iterative, 46
 regular iterative, 18
Min-max, 262
Mixed strategies, 243
Mixture, 36

N

Nonlinear algorithm, 22

O

Operating expenditures, 215
Operations research, 229
Optimal planning, 233
 productivity, 223
 reserves, 235
 strategies, 253
 supply, 234
Ordinary approach, 11

P

Planning of reserves, 230
Portrait, generalized, 111
Problem of optimality, 5
Process, deterministic, 10
  stochastic, 10

Q

Quantization, 241
Queuing theory, 226

R

Realizability of Boolean functions on a single threshold element, 259
Recurrent scheme, 39
Relaxation algorithm, 22
Reliability indices, 213
Repairs, 227
Risk, average, 6, 100

S

Self-learning, 45, 95, 99
Separability, 134
Simplex-method, 28
Slater's regularity condition, 27

Statistical theory of signal reception, 158
  decision theory, 158
Stochastic process, 11
Strategy, 249
Synthesis of optimal systems, 197, 206
  in the presence of noise, 199
System of automatic control, 176
  adaptive, 181
    simplified (with the model), 187
  complex, design, 223
  control by perturbation, 187
  discrete with the constant gain coefficient, 22
    with the non-linear gain coefficient, 22
    with time-varying gain, 22
  extremal, 191

T

"Teacher," 45
Training
  of automata, 272
  of threshold elements, 264
Transition diagram, 268
  table, 269

V

Variation of the average risk, 101

Mathematics in Science and Engineering

A Series of Monographs and Textbooks

Edited by RICHARD BELLMAN, *University of Southern California*

1. T. Y. Thomas. Concepts from Tensor Analysis and Differential Geometry. Second Edition. 1965
2. T. Y. Thomas. Plastic Flow and Fracture in Solids. 1961
3. R. Aris. The Optimal Design of Chemical Reactors: A Study in Dynamic Programming. 1961
4. J. LaSalle and S. Lefschetz. Stability by by Liapunov's Direct Method with Applications. 1961
5. G. Leitmann (ed.). Optimization Techniques: With Applications to Aerospace Systems. 1962
6. R. Bellman and K. L. Cooke. Differential-Difference Equations. 1963
7. F. A. Haight. Mathematical Theories of Traffic Flow. 1963
8. F. V. Atkinson. Discrete and Continuous Boundary Problems. 1964
9. A. Jeffrey and T. Taniuti. Non-Linear Wave Propagation: With Applications to Physics and Magnetohydrodynamics. 1964
10. J. T. Tou. Optimum Design of Digital Control Systems. 1963.
11. H. Flanders. Differential Forms: With Applications to the Physical Sciences. 1963
12. S. M. Roberts. Dynamic Programming in Chemical Engineering and Process Control. 1964
13. S. Lefschetz. Stability of Nonlinear Control Systems. 1965
14. D. N. Chorafas. Systems and Simulation. 1965
15. A. A. Pervozvanskii. Random Processes in Nonlinear Control Systems. 1965
16. M. C. Pease, III. Methods of Matrix Algebra. 1965
17. V. E. Benes. Mathematical Theory of Connecting Networks and Telephone Traffic. 1965
18. W. F. Ames. Nonlinear Partial Differential Equations in Engineering. 1965
19. J. Aczel. Lectures on Functional Equations and Their Applications. 1966
20. R. E. Murphy. Adaptive Processes in Economic Systems. 1965
21. S. E. Dreyfus. Dynamic Programming and the Calculus of Variations. 1965
22. A. A. Fel'dbaum. Optimal Control Systems. 1965
23. A. Halanay. Differential Equations: Stability, Oscillations, Time Lags. 1966
24. M. N. Oguztoreli. Time-Lag Control Systems. 1966
25. D. Sworder. Optimal Adaptive Control Systems. 1966
26. M. Ash. Optimal Shutdown Control of Nuclear Reactors. 1966
27. D. N. Chorafas. Control System Functions and Programming Approaches (In Two Volumes). 1966
28. N. P. Erugin. Linear Systems of Ordinary Differential Equations. 1966
29. S. Marcus. Algebraic Linguistics; Analytical Models. 1967
30. A. M. Liapunov. Stability of Motion. 1966
31. G. Leitmann (ed.). Topics in Optimization. 1967
32. M. Aoki. Optimization of Stochastic Systems. 1967
33. H. J. Kushner. Stochastic Stability and control. 1967
34. M. Urabe. Nonlinear Autonomous Oscillations. 1967
35. F. Calogero. Variable Phase Approach to Potential Scattering. 1967
36. A. Kaufmann. Graphs, Dynamic Programming, and Finite Games. 1967
37. A. Kaufmann and R. Cruon. Dynamic Programming: Sequential Scientific Management. 1967
38. J. H. Ahlberg, E. N. Nilson, and J. L. Walsh. The Theory of Splines and Their Applications. 1967

39. Y. Sawaragi, Y. Sunahara, and T. Nakamizo. Statistical Decision Theory in Adaptive Control Systems. 1967

40. R. Bellman. Introduction to the Mathematical Theory of Control Processes, Volume I. 1967; Volume II. 1971 (Volume III in preparation)

41. E. S. Lee. Quasilinearization and Invariant Imbedding. 1968

42. W. Ames. Nonlinear Ordinary Differential Equations in Transport Processes. 1968

43. W. Miller, Jr. Lie Theory and Special Functions. 1968

44. P. B. Bailey, L. F. Shampine, and P. E. Waltman. Nonlinear Two Point Boundary Value Problems. 1968

45. Iu. P. Petrov. Variational Methods in Optimum Control Theory. 1968

46. O. A. Ladyzhenskaya and N. N. Ural'tseva. Linear and Quasilinear Elliptic Equations. 1968

47. A. Kaufmann and R. Faure. Introduction to Operations Research. 1968

48. C. A. Swanson. Comparison and Oscillation Theory of Linear Differential Equations. 1968

49. R. Hermann. Differential Geometry and the Calculus of Variations. 1968

50. N. K. Jaiswal. Priority Queues. 1968

51. H. Nikaido. Convex Structures and Economic Theory. 1968

52. K. S. Fu. Sequential Methods in Pattern Recognition and Machine Learning. 1968

53. Y. L. Luke. The Special Functions and Their Approximations (In Two Volumes). 1969

54. R. P. Gilbert. Function Theoretic Methods in Partial Differential Equations. 1969

55. V. Lakshmikantham and S. Leela. Differential and Integral Inequalities (In Two Volumes). 1969

56. S. H. Hermes and J. P. LaSalle. Functional Analysis and Time Optimal Control. 1969

57. M. Iri. Network Flow, Transportation, and Scheduling: Theory and Algorithms. 1969

58. A. Blaquiere, F. Gerard, and G. Leitmann. Quantitative and Qualitative Games. 1969

59. P. L. Falb and J. L. de Jong. Successive Approximation Methods in Control and Oscillation Theory. 1969

60. G. Rosen. Formulations of Classical and Quantum Dynamical Theory. 1969

61. R. Bellman. Methods of Nonlinear Analysis, Volume I. 1970

62. R. Bellman, K. L. Cooke, and J. A. Lockett. Algorithms, Graphs, and Computers. 1970

63. E. J. Beltrami. An Algorithmic Approach to Nonlinear Analysis and Optimization. 1970

64. A. H. Jazwinski. Stochastic Processes and Filtering Theory. 1970

65. P. Dyer and S. R. McReynolds. The Computation and Theory of Optimal Control. 1970

66. J. M. Mendel and K. S. Fu (eds.). Adaptive, Learning, and Pattern Recognition Systems: Theory and Applications. 1970

67. C. Derman. Finite State Markovian Decision Processes. 1970

68. M. Mesarovic, D. Macko, and Y. Takahara. Theory of Hierarchial Multilevel Systems. 1970

69. H. H. Happ. Diakoptics and Networks. 1971

70. Karl Astrom. Introduction to Stochastic Control Theory. 1970

71. G. A. Baker, Jr. and J. L. Gammel (eds.). The Padé Approximant in Theoretical Physics. 1970

72. C. Berge. Principles of Combinatorics. 1971

73. Ya. Z. Tsypkin. Adaptation and Learning in Automatic Systems. 1971

74. Leon Lapidus and John H. Seinfeld. Numerical Solution of Ordinary Differential Equations. 1971

75. L. Mirsky. Transversal Theory, 1971

76. Harold Greenberg. Integer Programming, 1971

77. E. Polak. Computational Methods in Optimization: A Unified Approach, 1971

78. Thomas G. Windeknecht. A Mathematical Introduction to General Dynamical Processes, 1971

79. M. A. Aiserman, L. A. Gusev, L. I. Rozonoer, I. M. Smirnova, and A. A. Tal'. Logic, Automata, and Algorithms, 1971

In preparation

Andrew P. Sage and James L. Melsa. System Identification

R. Boudarel, J. Delmas, and P. Guichet. Dynamic Programming and Its Application to Optimal Control

William Stenger and Alexander Weinstein. Methods of Intermediate Problems for Eigenvalues Theory and Ramifications

TJ
217
T713

DEC 4 1972